Fundamentals of TRIBOLOGY

Fundamentals of TRIBOLOGY

by

Aruna Chandra | Sarah Parmar | Ekta Jain

Wordpen Academics
Publishing Science That Matter

ISBN: 978-93-6084-093-8

Published by : WORDPEN ACADEMICS
4736/23, Ansari Road, Darya Ganj,
New Delhi – 110 002
Phone: +91-011-35004336
E-mail: info@wordpenacademics.com

***Printed at*: Replika Press Pvt. Ltd.**

Exclusive Distributor: ***Bio-Green Books, New Delhi***

Preface

"Fundamentals of Tribology" serves as a comprehensive guide to understanding the intricate science of friction, wear, and lubrication in mechanical systems. Tribology, the study of interacting surfaces in relative motion, plays a vital role in a wide range of industries, from automotive and aerospace engineering to manufacturing and biomedical applications. This book is designed to provide students, engineers, and researchers with a solid foundation in the principles, theories, and applications of tribology.

In a world where efficiency, reliability, and sustainability are paramount, tribology emerges as a crucial discipline for optimizing the performance and longevity of mechanical systems. By understanding the complex interplay between surfaces, lubricants, and operating conditions, engineers can design more durable, efficient, and environmentally friendly machines and components.

As editors of this volume, we have endeavored to compile a diverse array of perspectives from leading experts and practitioners in the field of tribology. Through a blend of theoretical insights, practical applications, and cutting-edge research, this book covers topics ranging from surface characterization and friction mechanisms to lubrication strategies and wear prevention techniques.

Our overarching goal in presenting this book is to foster a deeper understanding and appreciation of tribology as a cornerstone of modern engineering. By providing readers with the knowledge and tools needed to tackle real-world tribological challenges, we aim to empower individuals and organizations to drive innovation, enhance performance, and advance sustainability across diverse industries and applications.

We extend our heartfelt gratitude to all the contributors who have generously shared their expertise, experiences, and insights, as well as to the readers who embark on this journey of discovery with us. May this volume serve as a valuable resource and inspiration for all those who seek to unlock the full potential of tribology in the pursuit of excellence and progress.

Aruna Chandra | Sarah Parmar | Ekta Jain

Contents

Part 1

Hydrodynamic Lubrication

1

Theory of Tribo-Systems

1. Introduction

What is tribology? Why people need tribology?

Some people say that tribology is friction, wear or lubrication. Others say that it is friction plus wear and plus lubrication. However both of them are not accurate enough. Tribology takes all theoretical and applied results from friction, wear and lubrication obtained in the past, inputs into them with much more new senses and contents based on the development of science and technology. It is striving to constitute a theoretical and technical platform to meet the future requirement. Tribology cannot be looked simply as equal only to friction, wear, lubrication or any other technique related.

The early stage of applying knowledge of friction, wear and lubrication in human productive and living practice can be traced back to 3000 BC or earlier (Dowson, 1979). This multi-disciplinary branch of science and technology and its application in comprehensive areas were studied in many different sub-subjects independently from very different points of view over a long period. A suggestion from H Peter Jost gave this old field a powerful impact and poured into it youthful vigor (Her Majesty's Stationery Office, 1966). It developed quicker and quicker thereafter. Tribology is a both old and young discipline.

In the first phase of development of tribology since 1965, due to its universal existence in nature according to the definition given by Jost on one side and the belief of tribologists in many countries that they could make huge benefit for industry on another side, the influence of tribology increased dramatically fast in the seventies and eighties of the last century. Promise of saving 5 billion pounds per year in UK in the Jost Report pushed forward tribologists working on applying existed knowledge of friction, wear and lubrication to solve engineering problems.

New techniques related to friction, wear and lubrication developed then rapidly in the following phase even though some people they did not like the name "tribology". Many books published in this stage with the title "Tribology" but no one discussed on the questions that what was tribology and why they used the word "tribology" except Jost did in his famous Report.

The later situation has shown that to achieve the potential benefit is not so easy (Xie, 1986; Xie & Zhang, 2009). A name, a definition and simply putting all knowledge components together subjectively are not enough. A concept system, theory system and method system, which can match the name, definition and nature of tribology and then can promote an independent development and application of tribology, are expected.

It is valuable to mention that "Tribology" was defined as one of the four major disciplines of Mechanical Systems by a Committee of NSF of US in 1983 (The Panel Steering Committee for the Mechanical Engineering and Applied Mechanics Division of the NSF, 1984) and then the "Journal of Lubrication Technology" was renamed as "Journal of Tribology" of Transaction of ASME. Only ten years later, a gentleman from US indicated in an informal speech in Beijing that a change under way was the gradual disappearance of the term "tribology" from programs and projects of NSF in US. In this period fewer papers which dealt with the relation between tribology and mechanical systems could be found in the journal. It implies that no enough effort has been made to carry out the original intention of the committee. Many famous tribologists prophesied that tribology would become or be replaced by surface engineering.

It shows some undesired situation in the development of tribology. There are at least three problems with it. Firstly tribology was born on the foundation of known appearances of friction, wear and lubrication but the difference between tribology and friction, wear and lubrication has not been paid attention to investigate into. Naturally the traditional way of studying friction, wear and lubrication independently is still having its visible influence on tribology. Secondly as tribology is so universal and so important to engineering and industry, much attention has been paid to the tribology-based applied techniques and a very fast development of the techniques has been achieved. Due to the nature of tribology, which will be discussed later, most of the techniques can be applied only to a specific branch of field for a specific target. Many people they work in the field of tribology but they don't think they are tribologists. Some of them think they are chemists, material scientists, biologists or mechanical engineers. Therefore the theoretical study of tribology, especially the efforts on finding a systematic framework for tribology cannot benefit further from such a fast development of technique. Thirdly people don't know how to use the results obtained under one condition to another condition and how to compare the results from one kind of test machines with what of another kind of test machines, in other words, there is no general model for tribology and almost no modern mathematic tool can be used in tribology. Engineers have to make each decision individually in design depending on experience or experiment. They cannot construct a tribological design in true sense for their products because no model can be found in simulating the behaviors of tribology other than friction, wear or lubrication individual. Therefore there is no strong enough attraction to take tribology as an independent discipline in industry further.

Tribology has been defined in 1965 as "the science and technology of interacting surfaces in relative motion and of the practices related thereto" (Her Majesty's Stationery Office, 1966) and modified later as "the science of behaviors of interaction surfaces in relative motion together with the active medium concerned (each of them is a tribo-element) in natural systems, their results and the technology related thereto" (Xie, 1996). A question arises then that why people need *the interacting surfaces in relative motion*? Both definitions deal with appearance aspects rather than functional aspects of tribology and cannot answer the question. Obviously any surface cannot exist independently and must be a part of a component. The relative motion of surfaces is defined by the relative motion of components and where the surfaces reside on. The interactions transmitted between surfaces are from the components in contact on the surfaces as well. In most (not all) cases two interacting surfaces in relative motion function as *a joint* which permits only some kinds of relative motion and prevents other kinds of relative motion between two components in contact. Such joints are named kinematic pairs in mechanisms. The interacting surfaces in relative motion must function with other elements in a system or function with other elements for a system.

Therefore the problems with tribology are problems of systems science and systems engineering. In a sense, without system there would be no tribology.

In the very early stage of tribology people have begun to think about system problems (Fleischer, 1970; Czichos, 1974; Salomon, 1974). A comprehensive study on applying system concepts to friction, wear and lubrication was given by Czichos which described how to use general systems theory and engineering system analysis in treating tribological problems (Czichos, 1978). Without an effective way for mathematic computation limited its application. Dai and Xue (Dai & Xue, 2003) tried to evaluate tribological behaviors with an entropy calculation in tribo-systems while Ge and Zhu (Ge & Zhu, 2005) worked out through a fractal analysis for a similar attempt. Either entropy calculation or fractal analysis cannot describe explicitly and quantitatively the character of movement of interacting surfaces in relative motion. Since they deal only with entropy or fractal parameters, transforming all other physical and geometric behaviors into an entropy or fractal change in calculation is unavoidable. It involves the transformation of a large amount of knowledge concerning with tribology getting together in the past into the thermodynamic or fractal knowledge and is almost impossible in practice.

Considering that relative motion is the first important character of tribology and is also a basic behavior studied in mechanisms, some concepts in mechanisms should be discussed before going to construct a *function based* systems theory for tribology.

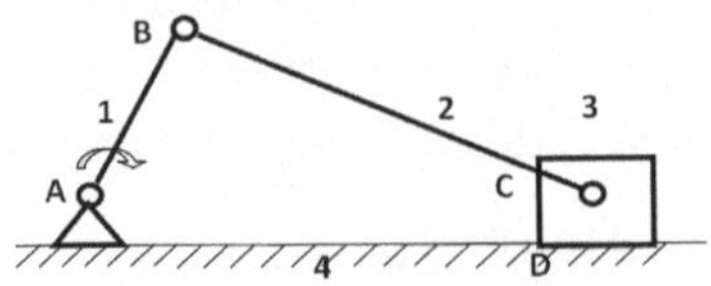

Fig. 1. A crank-slider mechanism

When several components are joined together by kinematic pairs it constructs a kinematic chain. The necessary condition that a kinematic chain becomes a mechanism, in other words a mechanical system is all components in the chain having definitive relative motions. This condition can be rewritten as that the number of motion conditions given (input) outside of the system equals to the number of residual degrees of freedom of the chain. For example, as shown in Figure 1 there is a plane kinematic chain of four components (1 - 4) with one fixed component (4, chassis), three revolute pairs (A – C) and one prismatic pair (D). Each movable component has three degrees of freedom while each revolute pair or each prismatic pair cancels two degrees of freedom. Revolute pairs and prismatic pairs all formed with surface contact are known as lower pairs and pairs formed with point contact or line contact are known as higher pairs in mechanisms. Each higher pair cancels one degree of freedom in plane analysis. Then the residual degrees of freedom of the plane chain can be calculated as

$$RDOF = 3MC - 2L - H = 3(4-1) - 2 \cdot 4 = 1 \tag{1}$$

In formula (1) *MC*, *L* and *H* are the number of movable components, the number of lower pairs and the number of higher pairs respectively. The result shows 1 motion condition input is in need of becoming the chain to a crank-slider mechanism. In this example when a rotating speed of the crank (1) is given the relative motions of all other movable components are defined and can be derived from the crank rotating speed. In the derivation each pair (interacting surfaces in relative motion) functions to permit some kinds of relative motion and prevent the others between two components joined by the pair, and each component functions to keep the surfaces of pairs having fixed positions on the component. Then the mechanism can be looked as consisting of two sub-systems: a component system and a pair system. The component system and the pair system work together to guarantee the mechanism with a definite motion when the number of motion conditions input is enough. Such function is a motion guarantee function.

Back to tribology, a tribo-pair is the physical realization of a kinematic pair and the tribo- pairs in total in a machine system constitutes the main part of a tribo-system.

Tribo-systems can be understood in another way. A system of higher rank can be divided into several systems of lower rank or sub-systems and vice versa. The division can be implemented in different ways. For example, a machine as a system can be divided into assembles, such as a rotor assemble, a chassis assemble, or can be divided according to the function of the sub-systems as well, such as a coolant circulation system, a brake system etc. A machine system can also be divided according to the character of behaviors of the sub- systems, for example, dividing the machine into a mechanical system, a thermodynamic system, an electric system and what will be discussed in detail in tribology, a tribo-system, etc.

Such a division is making an abstraction of machine systems. It does not limit the division to groups of elements or kinds of functions but keep the investigation into a given category of behaviors and their results. For example, in general there is an electric circuit diagram for a machine and the diagram is just a description of the electric system abstracted from the machine.

When a machine system or another natural system is abstracted into a system consisting of tribo-elements and some supporting auxiliary sub-systems for studying behaviors on or between the interacting surfaces in relative motion, results of the behaviors and technology related to, a tribo-system is then constructed.

In most cases there will be a liquid, a gas or a fat lubricant film kept between the interacting surfaces in relative motion to reduce friction and wear. Solid lubricant films will not be included in this discussion and looked as parts of the surfaces with a motion similar to the surfaces. The auxiliary sub-system for fluid lubrication including at first a cycling sub- system, a cooling sub-system and a filtering sub-system operates to keep the fluid film staying between the interacting surfaces in relative motion and to make the surfaces work efficiently, reliably and friendly to human and environment.

Due to there is a time variable character with the tribo-systems which will be discussed later, the change in structure, behavior and function or the change of working condition in short should be monitored (necessary) and be adaptively controlled (suggested) to avoid low efficiency work, abnormal wear, environment pollution or catastrophic damage. Therefore the condition monitoring sub-system or the condition controlling sub-system usually takes place in the auxiliary sub-system of tribo-systems. Figure 2 gives a general construction diagram for tribo-systems (Xie, 1996).

It can be concluded that a machine system is consisted of a component system and a tribo- system from the view point of motion. The tribo-system together with the component system plays a motion guarantee function which keeps each part of the machine system with a definite motion when the number of motion conditions input outside of the system is enough.

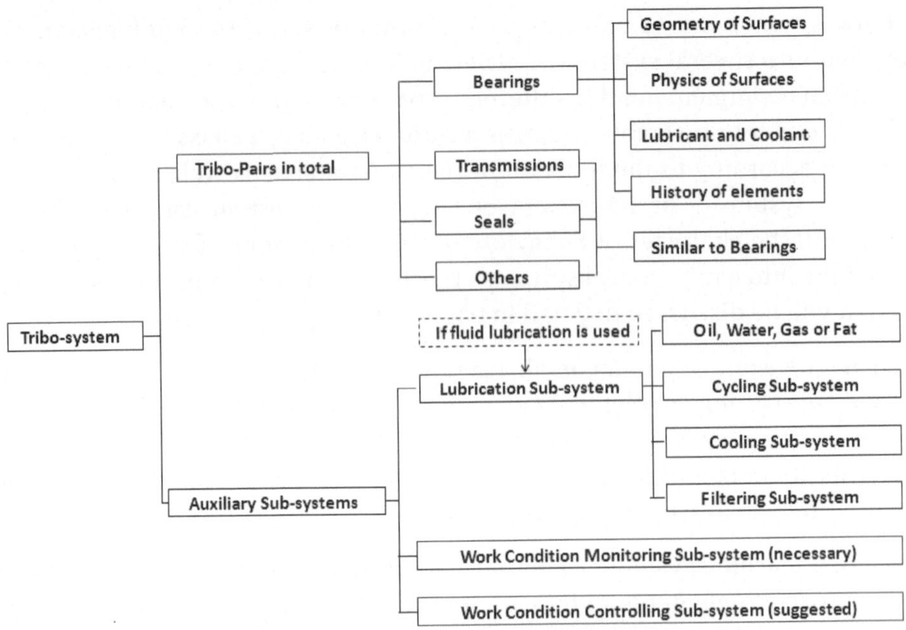

Fig. 2. The block diagram of a typical tribo-system

Tribology science and technology is very important in obtaining *the best way* (theory and application) to complete the motion guarantee function of tribo-systems. Tribology exists universally. Where there is relative motion there is tribology. Tribo-systems play sometimes very critical roles in machine systems and work sometimes under extreme severe condition. It implies that the motion guarantee function must be implemented with high reliability, low energy consumption, low cost, low pollution, human and environment friendship etc.

The study of friction, wear, lubrication and other tribological techniques are part of the efforts in finding the best way. Ignoring the fundamental function of tribo-systems and appreciating an *appearance based* study of friction, wear, lubrication or techniques from surface engineering, nanotechnology or biology etc, even though some physical or geometric results can be obtained, the study cannot give a clear overview on the relation between the results. Putting the results together in practice usually throws engineers into confusion. It will also increase the difficulty in tribo-system modeling and in looking for mathematic tools for an overall system and life cycle behavior simulation. Lake of overall system model and mathematic tool for simulation makes that the results from one working condition cannot be used in another condition, from one period of the life cannot be used in another period and from the study of friction cannot be used in the study of wear or the study of lubrication etc. Furthermore such a situation makes almost impossible to implement a tribological design since the tribo-design is the design of tribo-systems. It is well known that tribological design is a main channel for embedding tribology knowledge into products.

Therefore different from the independently study of friction, wear, lubrication or any technique related, tribology should be studied with a system viewpoint and cannot overlook the basic function of tribo-systems, the motion guarantee function for a machine system.

Tribology study is to find the best way (theory and application) to complete the motion guarantee function. It is no doubt that all results from the study of friction, wear, lubrication and other technique related are indispensable in reaching the goal.

2. How tribo-systems behave?

After a detailed discussion on the basic function of tribo-systems the question arises that how a tribo-system behaves to complete the function?

Some basic knowledge about the character and pattern of change of the systems is necessary in describing their behaviors. For example, one character of a mechanical system or a mechanism is that all components in the system have definitive relative motion and it can be checked with formula (1). The pattern of change of the mechanism is governed by principles in kinematics from which the behaviors can be derived. For the crank-slider mechanism shown in Figure 1 the relative motions of all movable components can be derived with a given rotating speed of the crank. Such a requirement will be similar for thermodynamic systems, electric systems and all other systems abstracted from a system of higher rank according to a given category of behaviors. Tribology is a multi-disciplinary area. Any principle in other disciplines cannot describe and govern the character and pattern of change of tribo-systems accurately. A task of top priority is to organize the basic knowledge which matches the name, the definition and the nature of tribology rather than matches what concerning with only friction, wear, lubrication or any individual technique related.

After the investigation in many years the author suggests that there are three axioms in tribology and they can be used as a base or a start point to study into the character and pattern of change of tribo-systems. So-called an axiom means what people cannot find any opposite example with the axiom even though they cannot prove it theoretically (Suh, 1990). The three axioms in tribology (Xie, 2001) are: (1) The first Axiom: Tribological behaviors are system dependent. (2) The second Axiom: The property of tribo-elements and then the systems containing tribo-elements are time dependent. (3) The third Axiom: The results of tribological behaviors are the results of mutual action and strong coupling of many behaviors of other disciplines under a tribological condition consisted of interacting surfaces in relating motion. The three axioms will be discussed in more detail in the following.

2.1 The first axiom: Tribological behaviors are system dependent

Changes taking place on or between the interacting surfaces in relative motion are what to be investigated into in tribology and called tribological behaviors.

Interactions and relative motion are causes of the behaviors. Results of the behaviors include a recoverable and irrecoverable change of intrinsic property of elements, a change of the state of the system consisted of the elements and a material, energy and information exchange with the environment in the forms of input and output. The intrinsic property includes geometric, physical and historic aspects and will be discussed later. A single surface or medium substance cannot implement any tribological behavior. In Fig. 3 there is a simplest tribo-system including three elements. The system is enveloped with a system block and exchanges material, energy and information via input and output with environment. The system dependent character governs not only the behaviors of simplest systems but also any more complex system consisted of simplest systems and their supporting auxiliary sub- systems (Xie, 2010). The first axiom focuses on the relationship of structures.

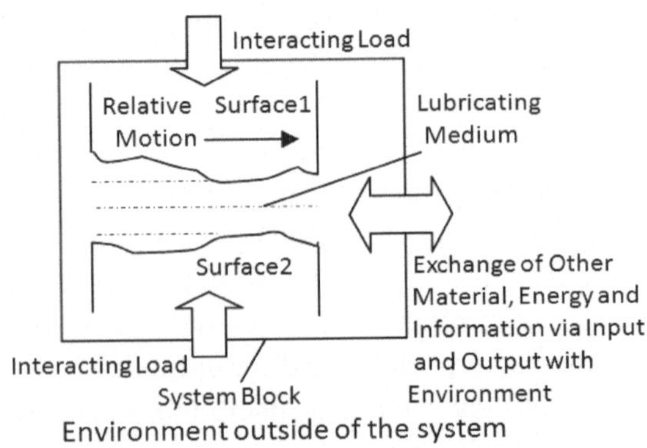

Fig. 3. A simplest tribo-system

2.2 The second axiom: The property of tribo-elements and then the systems containing tribo-elements are time dependent

In comparison with the material in the body of a component, the material of any element in a tribo-pair bears much more intensive load and works under a much more severe condition. As shown in Figure 4 when the roughness of surfaces in contact is considered, the real contact area is much smaller than the nominative contact area. Through the much smaller contact area it transmits a load equal to what transmitted by the body of the component with an area of the body section. The load density is then very high at the real contact area. On the other hand the transmission is implemented between different materials in a pair while it is through the same material in the case of the body of a component. Additional physical or chemical reaction between different materials may occur under such a condition. Furthermore there is a relative motion. It accelerates the change of their physical property, chemical composition and geometric configuration, especially due to the relative motion the change is continuously repeated, sometimes with very high frequency. The relative motion produces heat and then high temperature and other kinds of active energy. They will no doubt promote the change in physical aspects and chemical aspects. All of them make the change of the property of each element in the tribo-pair much more fast in comparison with what in the body of a component. Due to the severity in most cases the change is irrecoverable. Therefore as in performance analysis or design people usually consider what they deal with is a time-invariable system they must consider it as a time-variable system in tribological analysis and tribological design. As shown in Fig. 5 the speed of change of performance is variable in a life cycle. In the earlier stage of work of a new system the speed of change is high and this is a running in stage. Afterwards the speed of change will be slow in a stable operation stage. At last the speed of change increases faster and it predicts the end of life. The variation of speed of change is very complex in many systems.

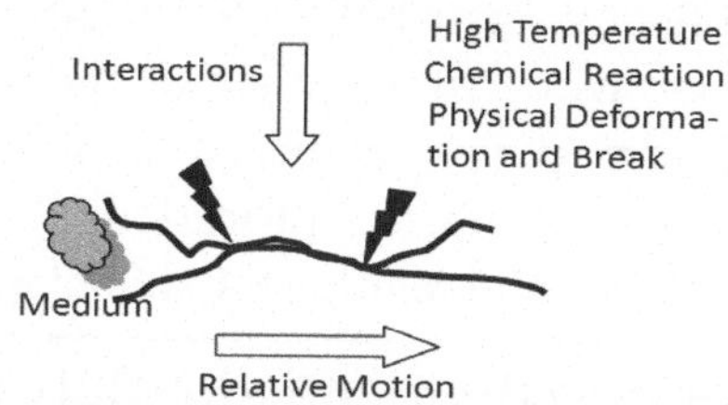

Fig. 4. The severity of work condition of a tribo-pair

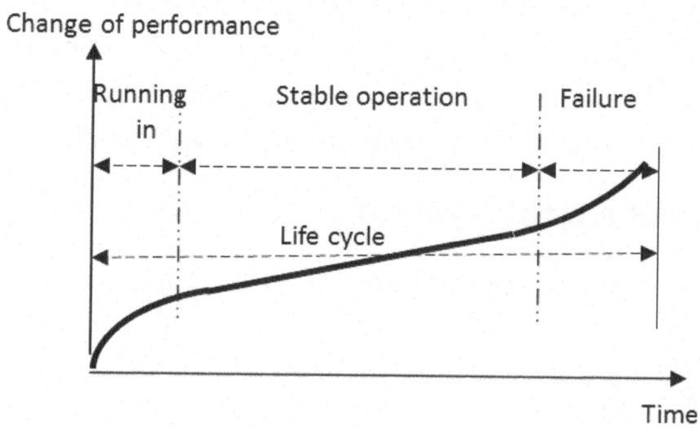

Fig. 5. Change of performance of a tribo-system in a life cylce

2.3 The third axiom in tribology: The results of tribological behaviors are the results of mutual action and strong coupling of many behaviors of other disciplines under a tribological condition consisted of interacting surfaces in relating motion

Obviously from the simplest tribo-system shown in Fig. 1 the force interaction, relative motion of the surfaces and the medium substance between the surfaces are a mechanical behavior. Transformation of the mechanical energy consumed in motion into heat energy and the diffusion of heat in surrounding, which makes a stable or unstable temperature field, are thermodynamic behaviors and heat transfer behaviors. The molecular interaction (including transferring) between surfaces and surfaces with medium is a physical or physical-chemical behavior. The reactions in ion level and atomic level are chemical behaviors. If there is any electric or magnetic field, which produces attractive or repelling interaction, or changes the arrangement of molecules in materials, or induces eddy current and heat, all of them are electric behaviors or magnetic behaviors, and so on. Most of them behavior simultaneously and inevitably change the structure of the system recoverably and irrecoverably. Then in turn they bring about the final results different from the results when they behavior singly. The results are different also from a simple addition of the results of individual behaviors. There is strong coupling between such behaviors. Tribology is the science and technology, which provides theories and techniques to describe and control the pattern of coupling between behaviors under tribological condition. No other discipline can meet such a need other than tribology, even though for any individual behavior there are principles from the discipline related which can predict its results. The difficulty for tribologists is that they have to know all the relative disciplines together with tribology simultaneously. Such a character of inter-disciplines and multi-disciplines requires a new methodology for tribology different from what for friction, wear or lubrication. In distinguish with the first axiom the third axiom focuses on the relationship of behaviors.

3. How to model a tribo-system and simulate its behaviors?

The structure is a description of intrinsic facts of a tribo-system while *the behaviors* are a description of change of the tribo-system. The structure of a system exists regardless whether there is an input. The behaviors must follow an input and can be derived from the input for a given structure in principle.

3.1 The structure of a tribo-system

Czichos (Czichos, 1978) described the structure of a tribo-system with a parameter set as

$$S=\{E,P,R\} \tag{2}$$

Where $E=\{e_1,e_2,\ldots,e_N\}$ is a sub-set showing that there are N tribo-elements in total in the system, $P=\{p_{e1},p_{e2},\ldots\ldots,p_{eN}\}$ is a sub-set describing the property of each element in the system and $R=\{r_{e1e2},r_{e1e3},\ldots r_{e1eN},r_{e2e3},\ldots\}$ is a sub-set collecting all relations between elements in the system.

Such description carries out a problem. Since the relative motion is the first important character of tribology and then the relative displacements between elements changing with the motion condition input, therefore the relative displacements between elements cannot be treated as an intrinsic fact. Several other examples can be listed as well.

To avoid the problem the author modified the description of a structure as (Xie, 2010)

$$S=\{E,P,H\} \tag{3}$$

Where $H=\{h_w,h_{e1},h_{e2},\ldots\ldots h_{eN}\}$ is a sub-set including the history of the system as a whole and of each element.

Each element e_i, $i = 1....N$, in the sub-set E represents a surface or a medium substance, for example a journal surface, a bearing surface, a cylinder bore surface, a piston skirt surface or the lubricant film between the surfaces.

Each element p_{ei}, $i = 1....N$, in the sub-set P describes the property of element e_i. In more detail the contents of property of each element can be divided into two groups, i.e. $p_{ei} = \{pg, pp\}_i$, $i = 1....N$, where pg is the geometric parameter group of property of the element, for example the diameter and width of the bearing surface in macro scale and the roughness in micro scale, and pp is the physical parameter group of property of the element. pp should be understood in a generalized sense including all physical, chemical and biological features besides geometric. It usually can be described by a group of physical, chemical and biological parameters, such as hardness, viscosity, acidity, activity etc. but there are some exceptions. Such features are affected by material composition, manufacturing process, service history, surrounding temperature, atmosphere, etc.

According to the second axiom in tribology, the property of elements and systems is time depended. The structure is a description of intrinsic facts but it is not invariable for a tribo- system. There are recoverable changes and irrecoverable changes in the structure due to the interaction and relative motion of surfaces. As described in formula (3), E is obvious invariable, the only variable things in S are P and H. Each element h_w, h_{ei}, $i = 1....N$, in the sub-set H are too complex to be described with parameters, usually they are a series of records in natural language. Using H rather than using a time parameter t here is because of that t notes only a time scale but what happened at t is more important for understanding the change of the structure. The elements of H do not act directly upon the structure but affect the values of parameters in pg and pp. For each effect some principles which govern the progress of effect can be found in related discipline. For example an elastic deformation of the surfaces is a recoverable change of pg which follows the change of interacting load on the surfaces governed by principles in the theory of elasticity, while a plastic deformation or wear of the surfaces is an irrecoverable change of pg, it is defined by what happened in the history and governed by principles in the theory of plasticity and tribology.

3.2 The behavior simulation of a tribo-system

Different from what used in references (Dai & Xue, 2003; Ge & Zhu, 2005), a state space method is applied here to simulate the behaviors. The state space method is a combination of general systems theory with engineering systems analysis and has wide application in dynamic system analysis, control engineering and many non-engineering analysis (Ogata, 1970, 1987). It takes a vector quantity called *state* as a scale to coordinate and evaluate the results of behaviors. When an input is applied upon a system, the system behaves from one state to another state and gives an output. For a time-invariable linear system a state equation (4) and an output equation (5) can be used to describe the results of behaviors:

$$\dot{X} = AX + BU \tag{4}$$

$$Y = CX + DU \tag{5}$$

where *X, U, Y* are the state vector, input vector and output vector of the system respectively. *A* and *B* are the system matrix, input matrix for equation (4) while *C* and *D* the output matrix for equation (5) respectively. All of them consist of the elements of structure of the system. *A, B, C* and *D* are constant for a time-invariable linear system.

In general the elements in a state vector are what concerned with the results of behaviors. As discussed previously, the first important behavior to be studied in tribo-systems is the relative motion. Any surface cannot exist independently and must be a part of a component of the machine system from which the tribo-system abstracted. The relative motion of surfaces is defined by the relative motion of components and where the surfaces reside on. Therefore for tribo-systems in the state vectors there are usually the parameters of displacements and time derivatives of displacements of components. For example the state of a single mass moving horizontally can be written as

$$X = [x, \dot{x}]^T$$

in which, *x* is the coordinate of the mass in *x* direction. When there are behaviors besides mechanics to be studied, parameters of related disciplines may emerge in the state vector, for example the electric current *i* in the coil of the electric magnet of an adaptive magnetic bearing.

For tribo-systems the situation will be complex. There are three possible ways to be selected.

1. If in behavior simulation the change of structure is not considered there will be a time- invariable linear system, i.e.

   $$S = const \tag{6}$$

   and simultaneously

   $$A = const, B = const, C = const, D = const \tag{6a}$$

2. If in behavior simulation the recoverable change of structure is considered only there will be a time-invariable non-linear system, i.e.

   $$S = S(X), A = A(X), B = B(X), C = C(X), D = D(X) \tag{7}$$

   Simultaneously there will be also

   $$P = \{pg, pp\} = P(X) = \{pg(X), pp(X)\} \tag{7a}$$

For any artifact system a requirement of behavior repeatability in an observation of short period is obviously necessary for reuse. Therefore the state *X* is repeatable The recoverable change of structure implies that the structure is a function of the state and independent to time. Whenever a similar input applied on a system with a similar state the system will have a similar state change and similar output. In other words the system behaves similarly. In an observation of short period the irrecoverable change due to very small in value in comparison with the recoverable change is negligible.

In an observation of short period, *pg* or *pp* changes with *X* due to many causes under the tribological condition, i.e. on or between the interacting surfaces in relative motion. Because *X* is repeatable and *pg* or *pp* is a function of *X* only, the patterns of change of *pg* or *pp* are relative simple. For each cause there will be some principles dealing with how the cause affects the change of parameters of *pg* or *pp*. These principles are in general relative to a discipline independent to tribology. Meanwhile a governing equation system, which may be a theoretical, experimental or statistical one, can be found in the discipline to describe the patterns of change of parameters of *pg* or *pp* under the tribological condition. As discussed before, for an elastic deformation the governing equation system can be found in the theory of elasticity and dynamics for a temperature distribution change the governing equation system can be found in the thermodynamics and heat transfer, for a change of viscosity of lubricants in terms of relative motion the governing equation system can be found in rheology, etc.

3. Irrecoverable changes are performed in entire processes of manufacturing, assembling, packaging, storing and transporting and will accumulate with service time and reach a comparable extent at last. It is history depended. In behavior simulation a time-variable non-linear system have to be treated, i.e.

$$S = S(X,t) \text{ or more accurate that } S = S(X,H)$$
$$\text{and } A = A(X,H), B = B(X,H), C = C(X,H), D = D(X,H) \quad (8)$$

Since formula (3) and that the elements of *H* do not act directly upon the structure but affect the values of parameters in *pg* and *pp*, the following formula can be established

$$P = \{pg, pp\} = P(X,H) = \{pg(X,H), pp(X,H)\} \quad (8a)$$

It shows that the property of a tribo-system changes with the system state and the history of the system.

In an observation of long period, *pg* or *pp* changes not only with *X* but also with *H*. There are many issues concerning with irrecoverable changes of the structure of machine systems. Wear, fatigue, plastic flow, creep, aging and corrosion are the most important irrecoverable changes. It is no doubt that wear is one of the issues studied in tribology. Fatigue takes place on the surfaces bringing forth a kind of fatigue wear. Plastic flow or creep carries out a permanent deformation of surfaces in macro scale which harms the motion guarantee function. Plastic flow in micro scale makes a change of elastic contact to plastic contact and will generate origins of surface fatigue after a number of cycles of repeat. Aging changes parameters in *pp* for solid surface materials and makes them inclining to failure. Aging spoils the performance of lubricants, increases corrosiveness and decreases the capability of lubrication. Corrosion of interacting surfaces in relative motion is also a kind of wear due to the chemical reaction of some compositions in lubricant or atmosphere with the materials of surfaces or due to the mechanical effect of break of air bubbles in the lubricant film. Obviously most issues concerning with irrecoverable changes are taken place in tribo- systems and studied in tribology.

According to the third axiom in tribology, the results of tribological behaviors are the results of mutual action and strong coupling of behaviors of many disciplines under a tribological condition constituted by interacting surfaces in relative motion. Because of that history or time is unrepeatable, the irrecoverable change is more complex in description than the recoverable change and almost no simple equation system can be found in any discipline. The different causes occurred singly or jointly at different moment in the history and their results were accumulated or coupled each other and result an irrecoverable change of the structure at a given time. In other words the structure is a carrier of mutual action and strong coupling of behaviors of many disciplines and gives a structure change in total at last as the results.

3.3 How to solve the state equations and output equations

In the behavior simulation of tribo-systems a time-variable non-linear system must be faced. The state equations and output equations will be as

$$\dot{X} = A(X,H)\cdot X + B(X,H)\cdot U(t) \tag{9}$$

$$Y = C(X,H)\cdot X + D(X,H)\cdot U(t) \tag{10}$$

Solving state equations is an initial value problem.

For a time-invariable linear system formula (4) can be integrated analytically when in formula (6a) *A* and *B* are constant. At any instant t_1 an input *U* (*t*) is applied to a system in an initial state X_1, then the system behaves to a state X_2 at an instant $t_2 = t_1 + \Delta t$ and give an output *Y* based on formula (5). It implies that similar initial state and similar input result similar change of state and similar output after a similar time interval Δt. After obtaining a new X_2 the new output Y_2 can be computed accordingly with formula (5) and constant matrixes *C* and *D*.

For time-variable non-linear systems the situation will be a little complex. Since matrix *A, B, C* or *D* is a function of the state and time (history related), integrating formula (9) and (10) analytically is in general impossible. The problem is similar with time-invariable non-linear systems when the matrixes *A, B, C* and *D* are functions of state *X* as shown in formula (7) and (7a) and will not be discussed separately in the following.

Numerical method is used for solving formula (9) and (10) for a time-variable non-linear system. The equations are discretized and integrated in a small time increment Δt step by step. When the Δt is small enough one can suppose that matrix *A, B, C* or *D* is independent to *X* and *t* and is constant in the time interval Δt, i.e. the system becomes a time-invariable linear system. In the integration, matrix *A, B, C* or *D* as a constant matrix and the values of their elements are calculated base on the results of last step with state X_1 and time t_1. After integration, there will be a change for both state and time, i.e. $X_2 = X_1 + \Delta X$ and $t_2 = t_1 + \Delta t$. Afterwards the elements in matrixes *A, B, C* and *D* should be recalculated according to X_2 and t_2 for the next step of integration if any of them is state and time related. Similar to the time-invariable and linear assumption made in the integration, a decoupling assumption is made also that the effect of any behavior on the values of elements in matrixes *A, B, C* and *D* can be calculated independently with the governing equations of related discipline or obtained from an experiment under a condition considering only the change of *X* and *t* ignoring other coupling effects. For example, in the simulation of the lubrication behavior in a piston skirt – cylinder bore pair, the lubricant film between the skirt surface and the bore surface undergoes a viscosity change when the piston changes its position along the bore due to a non-uniform distribution of temperature. The viscosity is a parameter in *pp* and its change may affect some elements in matrix *A, B, C* or *D*. A viscosity η_1 corresponding to temperature T_1 at y_1, the coordinate of shirt in the bore, is used for obtaining matrix *A, B, C* or *D*. After integrating over a Δt, y_1 becomes to y_2, T_1 becomes to T_2, η_1 becomes to η_2 and the matrix *A, B, C* or *D* will be recalculated with η_2 for the next integration. For recoverable change in an observation of short period the function $\eta(T)$ can be obtained by fitting experiment data and accurate enough. For irrecoverable change in an observation of long period a function in the form of $\eta(T,H)$ is necessary. In the history, many causes of very different kinds can affect the relation between η and *T* and make the lubricant aging. The causes before service include the kind of base oil, the technology and process of refining, the additive used etc. while the causes after service include the service temperature, service atmosphere, pollution condition and filtration efficiency in service etc. Knowledge of $\eta(T,H)$ have to be acquired for each application. Aging is a long period change and progresses very slowly. In numerical integration one can use a relative long time interval for such kinds of irrecoverable change other than recoverable change while a small time interval has to be used to keep the accuracy of simulation for recoverable change in time-invariable non-linear system.

There are many mathematic tools which make such an application available, for example, the Runge-Kutta Procedure (Chen, 1982). The difficulty in solving the problem is to find a balance between time consuming while a smaller time step (Δt) is used and low precision while a larger time step is used in integration (Xu, 2007).

4. Examples of modeling and simulation

4.1 Example1

The cylinder – piston – conrod – crank system of a single cylinder internal combustion engine is shown in Fig. 6. The system can be abstracted into a tribo-system with following tribo-pairs: piston skirt – cylinder bore, wrist pin – small end bearing, journal of crank – big end bearing and journal of crankshaft and main bearing, i.e. one prismatic pair and three revolute pairs in the system totally.

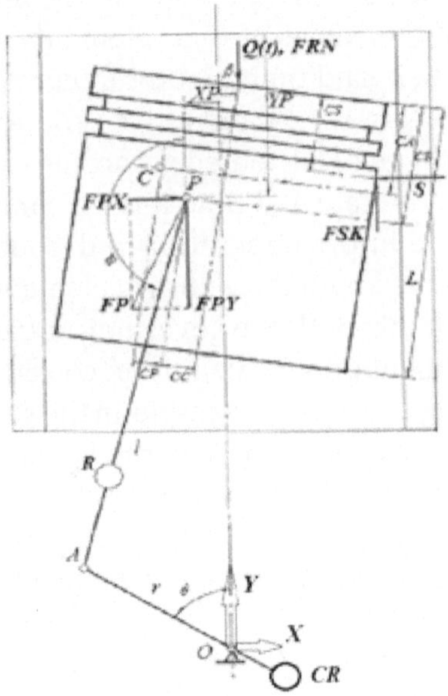

Fig. 6. Cylinder-piston-conrod-crank mechanism

A system block diagram for the piston skirt – cylinder bore pair is shown in Fig. 7 which gives a survey on the relationship between the skirt-bore tribo-pair and environment.

In the simulation the secondary motion of the piston and the change of inertia of the conrod are considered. The influence of the offset of the wrist pin can be considered as well and is taken as zero, positive or negative for comparison. The behaviors of lubricant film between the skirt surface and bore surface are treated according to the theory of thermal- hydrodynamic lubrication. The configuration of the skirt and bore can be given in simulation and a thermal distortion, force deformation and wear process can be calculated separately with the theory of heat transfer, theory of elasticity and regression of measured wear data. Their effects will couple between each other and with what of other behaviors in the iteration but a rigid skirt and a rigid bore without wear are supposed in the example. All of the eccentricities of journals in bearings and all of the elastic deformation of other components in the system are neglect. Their behaviors can be simulated separately and is decoupling with other behaviors in the global simulation.

The tribological behavior concerned in the system is a hydrodynamic lubrication behavior between the skirt surface and the bore surface. It results a thrust force *S* which balances the interacting load on the lubricant film and a shear resistant (friction) force *FSK* against the relative motion. In general there are two ways to treat the hydrodynamic behavior. One is looking the forces produced in the film like the inputs of the system. The other is taking the lubricant film as a structure element between surfaces. In this example the first way of treatment is applied.

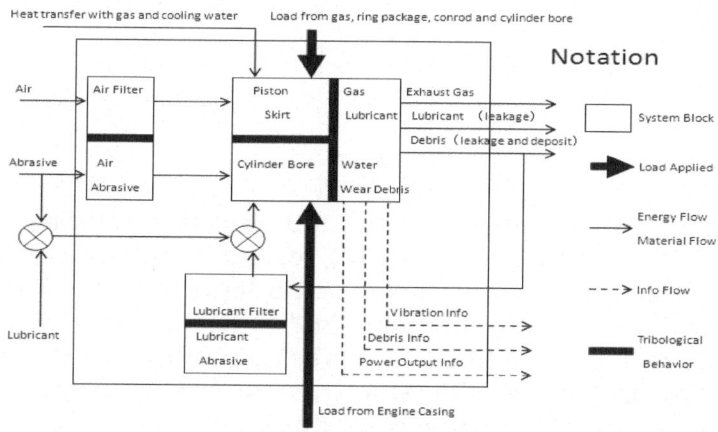

Fig. 7. The system block diagram of a cylinder bore-piston skirt

Piston ring package is considered separately also and the friction force between ring surfaces and cylinder bore is treated as an input (*FRN* in Fig. 6) applied on the piston.

Other inputs are the gas pressure $Q(t)$ on the top of the piston, the thrust force from the cylinder bore surface on the piston skirt surface *S*, the force on the wrist pin *FP*. All of them are balanced by a resistant torque moment (load) on the crankshaft.

The output can be selected according to what one wants to know in the simulation.

The state matrix equation of the system and the output matrix equation can be written as follows.

$$\begin{bmatrix} X_P \\ \dot{X}_P \\ \beta \\ \dot{\beta} \\ \theta \\ \dot{\theta} \end{bmatrix}' = \begin{bmatrix} 0 & 1 & 0 & 0 & 0 & 0 \\ 0 & 0 & 0 & 0 & 0 & A_{26} \\ 0 & 0 & 0 & 1 & 0 & 0 \\ 0 & 0 & 0 & 0 & 0 & A_{46} \\ 0 & 0 & 0 & 0 & 0 & 1 \\ 0 & 0 & 0 & 0 & 0 & A_{66} \end{bmatrix} \begin{bmatrix} X_P \\ \dot{X}_P \\ \beta \\ \dot{\beta} \\ \theta \\ \dot{\theta} \end{bmatrix} + \begin{bmatrix} 0 & 0 & 0 & 0 & 0 & 0 \\ 0 & 1 & 0 & 0 & 0 & 0 \\ 0 & 0 & 0 & 0 & 0 & 0 \\ 0 & 0 & 0 & 1 & 0 & 0 \\ 0 & 0 & 0 & 0 & 0 & 0 \\ 0 & 0 & 0 & 0 & 0 & 1 \end{bmatrix} \begin{bmatrix} 0 \\ U_2 \\ 0 \\ U_4 \\ 0 \\ U_6 \end{bmatrix} \quad (11)$$

When the hydrodynamic behavior between the skirt surface and bore surface is looked as an input applied on the system (via skirt surface), the resultant force of the hydrodynamic film pressure *S* and the resultant force of the resistant shear stress *FSK* will be the elements in U_2 and U_4. The hydrodynamic behavior depends on the gap geometry, the relative motion of surfaces and the lubricant viscosity. The gap geometry is changed with the wrist pin center displacement X_p and the piston tilting angle β in this case. The relative motion includes a tangential and normal component. The lubricant viscosity changes with temperature which has a distribution along the cylinder wall in *y* direction. The temperature distribution changes with the engine working condition but keeps unchanged in the example. All of them will be calculated in a separate program based on Reynolds Equation (Pinkus & Sternlicht, 1961).

$$\begin{bmatrix} \dot{\theta} \\ P_{LOSS} \\ X_P \\ \beta \\ F_{RHT} \\ F_{LFT} \end{bmatrix} = \begin{bmatrix} 0 & 0 & 0 & 0 & 0 & C_{16} \\ 0 & 0 & 0 & 0 & 0 & C_{26} \\ 0 & 0 & 0 & 0 & 0 & C_{36} \\ 0 & 0 & 0 & 0 & 0 & C_{46} \\ 0 & 0 & 0 & 0 & 0 & C_{56} \\ 0 & 0 & 0 & 0 & 0 & C_{66} \end{bmatrix} \begin{bmatrix} X_P \\ \dot{X}_P \\ \beta \\ \dot{\beta} \\ \theta \\ \dot{\theta} \end{bmatrix} \tag{12}$$

Fig. 8 gives the change of output in 7200 crankshaft rotating angle by formula (12) , where (a), (b), (c), (d), (e) and (f) are the deviation of crankshaft speed q˙ , change of friction power loss P_{LOSS} in the skirt-bore pair, displacement X_p of the wrist pin center in *X* direction, tilting angle β around the wrist pin center, thrust force F_{RHT} on the right side of the skirt and thrust force *FLFT* on the left side of the skirt from the hydrodynamic lubrication film respectively.

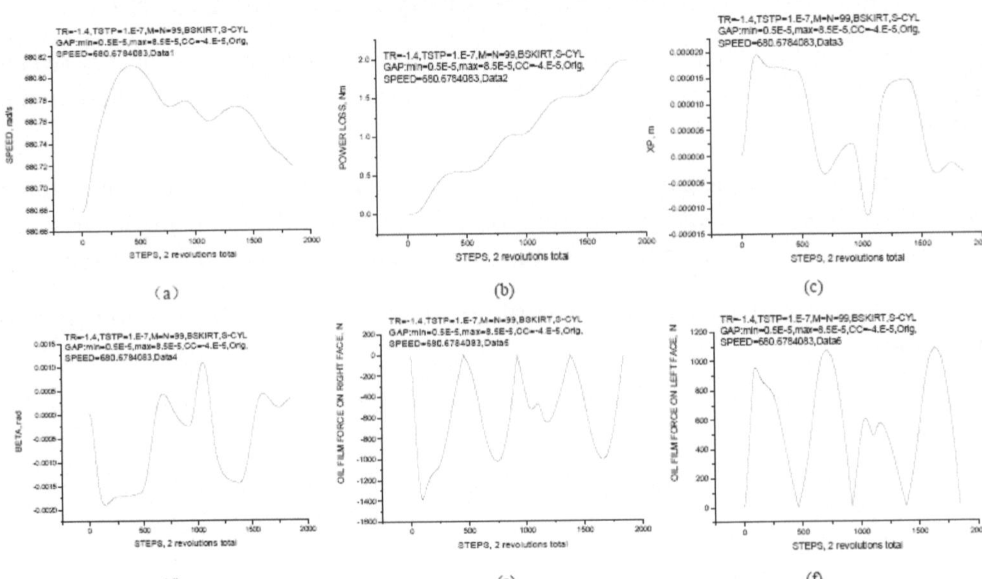

Fig. 8. Output of the system in 720° rotating angle of crankshaft

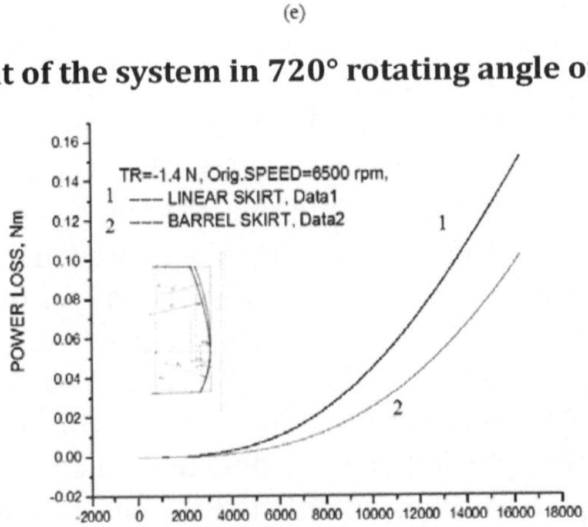

Fig. 9. Influence of skirt configuration on the friction power loss

Fig. 9 gives a comparison on the friction power loss when different skirt configurations are used. The geometry of skirt influences the gap between surfaces and then changes the hydrodynamic film pressure in values and distribution and changes the shear stress. It shows that the barrel skirt has a smaller friction loss.

Table 1 shows a comparison on the friction power loss between different values of wrist pin offset. The linear skirt is more sensitive to the offset than the barrel skirt is.

Table 1. Effects of wrist pin offset and skirt profile on piston skirt friction power loss

Computation number	Wrist Pin Offset	Friction Power Loss in 720o
Linear Skirt LS99-2-C-1	Left Offset CC=+4.E-5 m	2.32121 Nm
Linear Skirt LS99-2-C-0	Zero Offset CC=0.m	2.31236 Nm
Linear Skirt LS99-2-C-2	Right Offset CC=-4.E-5 m	2.30477 Nm
Barrel Skirt BS99-2-C-1	Left Offset CC=+4.E-5 m	1.97164 Nm
Barrel Skirt BS99-2-C-0	Zero Offset CC=0.m	1.97038 Nm
Barrel Skirt BS99-2-C-2	Right Offset CC=-4.E-5 m	1.96907 Nm

If the forces transmitted in the pairs *P, A* and *O* are interesting **there** will be another output matrix equation as

$$\begin{bmatrix} F_{PX} \\ F_{PY} \\ F_{AX} \\ F_{AY} \\ F_{OX} \\ F_{OY} \end{bmatrix} = \begin{bmatrix} 0 & 0 & 0 & 0 & 0 & C'_{16} \\ 0 & 0 & 0 & 0 & 0 & C'_{26} \\ 0 & 0 & 0 & 0 & 0 & C'_{36} \\ 0 & 0 & 0 & 0 & 0 & C'_{46} \\ 0 & 0 & 0 & 0 & 0 & C'_{56} \\ 0 & 0 & 0 & 0 & 0 & C'_{66} \end{bmatrix} \begin{bmatrix} X_P \\ \dot{X}_P \\ \beta \\ \dot{\beta} \\ \theta \\ \dot{\theta} \end{bmatrix} \tag{13}$$

Where F_{PX}, F_{PY}, F_{AX}, F_{AY}, F_{OX} and F_{OY} are the force components transmitted in the small end bearing of conrod, in the big end bearing of conrod and in the main bearings (in total) of crankshaft respectively of the IC engine in discussion. The change of such forces in 7200 crankshaft rotating angle is shown in Fig. 10.

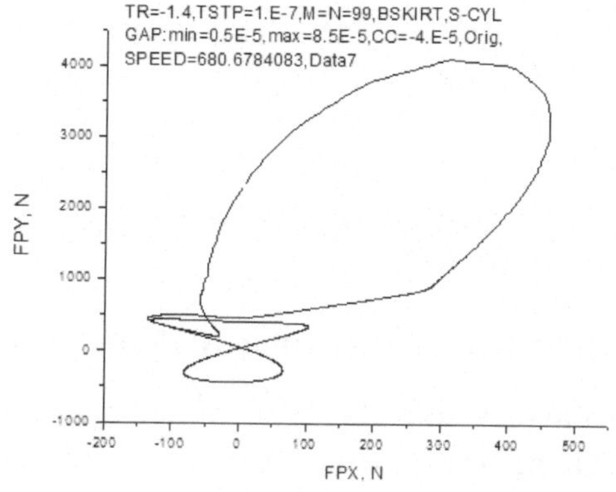

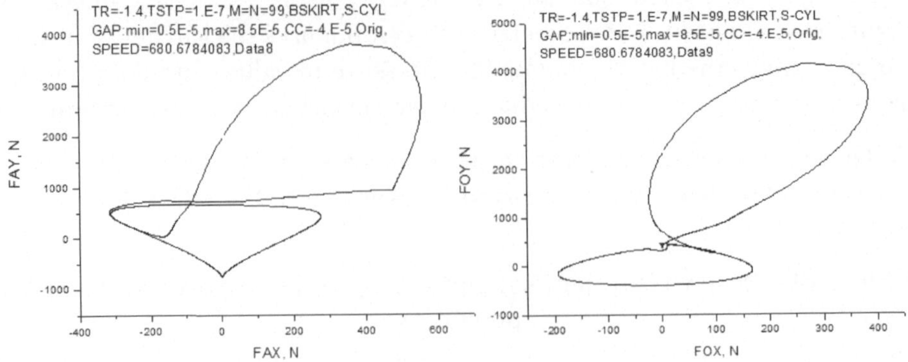

Fig. 10. Forces transmitted in the bearing of an IC engine. (a) Small end bearing of conrod. (b) Big end bearing of conrod. (c) Main bearing of crankshaft

The derivation of elements *A*16 to *A*66 , *C*16 to *C*66 , *U*2 to *U*6 and *C*1¢ 6 to *C*6¢ 6 in formulas (11), (12) and (13) can be found **in** Appendix.

4.2 Example 2

As shown in Fig. 11 there is a rotor-bearing system of a 300MW turbo-generator set consisted of the rotor of a high pressure cylinder (HP), an intermediate pressure cylinder (IP), a low pressure cylinder (LP), a generator, an exciter and eight hydrodynamic bearings (1# - 8#) on pedestals. A simplification is made in the example that the eight bearings are all plane bearings to reduce the amount of computation. The rotor in total is an elastic component supported by the bearings and can vibrate laterally. Obviously it is a statically indeterminate problem. The load on each bearing is determined by the relationship between the elevations of journal centers which are controlled by a camber curve checked at last in installation. There are many reasons which can change the relationship, for example the journals may float with different eccentricity *e* (Fig. 17) on the hydrodynamic film and the pedestals may change their heights due to the changes of working temperatures during different turbine output and then change the bearing loads under a statically indeterminate condition.

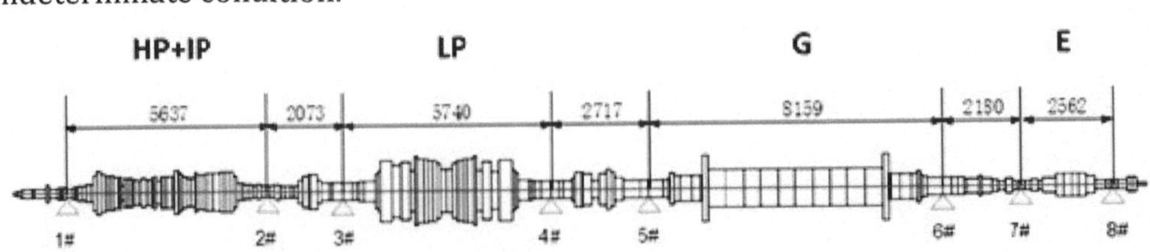

Fig. 11. The rotor-bearing system of a 300MW turbo-generator set

The tribological behaviors considered in the example are the hydrodynamic behaviors in bearings. There are three points to be considered.

1. For a hydrodynamic bearing the rotating journal is floating on the hydrodynamic film and there is an eccentricity between the journal center and the bearing center. During installation the journal is dropped upon the bottom surface of the bearing bore. The eccentricity changes with the load on the bearing.

2. The change of the load or eccentricity changes the geometric property and physical property (*pg, pp* – see section 3.1) of the film when taking it as a structure element between surfaces.

3. If the change of *pp* approaching to some extent the film will excite a kind of severe vibration of the system called oil whirl or oil resonance (Hori, 2002) and may result a catastrophic damage of the turbo-generator set.

In general it is recognized that the oil whirl begins at the threshold of instability of the rotor- bearing system and usually has a frequency half the rotor speed. It is a tribological behavior induced vibration and indicates a decrease or loss of motion guarantee function.

The treatment of the hydrodynamic behavior in the film looks like inserting a structure element between surfaces and is different from what has done in example 1 (see section 4.1). In this case the film is a linearized spring-damper in time interval Δt and its *pp* can be represented by four constant stiffness coefficients k_{xx}, k_{xy}, k_{yx}, k_{yy} and four constant damping coefficients d_{xx}, d_{xy}, d_{yx}, d_{yy}. It implies an assumption of using *pp=const* instead of *pp=pp(X)* during integration in time interval Δt. The eight coefficients can be calculated before integration with a separate program for a given film configuration (bearing bore geometry, eccentricity and attitude angle) and relative motion (tangential and normal) between journal surface and bearing surface (Pinkus & Sternlicht, 1961). The eight spring-dampers together with the distributed mass-stiffness-damping of the rotor defines the threshold of instability. To constitute the state space equation the rotor is discretized into 194 sections (Fig. 12) according to a concentrated mass treatment which can be found in rotor-bearing system dynamics (Glienicke, 1972) and its detail is omitted in the example.

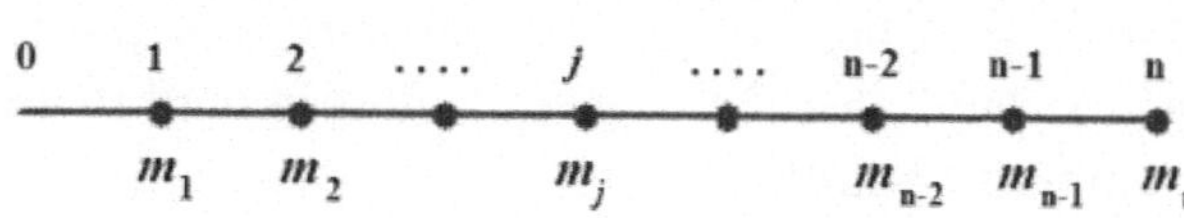

Fig. 12. A discretized model of the rotor

Each section (Fig. 13) consists of a field of length *l* with stiffness but without mass and a station with mass, inertia moment but without length.

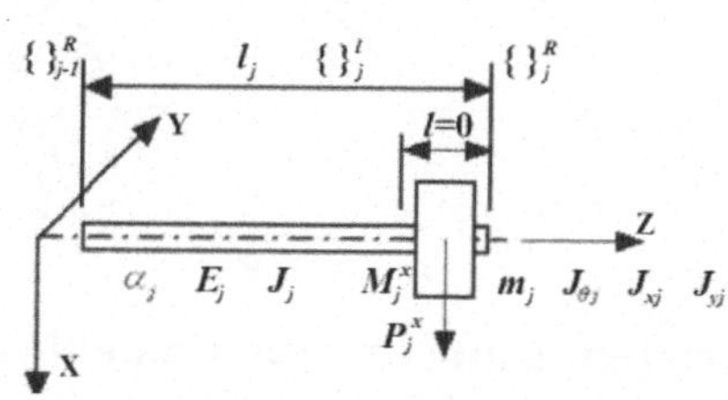

Fig. 13. A section of the rotor with a field and a station

The forces and moments applied on both side of a field and the related deformations are shown in Fig. 14 and Fig. 15.

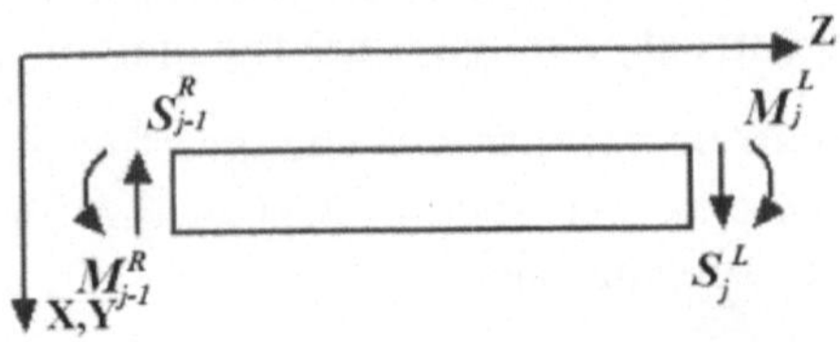

Fig. 14. The forces and moments on a field

The angular displacements and inertia monents of a station are described in Fig. 16. All of the inputs (forces and moments) apply only on the station. They make a balance between the forces and moments appling by the fields (right and left) and the inertia forces and moments. If there is a bearing attached to a section then the station is looked like supported by a linearized spring-damper with four direct stiffness and damping coefficients k_{xx}, k_{yy}, d_{xx}, d_{yy} and four cross stiffness and damping coefficients k_{xy}, k_{yx}, d_{xy}, d_{yx} as shown in Fig. 17. The cross stiffness and damping coefficients show an important difference between the

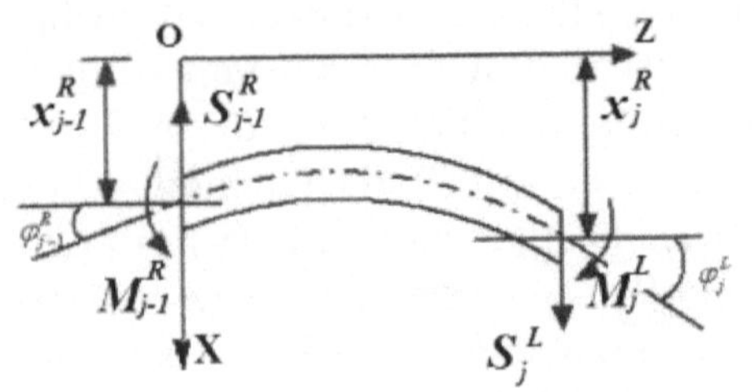

Fig. 15. The lateral deformation of a field

hydrodynamic film and isotropic solid material. The hydrodynamic film then plays the role of a component of the system. It should be emphsized that the height of the journal center is determined by the sum of the height of bearing center controlled by pedestal and the project of ecentricity *e* of the journal center on ordinate axis while the load on each bearing is determined by the journal height under a static inderminate condition.

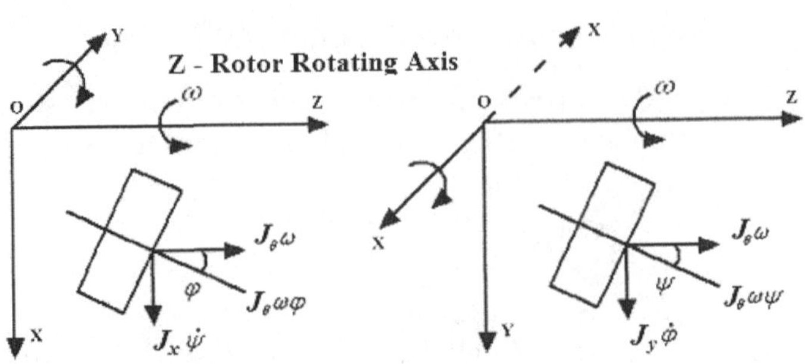

Fig. 16. Angular displacements and inertia moments of a station in X-Z and Y-Z plane

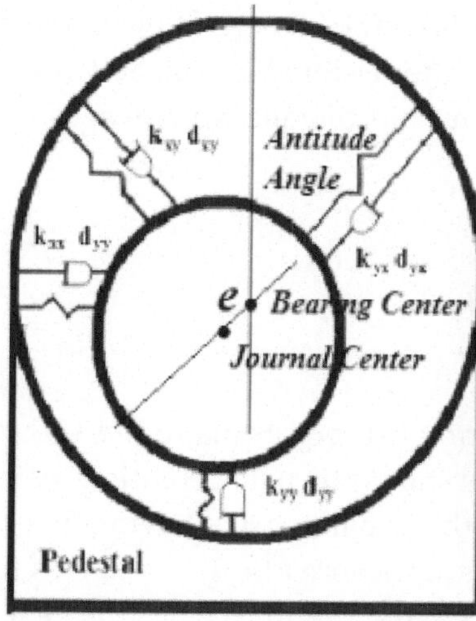

Fig. 17. A linearized model of the hydrodynamic film

Another form of formula (4) for one section, for example for section j, can be written as

$$\begin{bmatrix} m\ddot{x} \\ m\ddot{y} \\ J_x\ddot{\varphi} \\ J_y\ddot{\psi} \end{bmatrix}_j + \begin{bmatrix} d_{xx} & d_{xy} & 0 & 0 \\ d_{yx} & d_{yy} & 0 & 0 \\ 0 & 0 & 0 & -J_\theta\omega \\ 0 & 0 & J_\theta\omega & 0 \end{bmatrix}_j \begin{bmatrix} \dot{x} \\ \dot{y} \\ \dot{\varphi} \\ \dot{\psi} \end{bmatrix}_j + \begin{bmatrix} k_{xx} & k_{xy} & 0 & 0 \\ k_{yx} & k_{yy} & 0 & 0 \\ 0 & 0 & 0 & 0 \\ 0 & 0 & 0 & 0 \end{bmatrix}_j \begin{bmatrix} x \\ y \\ \varphi \\ \psi \end{bmatrix}_j$$

$$+ \begin{bmatrix} -\frac{12EJ}{l^3} & 0 & \frac{6EJ}{l^2} & 0 \\ 0 & -\frac{12EJ}{l^3} & 0 & \frac{6EJ}{l^2} \\ -\frac{6EJ}{l^2} & 0 & \frac{2EJ}{l} & 0 \\ 0 & -\frac{6EJ}{l^2} & 0 & \frac{2EJ}{l} \end{bmatrix}_{j+1} \begin{bmatrix} x \\ y \\ \varphi \\ \psi \end{bmatrix}_{j+1} + \begin{bmatrix} \frac{12EJ}{l^3} & 0 & \frac{6EJ}{l^2} & 0 \\ 0 & \frac{12EJ}{l^3} & 0 & \frac{6EJ}{l^2} \\ \frac{6EJ}{l^2} & 0 & \frac{4EJ}{l} & 0 \\ 0 & \frac{6EJ}{l^2} & 0 & \frac{4EJ}{l} \end{bmatrix}_{j+1} \begin{bmatrix} x \\ y \\ \varphi \\ \psi \end{bmatrix}_j \tag{14}$$

$$+ \begin{bmatrix} \frac{12EJ}{l^3} & 0 & -\frac{6EJ}{l^2} & 0 \\ 0 & \frac{12EJ}{l^3} & 0 & -\frac{6EJ}{l^2} \\ -\frac{6EJ}{l^2} & 0 & \frac{4EJ}{l} & 0 \\ 0 & -\frac{6EJ}{l^2} & 0 & \frac{4EJ}{l} \end{bmatrix}_j \begin{bmatrix} x \\ y \\ \varphi \\ \psi \end{bmatrix}_j + \begin{bmatrix} -\frac{12EJ}{l^3} & 0 & -\frac{6EJ}{l^2} & 0 \\ 0 & -\frac{12EJ}{l^3} & 0 & -\frac{6EJ}{l^2} \\ \frac{6EJ}{l^2} & 0 & \frac{2EJ}{l} & 0 \\ 0 & \frac{6EJ}{l^2} & 0 & \frac{2EJ}{l} \end{bmatrix}_j \begin{bmatrix} x \\ y \\ \varphi \\ \psi \end{bmatrix}_{j-1} = \begin{bmatrix} P^x \\ P^y \\ M_k \\ N_k \end{bmatrix}_j$$

where E is the Young's module of the rotor material and J is the area moment inertia, other parameters can be found in Fig. 12 to 17. The state space equation for the rotor bearing system can be obtained by assembling formula (14) for j=1 to j=n with free boundary condition at the two terminal ends. The assembled result formula will not be presented in the example.

A question arises that how the change of elevation distribution influences the threshold of instability of the system? It can be transformed into an eigenvalue problem. In general the solutions of equation are as follows

$$x_i = x_{0i}e^{v_i t} = x_{0i}e^{-a_i t}e^{jb_i t},$$
$$y_i = y_{0i}e^{-a_i t}e^{jb_{ij} t},$$
$$\varphi_i = \varphi_{0i}e^{-a_i t}e^{jb_i t},$$
$$\psi_i = \psi_{0i}e^{-a_i t}e^{jb_i t}, i = 1 \sim N. \qquad (15)$$

N is defined by the practical requirement and the computational facility. Only some interesting solutions should be paid attention to, for example the solution *i* in this discussion to explain the tribological behavior. In formula (15) the item *e* , the virtual part of the solution where *j* = , gives b_i which is the frequency of vibration (oil whirl). Meanwhile the item $e^{-a}it$, the real part of the solution, gives a_i which is the system damping of the system and predicts a speed of changing the amplitude of vibration concerning with the solution. When a_i takes a negative value the amplitude of vibration will increase with time and the solution is then instable. Only when it is positive the solution can be stable. Therefore $a_i = 0$ is a condition of threshold of instability of the system.

Back to formula (14), if the input vector [*px, py*, M_x, M_k, N_k]*T* is constant, most structure parameters are constant in a short period of observation except the eight stiffness and damping coefficients which are defined by the relative motion (the rotating speed of the rotor) and the load on the bearing. Under a given elevation distribution the change of system damping can be expressed in another form, the logarithmic decrement

$$\Delta = 2\pi a_i / b_i$$

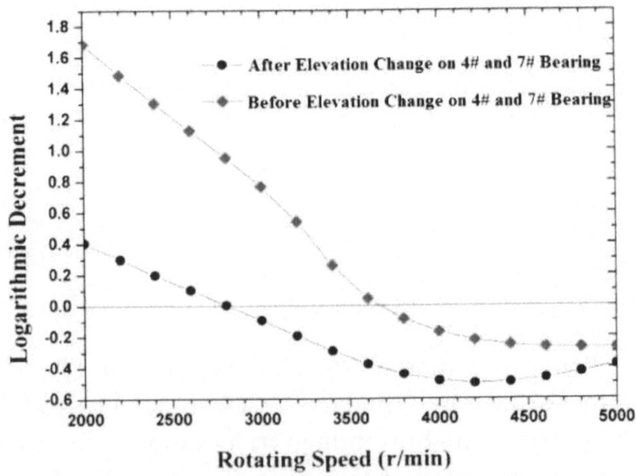

Fig. 18. Logarithmic decrement versus rotating speed for two different elevation distributions

Figure 18 gives two logarithmic decrement curves versus rotor rotating speed. The intersection point of each curve and abscissa ($\Delta = 0$) gives a margin of threshold of instability with related elevation distribution. The turbo-generator set in power plant must work under a speed of 3000 rpm. In Fig. 18 one can find that at a speed of 3000 rpm, before and after the change of elevation of 4# bearing (decreasing a value of 0.15 mm) and 7# bearing (increasing a value of 0.7 mm) the logarithmic decrement changes from 0.95 to - 0.05. It implies that the change makes the system becoming not stable. Some turbo-generator set works normally in full output but during low output in middle night a half frequency vibration component emerges. Elevation distribution change might be an important cause of such phenomena. Many efforts have been given to understand it (Li, 2001).

5. Conclusion

The problems with tribology are problems of systems science and systems engineering. In a sense, without system there would be no tribology. A machine system is consisted of a component system and a tribo-system from the view point of motion. The tribo-system is consisted of tribo-elements and some supporting auxiliary sub-systems abstracted from a machine system for studying behaviors on or between the interacting surfaces in relative motion, results of the behaviors and technology related to. The tribo-system together with the component system plays a motion guarantee function which keeps each part of the machine system with a definite motion. Tribology science and technology is very important in obtaining the best way (theory and application) to complete the motion guarantee function of tribo-systems.

Tribological behaviors are system dependent. The property of tribo-elements and then the systems containing tribo-elements are time dependent. The results of tribological behaviors are the results of mutual action and strong coupling of many behaviors of other disciplines under a tribological condition consisted of interacting surfaces in relating motion.

A state space method which is a combination of general systems theory with engineering systems analysis can be successfully applied to simulate the behaviors. Two examples are given to show how the system structure can be connected with the system behaviors via the state space method. With the state space method the structure is a carrier in realizing the mutual action and coupling. The structure can have a recoverable change and an irrecoverable change while the behaviors can be repeatable and unrepeatable in the simulation.

Appendix: Derivation of elements in the state space and output equations in example 1

In this example, the study will focus mainly on the skirt – bore tribo-pair of a cylinder – piston – conrod – crank system of an internal combustion engine.

As shown in Fig. 6 and in the following formulas, symbols Q – gas pressure on the top of piston, F – force or friction force, S –thrust force in total on piston skirt, T - torque moment load on the crankshaft, t – time, m – mass of a component, I – inertial moment of a component, P – center of small end pair of conrod, A – center of big end pair of conrod, O – center of crankshaft pair on casing, C – center of mass of piston assembly, R – center of mass of conrod, CR – center of mass of crankshaft, X,Y – coordinate directions, *PIS* – piston, *PIN* – wrist pin, *SK* – skirt, *RN* – piston ring package, R – conrod respectively and l — length of conrod, r — length of crank, jl – distance from A to R, hr – distance from O to CR.

Suppose that the influence of secondary motion of piston on the motion and equilibrium of conrod and crankshaft can be neglected. The following formulas yield the geometry and motion relationship between the conrod and crankshaft:

where, $C16 = 1$ and $C26$,$C36$,$C46$,$C56$,$C66$ concern the solution of Reynolds equation which governs the hydrodynamic lubrication behaviors between skirt and bore surfaces and cannot be presented explicitly. They will be computed numerically with a separate procedure before every integrating step from the value of elements in state vector obtained in last integrating.

If the forces transmitting in the pairs P, A and O are interesting the forces can be obtained with an equilibrium condition analysis for the piston on P, for the conrod on A and for the crankshaft on O. Replacing the first and second derivatives of displacements in formulas (22A) to (30A) and reordering yields

$$
\begin{aligned}
F_{PX} &= -\dot{\theta}^2 m_P (W4 + W4' \cdot W5) - m_P (W4' \cdot W5' - FY) - S \\
F_{PY} &= \dot{\theta}^2 m_P (W2 + W3 \cdot W5) + m_P (W3 \cdot W5' + g) + FY' \\
F_{AX} &= \dot{\theta}^2 \left[-m_P (W4 + W4' \cdot W5) + m_R r (1 - j)(\sin\theta - W5 \cos\theta) \right] \\
&\quad + \left[-m_P (W4' \cdot W5' - FY) - m_R r (1 - j) W5' \cos\theta - S \right] \\
F_{AY} &= \dot{\theta}^2 \left[m_P (W2 + W3 \cdot W5) + m_R (W2' + W3 \cdot W5) \right] \\
&\quad + \left[m_P (W3 \cdot W5' + g) + m_R (W3' \cdot W5' + g) + FY' \right] \\
F_{OX} &= \dot{\theta}^2 (W6 + W5 \cdot W6') + W5' \cdot W6' + m_P \cdot FY - S \\
F_{OY} &= \dot{\theta}^2 (W7 + W5 \cdot W7') + W5' \cdot W7' + FY' + (m_P + m_R + m_C) \cdot g
\end{aligned}
$$

Then the output matrix equation becomes

$$
\begin{bmatrix} F_{PX} \\ F_{PY} \\ F_{AX} \\ F_{AY} \\ F_{OX} \\ F_{OY} \end{bmatrix}
=
\begin{bmatrix}
0 & 0 & 0 & 0 & 0 & C'_{16} \\
0 & 0 & 0 & 0 & 0 & C'_{26} \\
0 & 0 & 0 & 0 & 0 & C'_{36} \\
0 & 0 & 0 & 0 & 0 & C'_{46} \\
0 & 0 & 0 & 0 & 0 & C'_{56} \\
0 & 0 & 0 & 0 & 0 & C'_{66}
\end{bmatrix}
\begin{bmatrix} X_P \\ \dot{X}_P \\ \beta \\ \dot{\beta} \\ \theta \\ \dot{\theta} \end{bmatrix}
$$

where

$$C'_{16} = -\dot{\theta} m_P (W4 + W4' \cdot W5) - \left[m_P (W4' \cdot W5' - FY) - S \right] / \dot{\theta}$$
$$C'_{26} = \dot{\theta} m_P (W2 + W3 \cdot W5) + \left[m_P (W3 \cdot W5' + g) + FY' \right] / \dot{\theta}$$

$$C'_{36} = \dot{\theta} \left[-m_P (W4 + W4' \cdot W5) + m_R r (1 - j)(\sin\theta - W5 \cdot \cos\theta) \right] + \left[-m_P (W4' \cdot W5' - FY) - m_R r (1 - j) \cdot W5' \cdot \cos\theta - S \right] / \dot{\theta}$$
$$C'_{46} = \dot{\theta} \left[m_P (W2 + W3 \cdot W5) + m_R (W2' + W3' \cdot W5) \right] + \left[m_P (W3 \cdot W5' + g) + m_R (W3' \cdot W5' + g) + FY' \right] / \dot{\theta}$$

$$C'_{56} = \dot{\theta} (W6 + W5 \cdot W6') + \left[W5' \cdot W6' + m_P \cdot FY - S \right] / \dot{\theta}$$
$$C'_{66} = \dot{\theta} (W7 + W5 \cdot W7') + \left[W5' \cdot W7' + FY' + g(m_P + m_R + m_C) \right] / \dot{\theta}$$

6. References

1. Chen, P. (1982). *State Space Methods and Application*. Publishing House of Electronics Industry, Beijing, China (In Chinese)
2. Czichos H. (1978). *Tribology: A Systems Approach to the Science and Technology of Friction, Lubrication and Wear*, ISBN 978-0444416766, Elsevier
3. Czichos H. The Principle of System Analysis and their Application to Tribology. *ASLE Trans*, Vol. 17, No. 4, (1974), pp. 300-306, ISSN 0569-8197
4. Dai, Z.; Xue, Q. Exploration Systematical Analysis and Quantitative Modeling of Tribo- System Based on Entropy Concept. *Journal of Nanjing University of Aeronautics & Astronautics*, Vol. 25, No.6, (2003), pp.585 ~ 589 ISSN 1005-2615 (In Chinese)
5. Dowson, D. (1979). *History of Tribology*. Wiley, ISBN 978-1860580703, London
6. Fleischer G. *Systembetrachtungen zur Tribologie*. Wiss. Z. TH Magdeburg, Vol. 14, (1970), pp.415-420
7. Ge, S.; Zhu, H. (2005). *Fractal in Tribology*. China Machine Press, ISBN 7-111-16014-2, Beijing, China (In Chinese)
8. Glienicke, J. (1972). Theoretische und experimentelle Ermittlung der Systemdaempfung gleitgelagerter Rotoren und ihre Erhoehung durch eine aeussere Lagerdaempfung. Fortschritt Berichte der VDI Zeitschriften, Reihe 11, Nr. 13, VDI-Verlag GmbH, Duesseldorf
9. Her Majesty's Stationery Office. (1966). Lubrication (Tribology) Education and Research: A Report on the Present Position and Industry's Needs. London
10. Hori, Y. (2005). *Hydrodynamic Lubrication*. Springer, ISBN 978-4431278986
11. Li, J. Analysis and Calculation of Influence of Steam Turbo-Generator Bearing Elevation Variation on Load. *North China Electric Power*, No. 11, (2001), pp.5 ~ 7, ISSN 1007- 2691 (In Chinese)
12. Ogata, K. (1970). *Modern Control Engineering*. Prentice-Hall, ISBN 9780135902325, New Jersey, USA

13. Ogata, K. (1987). *Discrete-time Control Systems*. Prentice-Hall, ISBN 9780132161022, New Jersey, USA
14. Pinkus, O.; Sternlicht, B. (1961). *Theory of Hydrodynamic Lubrication*. McGraw-Hill, New York, USA
15. Salomon G. Application of Systems Thinking to Tribology. *ASLE Trans*, Vol.17, No.4, (1974), pp.295-299, ISSN 0569-8197
16. Suh, N. (1990). *The Principle of Design*, Oxford University Press, ISBN 978-0195043457, USA The Panel Steering Committee for the Mechanical Engineering and Applied Mechanics
17. Division of the NSF. Research Needs in Mechanical Systems-Report of the Select Panel on Research Goals and Priorities in Mechanical Systems. *Trans ASME, Journal of Tribology*, Vol. 1, (1984), pp. 2~25, ISSN 0022-2305
18. Xie, Y. On the Systems Engineering of Tribo-Systems. *Chinese Journal of Mechanical Engineering* (English Edition), No 2, (1996), pp. 89-99, ISSN 1000-9345
19. Xie, Y. On the System Theory and Modeling of Tribo-Systems. *Tribology*, Vol.30, No.1, (2010), pp.1-8, ISSN 1004-0595 (In Chinese)
20. Xie, Y. *On the Tribological Database*. Lubrication Engineering, Vol.5, (1986), pp. 1-7, ISSN 0254-0150 (In Chinese)
21. Xie, Y. Three Axioms in Tribology. *Tribology*, Vol.21, No.3, (2001), pp.161-166, ISSN 1004- 0595 (In Chinese)
22. Xie, Y.; Zhang, S. (Eds.). (2009). *Status and Developing Strategy Investigation on Tribology Science and Engineering Application: A Consulting Report of the Chinese Academy of Engineering (CAE)*. Higher Education Press, ISBN 978-7-04-026378-7, Beijing, China (In Chinese)
23. Xu, S. (2007). *Digital Analysis and Methods*. China Machine Press, ISBN 978-7-111-20668-2, Beijing, China (In Chinese)

2

Tribological Aspects of Rolling Bearing Failures

1. Introduction

Rolling (element) bearings are referred to as *anti-friction* bearings due to the low friction and hence only slight energy loss they cause in service, especially compared to sliding or *friction* bearings. The minor wear occurring in proper operation superficially seems to suggest the question how rolling contact tribology should be of relevance to bearing failures. Satisfactorily proven throughout the 20th century primarily on small highly loaded ball bearings, the life prediction is actually based on material fatigue theories. Nonetheless, resulting subsurface spalling is usually called fatigue wear and therefore included in the discussion below. The influence of friction on the damage of rolling bearings, at first, is strikingly reflected, for instance, in foreign particle abrasion and smearing adhesion wear under improper running or lubrication conditions. On far less affected, visually intact raceways, however, temporary frictional forces can also initiate failure for common overall friction coefficients below 0.1. Larger size roller bearings with extended line contacts operating typically at low to moderate Hertzian pressure, generally speaking, are most susceptible to this surface loading. As large roller bearings are increasingly applied in the 21st century, e.g. in industrial gears, an attempt is made in the following to incorporate the rolling-sliding nature of the tribological contact into an extended bearing life model. By holding the established assumption that the stage of crack initiation still dominates the total lifetime, the consideration of the proposed competing normal stress hypothesis is deemed appropriate.

The present chapter opens with a general introduction of the subsurface and (near-) surface failure mode of rolling bearings. Due to its particular importance to the identification of the damage mechanisms, the measuring procedure and the evaluation method of the material response analysis, which is based on an X-ray diffraction residual stress determination, are described in detail. In section

4, a metal physics model of classical subsurface rolling contact fatigue is outlined. Recent experimental findings are reported that support this mechanistic approach. The accelerating effect of absorbed hydrogen on rolling contact fatigue is also in agreement with the new model and verified by applying tools of material response analysis. It uncovers a remarkable impact of serious high-frequency electric current passage through bearings in operation, previously unnoticed in the literature. Section 5 provides an overview of state-of-the-art research on mechanical and chemical damage mechanisms by tribological stressing in rolling-sliding contact. The combined action of mixed friction and corrosion in the complex loading regime is demonstrated. Mechanical vibrations in bearing service, e.g. from adjacent machines, increase sliding in the contact area. Typical depth distributions of residual stress and X-ray diffraction peak width, which indicate microplastic deformation and (low-cycle) fatigue, are reproduced on a special rolling bearing test rig. The effect of vibrationally increased sliding friction on near-surface mechanical loading is described by a tribological contact model. Temperature rise and chemical lubricant aging are observed as well. Gray staining is interpreted as corrosion rolling contact fatigue. Material weakening by operational surface embrittlement is proven. Three mechanisms of *tribocracking* on raceways are discussed: tribochemical dissolution of nonmetallic inclusions and crack initiation by either frictional tensile stresses or shear stresses. Deep branching crack growth is driven by another variant of corrosion fatigue in rolling contact.

2. Failure modes of rolling bearings

Bearings in operation, in simple terms, experience pure rolling in elastohydrodynamic lubrication (EHL) or superimposed surface loading. With respect to the differing initiation sites of fatigue damage, a distinction is made between the classical subsurface and the (near-) surface failure mode (Muro & Tsushima, 1970). In the following simplified analysis, the evaluation of material stressing due to rolling contact (RC) loading is based on an extended static yield criterion by means of the distribution of the equivalent stress. The more complex surface failure mode, which predominates in today's engineering practice also due to the improved steelmaking processes and the tendency to use energy saving lower viscosity lubricants, comprises several damage mechanisms. Raceway indentations or boundary lubrication, for instance, respectively add edge stresses on Hertzian micro contacts and frictional sliding loading to the ideal elastohydrodynamic operating conditions.

2.1 Subsurface failure mode

The Hertz theory of elastic contact deformation between two solid bodies, specifically a rolling element and a ring of a bearing, is used to analyze the spatial stress state (Johnson, 1985). Initial yielding and generation of compressive residual stresses (CRS) is governed by the distortion energy hypothesis. In a normalized representation, Figure 1 plots the distance distributions of the three principal normal stresses σ_x, σ_y and σ_z and the resulting v. Mises equivalent stress $\sigma^{v.Mises}$ below the center line of a purely radially loaded frictionless elastic line contact, where the maximum normal stress, i.e. the Hertzian pressure p_0, occurs. In the coordinate trihedral, *x*, *y* and *z* respectively indicate the axial (lateral), tangential (overrolling) and radial (depth) direction. The v. Mises equivalent stress reaches its maximum $s^{max} = 0.56\,'\,p$ in a distance $z^{v.Mises} = 0.71\,'\,a$ from the surface, which is valid in good approximation for roller and ball bearings (Hooke, 2003). The load is expressed as p_0 and *a* stands for the semiminor axis of the contact ellipse.

As illustrated in Figure 1 for a through hardened grade ($R_{p0.2}$=const.), the v. Mises equivalent stress can locally exceed the yield strength $R_{p0.2}$ of the steel that ranges between 1400 and 1800 MPa, depending, e.g., on the heat treatment and the degree of deformation of the material (segregations) or the operating temperature. From Hertzian pressures p_0 **of about 2500 to 3000** MPa, therefore, compressive residual stresses are built up. An example of a measured distance profile is shown in Figure 2a. By identifying the maximum position of the v. Mises and compressive residual stress, the Hertzian pressure is estimated to be 3500 MPa.

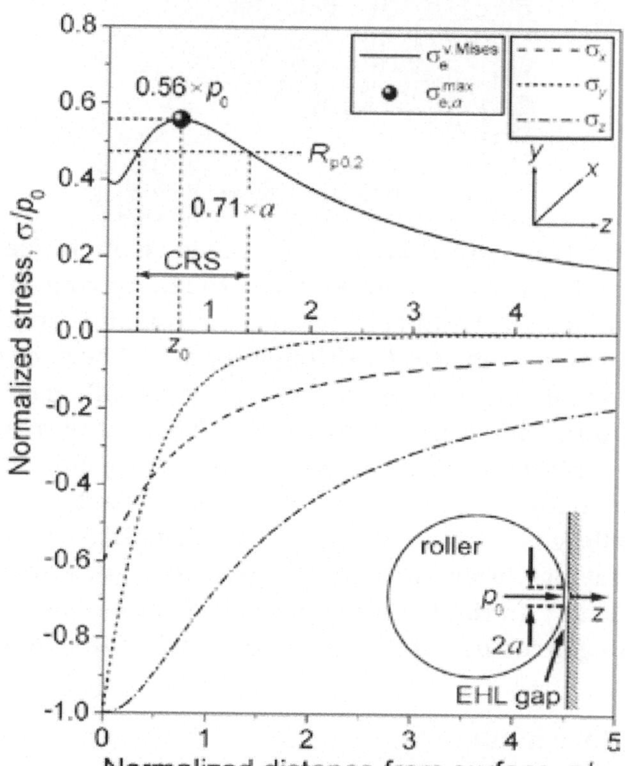

Fig. 1. Normalized plot of the depth distribution of the s_x, s_y, and s_z main normal and of the v. Mises equivalent stress below the center line of the Hertzian contact area

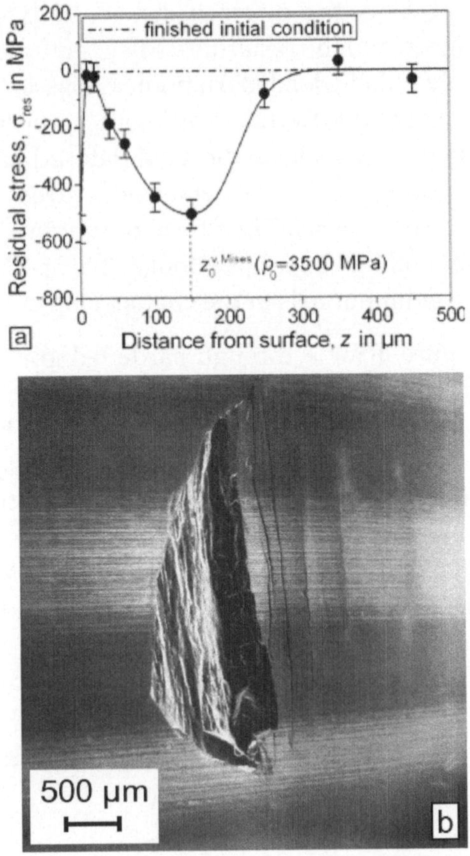

Fig. 2. Subsurface material loading and damage characterized, respectively, by (a) the residual stress distribution below the inner ring (IR) raceway of a deep groove ball bearing (DGBB) tested in an automobile gearbox rig, where the part is made of martensitically through hardened bearing steel and (b) a SEM image (secondary electron mode, SE) of fatigue spalling on the IR raceway of a rig tested DGBB with overrolling direction from left to right Up to a depth *z* of 20 µm, the indicated initial state after hardening and machining is not changed, which manifests good lubrication. The residual stress is denoted by s_{res}. Fatigue spalling is eventually caused by subsurface crack initiation and growth to the surface in overrolling direction (OD), as evident from Figure 2b (Voskamp, 1996). In the scanning electron microscope (SEM) image, the still intact honing structure of the raceway confirms the adjusted ideal EHL conditions.

2.2 Surface failure mode

Hard (ceramic) or metallic foreign particles contaminating the lubricating gap at the contact area, however, result in indentations on the raceway due to overrolling in bearing operation. The SEM images of Figures 3a and 3b, taken in the SE mode, show examples of both types:

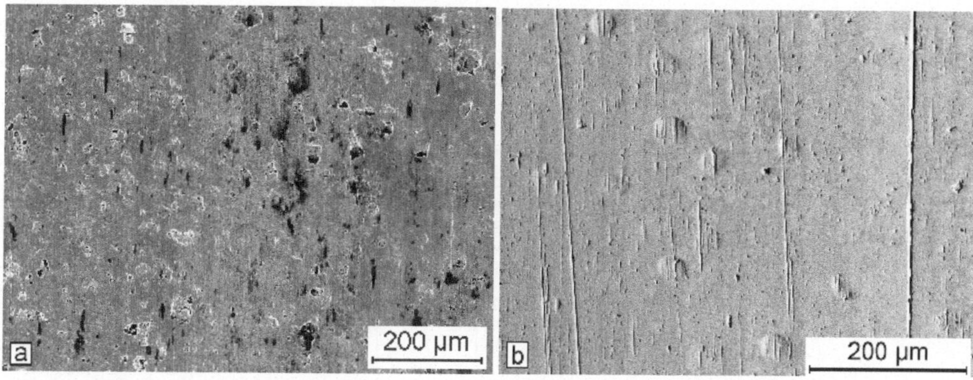

Fig. 3. SEM images (SE mode) of (a) randomly distributed dense hard particle raceway indentations (also track-like indentation patterns can occur, e.g. so-called frosty bands) from contaminated lubricant and (b) indentations of metallic particles on the smoothed IR raceway of a cylindrical roller bearing (CRB) that clearly reveal earlier surface conditions of better preserved honing structure

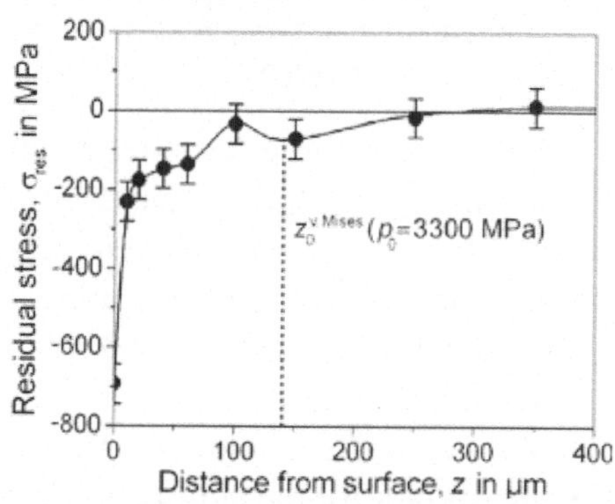

Fig. 4. Residual stress depth distribution of the martensitically hardened IR of a taper roller bearing (TRB) indicating foreign particle (e.g., wear debris) contamination of the lubricant

Cyclic loading of the Hertzian micro contacts induces continuously increasing compressive residual stresses near the surface up to a depth that is connected with the regular (e.g., lognormal) size distribution of the indentations. In the case of Figure 4, the superimposed profile modification by the basic macro contact is marginal, which means that the maximum Hertzian pressure of 3300 MPa is only applied for a short time. Compressive residual stresses in the edge zone are generated up to 60 µm depth. The high surface value reflects polishing of the raceway, associated with plastic deformation.

The stress analysis for evaluation of the v. Mises yield criterion in Figure 1 refers to the ideal undisturbed EHL rolling contact in a bearing with fully separating lubricating film, where (fluid) friction only occurs. In an extension of this scheme, the surface mode of rolling contact fatigue (RCF) is illustrated in Figure 5 on the example of indentations (size a_{micro}) that cover the raceway densely in the form of a statistical waviness at an early stage of operation:

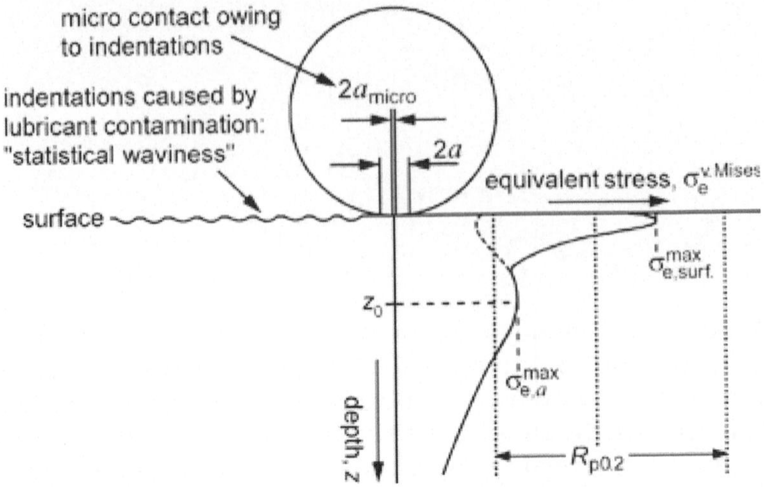

Fig. 5. Scheme of the v. Mises stress as a function of the distance from the Hertzian contact with and without raceway indentations (roller on a smaller scale)

The resulting peak of the v. Mises equivalent stress, max e,surf. , is influenced by the sharp- edged indentations of hard foreign particles (cf. Figure 3a). However, lubricant contamination by hardened steel acts most effectively because of the larger size. The contact area of the rolling elements also exhibits a statistical waviness of indentations. The stress concentrations on the edges of the Hertzian micro contacts promote material fatigue and damage initiation on or near the surface. Consequently, bearing life is reduced (Takemura & Murakami, 1998). It is shown in section 5.1 that, by creating tangential forces, additional sliding in frictional rolling contact can cause equivalent and hence residual stress distributions similar to Figures 5 and 4, respectively, on indentation-free raceways. The occurrence or dominance of the competing (near-) surface and subsurface failure mode depends on the magnitude of and the relative position of the (actually not varying) yield strength $R_{p0.2}$, as indicated in Figure 5.

The ground area of an indentation is unloaded. On the highly stressed edges, the lubricating film breaks down and metal-to-metal contact results in locally most pronounced smoothing of the honing marks. Figure 6a reveals the back end of a metal span indentation in overrolling direction. Strain hardening by severe plastic deformation leads to material embrittlement and subsequent crack initiation on the surface. Further failure development produces a so-called V pit of originally only several µm depth behind the indentation, as documented in Figure 6b. It is instructive to compare this shallow pit and the clearly smoothed raceway with the subsurface fatigue spall of Figure 2b that evolves from a depth of about 100 µm below an intact honing structure.

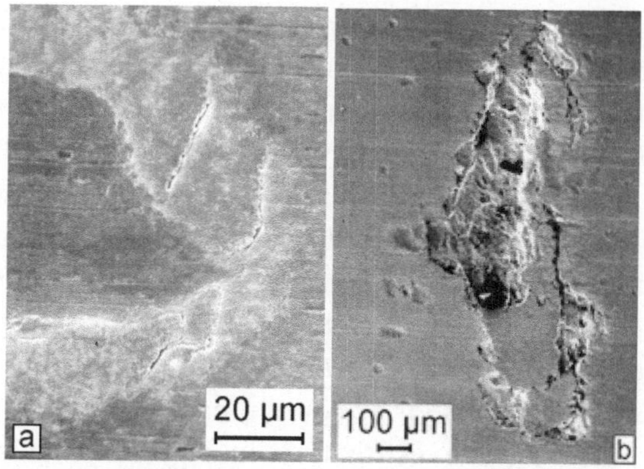

Fig. 6. SEM image (SE mode) of (a) incipient cracking and (b) beginning V pitting behind an indentation on the IR raceway of a TRB. Note the overrolling direction from left to right

3. Material based bearing performance analysis

Stressing, damage and eventually failure of a component occur due to a response of the material to the applied loading that generally acts as a combination of mechanical, chemical and thermal portions. The reliability of Hertzian contact machine elements, such as rolling bearings, gears, followers, cams or tappets, is of particular engineering significance. Advanced techniques of physical diagnostics permit the evaluation of the prevailing material condition on a microscopic scale. According to the collective impact of fatigue, friction, wear and corrosion and thus, for instance, depending on the type of lubrication, the degree of contamination, the roughness profile and the applied Hertzian pressure, failures are initiated on or below the raceway surface (see section 2). An operating rolling bearing represents a cyclically loaded tribological system. Depth resolved X-ray diffraction (XRD) measurements of macro and micro residual stresses provide an accurate estimation of the stage of material aging. The XRD material response analysis of rolling bearings is experimentally and methodologically most highly evolved. A quantitative evaluation of the changes in the residual stress distribution is proposed in the literature, for instance by integrating the depth profile to compute a characteristic deformation number (Böhmer et al., 1999). In the research reported in this chapter, however, the alternative XRD peak width based conception is used. The established procedure described in the following may be, due to its development to a powerful evaluation tool for scientific and routine engineering purposes in the SKF Material Physics laboratory under the guidance of *Wolfgang Nierlich*, referred to as the *Schweinfurt* methodology of XRD material response bearing performance analysis.

3.1 Intention and history of XRD material response analysis

The investigation aims at characterizing the response of the steel in the highly stressed edge zone to rolling contact loading. Plastification (local yielding) and material aging (defect accumulation) is estimated by the changes of the (macro) residual stresses and the XRD peak width, respectively. Failure is related to mechanical damage by fatigue and tribological loading, (tribo-) chemical and thermal exposure. Mixed friction or boundary lubrication in rolling-sliding contact is reflected, for instance, by polishing wear on the surface. The operating condition of cyclically Hertzian loaded machine parts shall be analyzed. The key focus is put on rolling bearings but also other components, like gears or camshafts, can be examined. XRD material response analysis permits the identification of the relevant failure mode. In the frequent case of surface rolling contact loading, the acting damage mechanism, such as vibrations, poor or contaminated lubrication, is also deducible. The quantitative remaining life estimation in rig test evaluation supports, for instance, product development or design optimization. This analysis option receives great interest especially in automotive engineering. Drawing a comparison with the calculated nominal life is of high significance. Also, not too heavily damaged (spalled) field returns can be investigated in the framework of failure analysis and research.

The practicable evaluation tools provided and applied in the following sections are derived from the basic research work of *Aat Voskamp* (Voskamp, 1985, 1996, 1998), who concentrates on residual stress evolution and microstructural alterations during classical subsurface rolling contact fatigue, and *Wolfgang Nierlich* (Nierlich et al., 1992; Nierlich & Gegner, 2002, 2008), who studies the surface failure mode and aligns the X-ray diffractometry technique from the 1970's on to meet industry needs. The application of the XRD line broadening for the characterization of material damage and the introduction of the peak width ratio as a quantitative measure represent the essential milestone in method development (Nierlich et al., 1992). The bearing life calibration curves for classical and surface rolling contact fatigue, deduced from rig test series, also make the connection to mechanical engineering failure analysis and design (Nierlich et al., 1992; Voskamp, 1998). The three stage model of material response allows the attribution of the residual stress and microstructure changes (Voskamp, 1985). With substantial modification on the surface (Nierlich & Gegner, 2002), this today accepted scheme proves applicable to both failure modes (Gegner, 2006a). The interdependent joint evaluation of residual stress and peak width depth profiles in the subsurface region of classical rolling contact fatigue completes the *Schweinfurt* methodology (Gegner, 2006a). Further developments of the XRD material response analysis, such as the application to other cyclically Hertzian loaded machine elements, are reported in the literature (Gegner et al., 2007; Nierlich & Gegner, 2006).

3.2 Residual stress measurement

To discuss the principles of material based bearing performance analysis, first a synopsis of the XRD measurement technique is provided. Data interpretation is subsequently described in section 3.3. The evaluation of a high number of measurements on run field and test bearings is necessary to create the appropriate scientific, engineering, and methodological foundations of XRD material response analysis. For efficient performance, the applied XRD technique must thus take into account the required fast specimen throughput at sufficient data accuracy. The rapid industrial-suited XRD measurement of residual stresses outlined below incorporates suggestions from the literature (Faninger & Wolfstieg, 1976). Usually, around ten depth positions are adequate for a profile determination. Residual stress free material removal with high precision occurs by electrochemical polishing. The spatial resolution is given by the low penetration power of the incident X-ray radiation to about 5 μm that is appropriate for the application.

XRD residual stress analysis is widely used in bearing engineering since the 1970's (Muro et al., 1973). In the investigations of the present chapter, computer controlled W goniometers with scintillation type counter tube are applied, which work on the principle of the focusing Bragg-Brentano coupled q–2q diffraction geometry (Bragg & Bragg, 1913; Hauk & Macherauch, 1984). The X-ray source is fixed and the detector gradually rotates with twice the angular velocity q˙ of the specimen to preserve a constant angle of 2q between the incident and reflected beam.

3.2.1 High intensity diffractometer

The positions of major modifications of the conventional goniometer design are numbered consecutively in Figure 7. The severe difficulties of XRD measurements of hardened steels in the past from the broad asymmetrical diffraction lines of martensite are well known (Macherauch, 1966; Marx, 1966). Exploiting the negligible instrumental broadening, however, these large peak widths of about 5° to 7.5° only permit the implementation of such fundamental interventions in the beam path to increase the intensity of the incident and emergent X-ray radiation by tailoring the required resolution. In position 1, the square instead of the line focal spot is used. Thus, the intensity loss by vertical masking at the beam defining slit is reduced. Position 2 is also labeled in Figure 7. The distance from the horizontally and vertically adjustable defining slit to the focal spot is extended to two-thirds of the diffractometer (or measuring) circle radius. Whereas the lower resolution is of no significance, the intensity of the primary beam is further enhanced. The aperture a is indicated. The depicted scattering and Soller slits limit peak width and divergence of the diffracted beam on the expense of intensity loss. Position 3 signifies that parallelization of the radiation is dispensed with. For the same purpose, the receiving slit is opened to a detection angle of 2° in position 4. The adapted diffractometer arrangement, optimized with the courage to problem-oriented simplification, eventually provides a 10 times higher recorded X-ray intensity without noticeable loss in accuracy for the broad interference lines of hardened bearing steels. The effect on determining peak position and width is negligible. The dispersion, defining the line shift relevant to residual stress evaluation, remains uninfluenced. The intensity corrections are not further discussed.

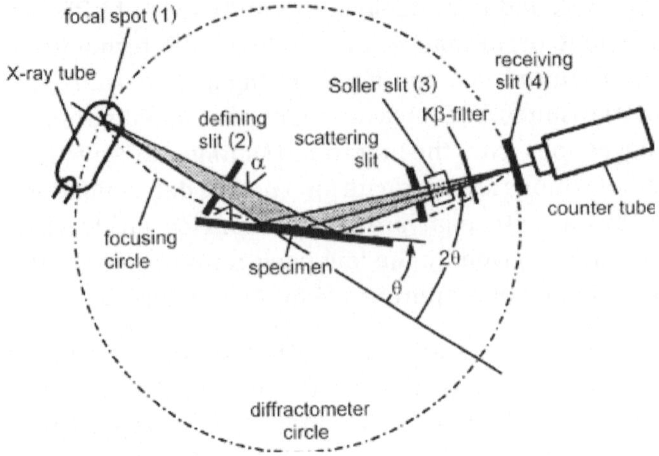

Fig. 7. Schematic diffractometer beam path with indicated modifications (1 to 4)

3.2.2 Implementation of the sin²ψ method

Stress determinations are generally based on the measurement of strain (Dally & Riley, 2005). The conversion occurs by elasticity theory. Residual stresses of the first kind result in lattice strains of the order of 1‰. This elastic distortion of the unit cell causes an anisotropic peak shift of interference lines, which is determinable by the XRD technique. The macro or volume residual stresses of a polycrystalline material are derived by measuring and evaluating the relative interplanar spacing for multiple specimen orientations bringing differently oriented sets of lattice plane into reflection. Larger line shifts preferable for a more sensitive strain determination occur in the backscattering region for 2q>130°, as obvious from the total differential of the Bragg condition for the monochromatic radiation:

$$\frac{dD}{D} = -\cot\theta \, d\theta \qquad (1)$$

According to Figure 7, q and 2q respectively denote the glancing Bragg and the diffraction angle. Such favorable lattice spacings *D* of the reflecting planes increase the measuring accuracy. For martensitic or bainitic microstructures, the recorded a-Fe (211) diffraction peak, excited by the long-wave Cr Ka radiation (wavelength l=0.229 nm), best meets this requirement with $2q_0$=156.1° and is analyzable even for incomplete line detection (see Figure 8) when the rotating detector would touch the X-ray tube at 2q>162°. Since its introduction in 1961 (Macherauch & Müller, 1961), the applied sin2y method is further developed and, for accuracy reasons (Macherauch, 1966), predominantly used for XRD macro residual stress measurements (Hauk, 1997; Noyan & Cohen, 1987). Due to the small penetration depth of the X-ray radiation of a few μm, a biaxial stress state exists in the specimen surface in good approximation. The strain can be measured from the line shift by

Eq. (1). Poisson's ratio and Young's modulus are denoted n and *E*, respectively. Applying Hooke's law, elemental geometry provides the fundamental equation of X-ray residual stress analysis:

$$-(\theta-\theta_0)\cot\theta_0 = \frac{D_{\varphi,\psi}-D_0}{D_0} = \varepsilon_{\varphi,\psi} = \frac{1+\nu}{E}\sigma_\phi \sin^2\psi - \frac{\nu}{E}(\sigma_1+\sigma_2) \qquad (2)$$

The azimuth and inclination Euler angles, j and y, characterize the direction of the interplanar spacing $D_{j,y}$ and the lattice strain $e_{j,y}$. Furthermore, s_1 and s_2 are the principle stresses parallel to the specimen surface (s_3=0). The values D_0 and q_0 refer to the strain-free undeformed lattice. For the surface stress component (y=90°) corresponding to j, a trigonometric relationship holds:

$$\sigma_\varphi = \sigma_1 \cos^2\varphi + \sigma_2 \sin^2\varphi \qquad (3)$$

Substituting the X-ray elastic constants (XEC):

$$\frac{1}{2}S_2 = \frac{1+\nu}{E}, \quad S_1 = -\frac{\nu}{E} \qquad (4)$$

into Eq. (2) ends up with the following expression:

$$\varepsilon_{\varphi,\psi} = \frac{1}{2}S_2\sigma_\varphi \sin^2\psi + S_1(\sigma_1+\sigma_2) \qquad (5)$$

For hardened steel, isotropic grain distribution is assumed. The measurement of seven specimen tilt angles y from -45° to +45° symmetric about y=0° in equidistant sin2y steps is sufficient to reliably derive the desired s_j value from the slope of the straight line fitted to the data of a $D_{j,y}$ or $e_{j,y}$ plot against sin2y for constant j (Nierlich & Gegner, 2008). High accuracy is already achieved by replacing D_0 with the experimental $D_{j,y}$ at y=0° (Voskamp, 1996). Recommendations for the X-ray elastic constants of the relevant steel microstructure are given in the literature (Hauk & Wolfstieg, 1976; Macherauch, 1966). For the XRD analyses reported in the present chapter, ½S_2=5.811´10^{-6} MPa^{-1} is applied.

3.2.3 Two stage diffraction line analysis and peak maximum method

Besides X-ray intensity gain in the beam path, the second major task of rapid macro residual stress measurement is thus an efficient routine for the involved line shift evaluations. Accelerated determination of the diffraction peak position 2q is achieved by an automated self-adjusting analysis technique tailored to the a-Fe (211) interference. The method is explained by means of Figure 8:

The pre-measurement at reduced counting statistics across the indicated fixed angular range of 5° provides the peak maximum with an error of ±0.2°. This rough localization suffices to define appropriate symmetric evaluation points in an interval of 3° around the identified center for the subsequent highly accurate pulse controlled main run. The peak position is deduced from a fitting polynomial regression. A significant additional saving in acquisition time of 60%, compared to the standard procedure, is achieved by this skillful analysis strategy with the modified arrangement of Figure 7, which equals the fastest up-to-date equipment also applied for the analyses in the present paper. Each individual residual stress determination on an irradiated area of 2´3 mm2 takes approximately 5 min. The single measured value scatter, expressing the measurement uncertainty by the standard deviation, is found to be about ±50 MPa, as correspondingly reported elsewhere in the literature (Voskamp, 1996).

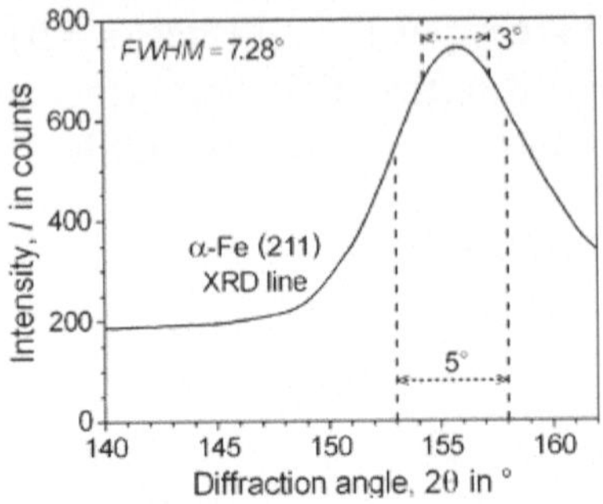

Fig. 8. Illustration of the self-developed peak finding procedure with a martensite diffraction reflex of full width at half maximum (*FWHM*) line breadth of 7.28°

Unlike, for instance, several production processes (e.g., milling), rolling contact loading usually leads to the formation of similar depth distributions of the circumferential and axial residual stresses (Voskamp, 1987). Aside from rare exceptions such as the additional impact of severe three-dimensional vibrations (Gegner & Nierlich, 2008), deviations of maximum 20% to 30% reflect experience. As also the course of the depth profile is more important for the XRD material response analysis than the actual values of the single measurements, in the following the residual stresses are only determined in the circumferential (i.e., overrolling) direction.

3.2.4 Automated XRD peak width evaluation

Due to the geometrical restrictions of the goniometer in Figure 7, the XRD line is only collected up to a diffraction angle of 162°. The peak width, expressed as *FWHM*, is measured at a specimen tilt of y=0° and provides information on the third kind (micro) residual stresses. For the extrapolation shown in Figure 9a, the background function is determined by a linear fit on the left of the line center. In the automated measurement procedure, the scintillation counter then moves to the onset of the diffraction peak. For the sake of simplicity, the background subtracted data of the subsequent line recoding in Figure 9b are fitted by an interpolating polynomial of high degree. The acquisition time per *FWHM* value and the measuring accuracy (one-fold standard deviation) amount to 3 to 5 min and 0.06° to 0.09°, respectively.

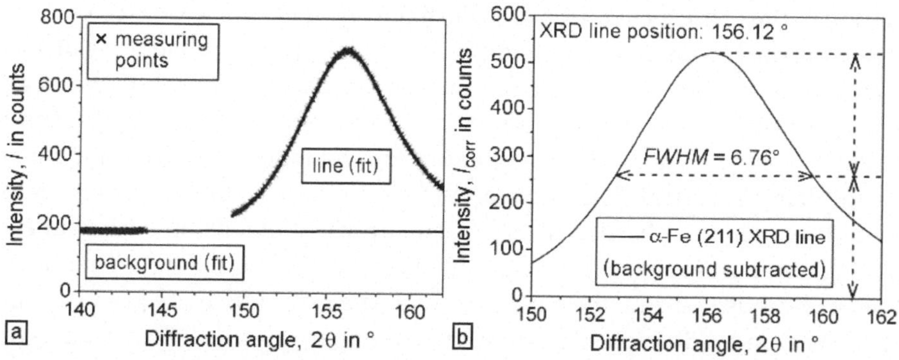

Fig. 9. Illustration of (a) the programmed XRD peak width analysis with intervals of data fitting and (b) the evaluation of the *FWHM* value for the diffraction line of Figure 9a

3.2.5 Completion of investigation methods for material response analysis

It becomes clear in the following that the reliable interpretation of the measured depth distributions of residual stresses and XRD peak width, aside from optional auxiliary retained austenite determinations to further characterize material aging (Gegner, 2006a; Jatckak & Larson, 1980), requires supportive investigation techniques for the condition of the raceway surface, microstructure, and oil or grease. Visual inspection, failure metallography, imaging and analytical scanning electron microscopy (SEM) and infrared spectroscopy of used lubricants are employed. Concrete examples of the application of these additional examination methods in the framework of XRD based material response bearing performance analysis are also discussed extensively in the literature (Gegner, 2006a; Gegner et al., 2007; Gegner & Nierlich, 2008, 2011a, 2011b, 2011c; Nierlich et al., 1992; Nierlich & Gegner, 2002, 2006, 2008).

3.3 Evaluation methodology of XRD material response analysis

The XRD peak width based *Schweinfurt* material response analysis (MRA) provides a powerful investigation tool for run rolling bearings. An actual life calibrated estimation of the loading conditions in the (near-) surface and subsurface failure mode represents the key feature of the evaluation conception (Nierlich et al., 1992; Voskamp, 1998).

The random nature of the effect of the large number of unpredictably distributed defects in the steel indicates a statistical risk evaluation of the failure of rolling bearings (Ioannides & Harris, 1985; Lundberg & Palmgren, 1947, 1952). The Weibull lifetime distribution is suitable for machine elements. The established mechanical engineering approach to RCF deals with stress field analyses on the basis, for instance, of tensor invariants or mean values (Böhmer et al., 1999; Desimone et al., 2006). On the microscopic level, however, the material experiences strain development when exposed to cyclic loading, which suggests a quantitative evaluation of the changes in XRD peak width during operation (Nierlich et al, 1992). Disregarding the intrinsic instrumental fraction, the physical broadening of an X-ray diffraction line is connected with the microstructural condition of the analyzed material (region) by several size and strain influences (Balzar, 1999). The peak width thus represents a measuring quantity for changing properties and densities of crystal defects. Lattice distortion provides the dominating contribution to the high line broadening of hardened steels. The average dimension of the coherently diffracting domains in martensite amounts to about 100 to 200 nm. Therefore, the XRD peak width is not directly correlated with the prior austenite grain size of few μm. The observed reduction of the line broadening by plastic deformation signifies a decrease of the lattice distortion. The minimum XRD peak width ratio, b/B, is the calibrated damage parameter of rolling contact fatigue that links materials to mechanical engineering (Weibull) failure analysis. The derived XRD equivalent values of the actual (experimental) L_{10} life at 90% survival probability (rating reliability) of a bearing population equal about 0.64 for the subsurface as well as 0.83 and 0.86, respectively for ball and roller bearings, for the surface mode of RCF (Gegner, 2006a; Gegner et al., 2007; Nierlich et al., 1992; Voskamp, 1998). Figures 10 and 11 display b/B data from calibrating rig tests. Here, b and B respectively denote the minimum *FWHM* in the depth region relevant to the considered (subsurface or near-surface) failure mode and the initial *FWHM* value. B is taken approximately in the core of the material or can be measured separately, e.g. below the shoulder of an examined bearing ring. The correlation between the statistical parameters representing a population of bearings under certain operation conditions and the state of aging damage (fatigue) of the steel matrix by the XRD peak width ratio measured on an accidentally selected part also reflects the intrinsic determinateness behind the randomness.

Based on Voskamp's three stage model for the subsurface and its extension to the surface failure mode (Gegner, 2006a; Voskamp, 1985, 1996), Figures 10 and 11 schematically illustrate the progress of material loading in rolling contact fatigue with running time, expressed by the number N of inner ring revolutions.

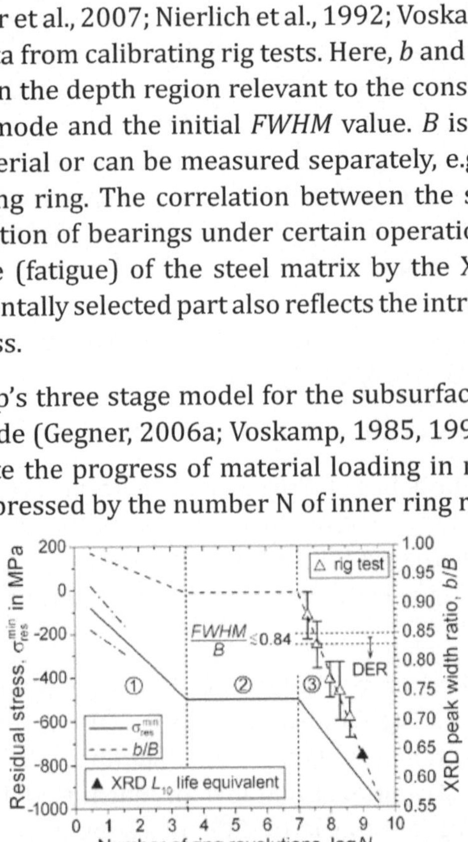

Fig. 10. Three stage model of subsurface RCF with XRD peak width ratio based indication of dark etching region (DER) formation in the microstructure and L_{10} life calibration (DGBB)

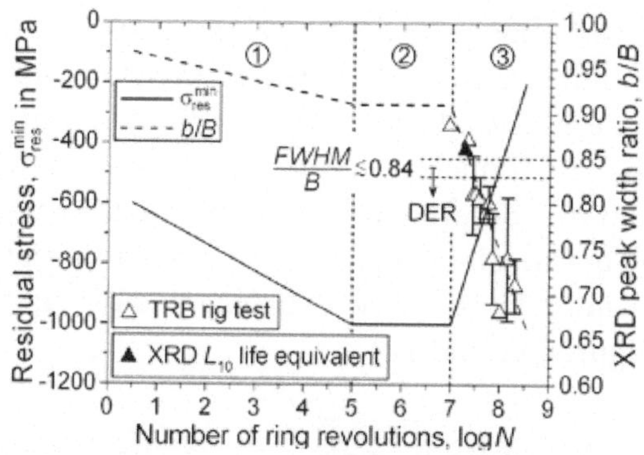

Fig. 11. Three stage model of surface RCF with XRD peak width ratio based DER indication and actual L_{10} life calibration (roller bearings) that refers to the higher loaded inner ring

The changes are best described by the development of the maximum compressive residual stress, s^{min} , and the RCF damage parameter, *b/B*, measured respectively **in** the depth and on (or near) the surface. The underlying alterations of the $s_{res}(z)$ and *FWHM*(*z*) distributions are demonstrated in Figure 12 for competing failure modes. The characteristic values are indicated in the profiles that in the subsurface region of classical RCF reveal an asymmetry towards higher depths (cf. Figure 1). The response of the steel to rolling contact loading is divided into the three stages of mechanical conditioning shakedown (1), damage incubation steady state (2), and material softening instability (3). Figures 10 to 12 provide schematic illustrations. The prevalently observed re-reduction of the compressive residual stresses in the instability phase of the surface mode, particularly typical of mixed friction running conditions, suggests relaxation processes. From experience, a residual stress limit of about −200 MPa is usually not exceeded, as included in the diagrams of Figures 11 and 12. The conventional logarithmic plot overemphasizes the differences in the slopes between the constant and the decreasing curves in the steady state and the instability stage of Figures 10 and 11. The existence of a third phase, however, is indicated by the reversal of the residual stress on the surface (cf. Figures 11 and 12) and also found in RCF component rig tests (Rollmann, 2000).

The first stage of shakedown is characterized by microplastic **deformation** and the quick build-up of compressive residual stresses when the yield strength, $R_{p0.2}$, of the hardened steel is locally exceeded by the v. Mises equivalent stress representing the triaxial stress field in rolling contact loading (cf. section 2). Short-cycle cold working processes of dislocation rearrangement with material alteration restricted to the higher fatigue endurance limit, in which carbon diffusion is not involved, cause rapid mechanical conditioning (Nierlich & Gegner, 2008). Further details are discussed in section 4.2. The second stage of steady state arises as long as the applied load falls below the shakedown limit so that ratcheting is avoided (Johnson & Jefferis, 1963; Voskamp, 1996; Yhland, 1983). In this period of fatigue damage incubation, no significant microstructure, residual stress and XRD peak width alterations are observed. Elastic behavior of the pre-conditioned microstrained material is assumed. In the extended final instability stage, gradual microstructure changes occur (Voskamp, 1996). The phase transformations require diffusive redistribution of carbon on a micro scale, which is assisted by plastification. From *FWHM*/*B* of about 0.83 to 0.85 downwards, a dark etching region (DER) occurs in the microstructure by martensite decay. Note that this is in the range of the XRD L_{10} value for the surface failure mode but well before this life equivalent is reached for subsurface RCF (cf. Figures 11 and 12).

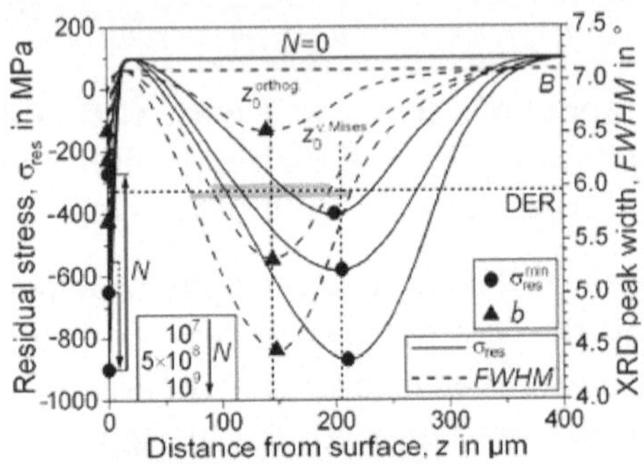

Fig. 12. Schematic residual stress and XRD peak width change with rising *N* during subsurface and surface RCF and prediction of the respective depth ranges (gray) of DER formation

Fatigue is damage (defect) accumulation under cyclic loading. Microplastic deformation is reflected in the XRD line broadening. The observed reduction of the peak width signifies a decrease of the lattice distortion. For describing subsurface RCF failure, the established Lundberg-Palmgren bearing life theory defines the risk volume of damage initiation on microstructural defects by the effect of an alternating load, thus referring to the depth of maximum orthogonal shear stress (Lundberg & Palmgren, 1947, 1952). However, the v. Mises equivalent stress, by which residual stress formation is governed, as well as each principal normal stress (cf. Figure 1) are pulsating in time. In the region of classical RCF below the raceway, the minimum XRD peak width occurs significantly closer to the surface than the maximum compressive residual stress (Gegner & Nierlich, 2011b; Schlicht et al., 1987). It is discussed in the literature which material failure hypothesis is best suited for predicting RCF loading (Gohar, 2001; Harris, 2001): Lundberg and Palmgren use the orthogonal shear stress approach but other authors prefer the Huber-von Mises-Hencky distortion or deformation energy hypothesis (Broszeit et al., 1986). The well-founded conclusion from the XRD material response analyses interconnects both views in a kind of paradox (Gegner, 2006a): whereas residual stress formation and the beginning of plastification conform to the distortion energy hypothesis, RCF material aging and damage evolution in the steel matrix, manifested in the XRD peak width reduction, responds to the alternating orthogonal shear stress.

The detected location of highest damage of the steel matrix agrees with the observation that under ideal EHL rolling contact loading most fatigue cracks are initiated near the depth (Lundberg & Palmgren, 1947). It is recently reported that the frequency of fracturing of sulfide inclusions in bearing operation due to the influence of the subsurface compressive stress field also correlates well with the distance distribution of the orthogonal shear stress below the raceway (Brückner et al., 2011). The three stages of the associated mechanism of butterfly formation, which occurs from a Hertzian pressure of about 1400 MPa, are documented in Figure 13: fracturing of the MnS inclusion (1), microcrack extension into the bulk material (2), development of a white etching wing microstructure along the crack (3). The light optical micrograph (LOM) and SEM image of Figures 14a and 14b, respectively, reveal in a radial (i.e., circumferential) microsection how the white etching area (WEA) of the butterfly wing virtually *emanates* from the matrix zone in contact with the pore like material separation of the initially fractured MnS inclusion into the surrounding steel microstructure.

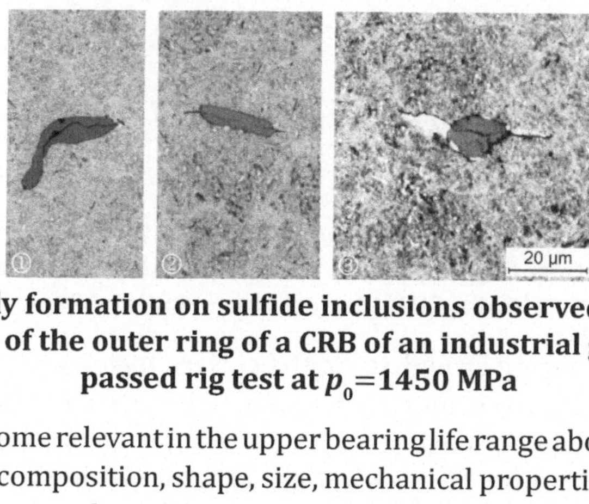

Fig. 13. Butterfly formation on sulfide inclusions observed in etched axial microsections of the outer ring of a CRB of an industrial gearbox after a passed rig test at p_0=1450 MPa

Butterflies become relevant in the upper bearing life range above L_{10}. Inclusions of different chemical composition, shape, size, mechanical properties and surrounding residual stresses are technically unavoidable in steels from the manufacturing process. The potential for their reduction is limited also from an economic viewpoint and virtually fully tapped in the today's high cleanliness bearing grades. Local peak stresses on nonmetallic inclusions, i.e. internal metallurgical notches, below the contact surface can cause the initiation of microcracks. Operational fracture of embedded MnS particles (see Figures 13, 14) is quite often observed and represents a potential butterfly formation mechanism besides, e.g., decohesion of the interphase (Brückner et al., 2011). Subsequent fatigue crack propagation is driven by the acting main shear stress (Schlicht et al., 1987, 1988; Takemura & Murakami, 1995). The growing butterfly wings thus follow the direction of ideally 45° to the raceway tangent. Figure 15 shows a textbook example from a weaving machine gearbox bearing at around the nominal L_{50} life. The overrolling direction in the micrograph is from right to left. The white etching constituents show an extreme hardness of about 75 HRC (1200 HV) and consist of carbide-free nearly amorphous to nano-granular ferrite with grain sizes up to 20 to 30 nm.

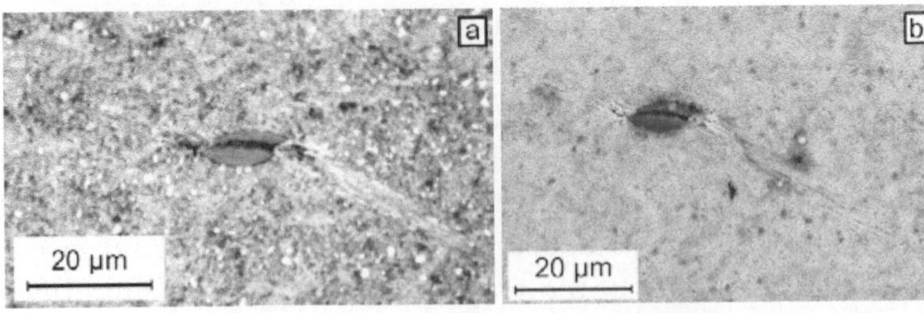

Fig. 14. LOM micrograph (a) and corresponding SEM-SE image (b) of butterfly development on a cracked MnS inclusion in the etched radial microsection of the stationary outer ring of a spherical roller bearing (SRB) after a passed rig test at a Hertzian pressure p_0 of 2400 MPa

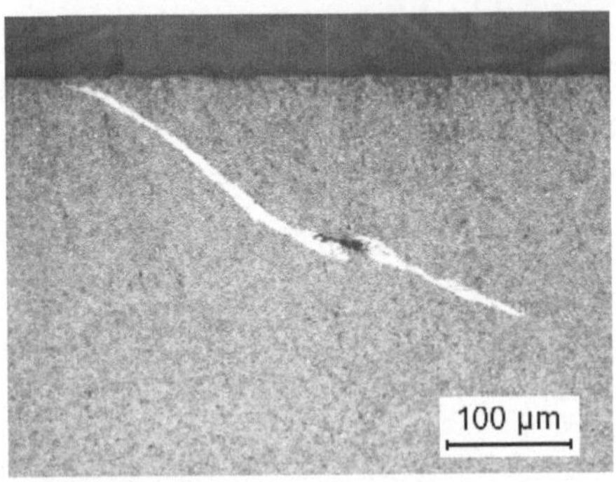

Fig. 15. Butterfly wing growth from the depth to the raceway surface in overrolling direction (right-to-left) in the etched radial microsection of the IR of a CRB loaded at p_0=1800 MPa

Critical butterfly wing growth up to the surface (see Figure 15), which leads to bearing failure by raceway spalling eventually, occurs very rarely (Schreiber, 1992). The metallurgically unweakened steel matrix in some distance to the inclusion can cause crack arrest. Multiple damage initiation, however, is found in the final stage of rolling contact fatigue. Subsurface cracks may then reach the raceway (Voskamp, 1996). Butterfly RCF damage develops by the microstructural transformation of low-temperature dynamic recrystallization of the highly strained regions along cracks rapidly initiated on stress raising nonmetallic inclusions in the steel (Böhm et al., 1975; Brückner et al., 2011; Furumura et al., 1993; Österlund et al., 1982; Schlicht et al., 1987; Voskamp, 1996), If this localized fatigue process occurs at Hertzian pressures below 2500 MPa (Brückner et al., 2011; Vincent et al., 1998), it is not recognizable alone by an XRD analysis that is sensitive to integral material loading (see section 3.2).

According to the Hertz theory, the depth zorthog. of the maximum of the alternating orthogonal shear stress and its double amplitude depend on the footprint ratio between the semiminor and the semimajor axis of the pressure ellipse (Harris, 2001; Palmgren, 1964): the values respectively amount to $0.5' a$ and $0.5' p_0$ in line contact and are slightly lower for ball bearings. From $z^{orthog.} = 0.5\ '\ a < z^{v.Mises}$ follows that the

FWHM distance curve reaches its minimum *b* significantly closer to the surface than the residual stresses, as it is illustrated in Figure 12 and apparent from the practical example of Figure 16a. This finding is exploited for XRD material response analysis (Gegner, 2006a). The residual stress and XRD peak width distributions are evaluated jointly in the subsurface region of classical rolling contact fatigue by applying the v. Mises and orthogonal shear stress interdependently. Data analysis is demonstrated in Figures 16a and 16b. Adjusting to the best fit improves the accuracy of deducing the Hertzian pressure p_0 from the measured profiles. Superposition with the load stresses results in a slight gradual shift of the residual stress and XRD peak width distribution to larger depths with run duration (Voskamp, 1996), which is neglected in the evaluation (see Figure 12). In the example of Figure 16a, material aging is within the scattering range of the L_{10} life equivalent value for both, thus in this case competing, failure modes of surface (b/B»0.83) and subsurface RCF (b/B»0.64): a relative XRD peak width reduction of $b/B^3$0.82 and b/B=0.67 is respectively taken from the diagram.

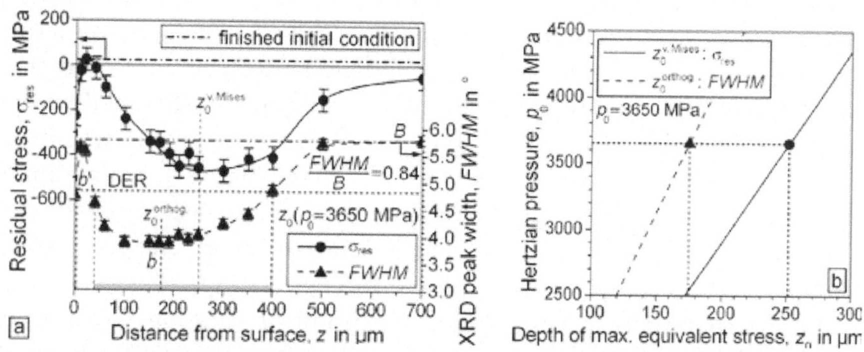

Fig. 16. Graphical representation of (a) the residual stress and XRD peak width depth distribution measured below the IR raceway of a DGBB tested in an automobile gearbox rig with indication of the initial as-delivered condition and (b) the joint subsurface profile evaluation

The greater-or- equal sign for the estimation of the surface failure mode considers the unknown small *FWHM* decrease on the raceway due to grinding and honing of the hardened steel in the as- finished condition (see Figures 12 and 16a) so that the alternatively used reference *B* in the core of the material or another uninfluenced region (e.g., below the shoulder of a bearing ring) exceeds the actual initial value at z=0 typically by about 0.02°. The original residual stress and XRD peak width level below the edge zone results from the heat treatment. The inner ring of Figure 16a, for instance, is made out of martensitically through hardened bearing steel.

The predicted dark etching regions at the surface and in a depth between 40 and 400 μm are well confirmed by failure metallography, as evident from a comparison of Figure 16a with Figures 17a and 17b. The DER-free intermediate layer is clearly visible in the overview micrograph. The dark etching region near the surface ranges to about 10 to 12 μm depth.

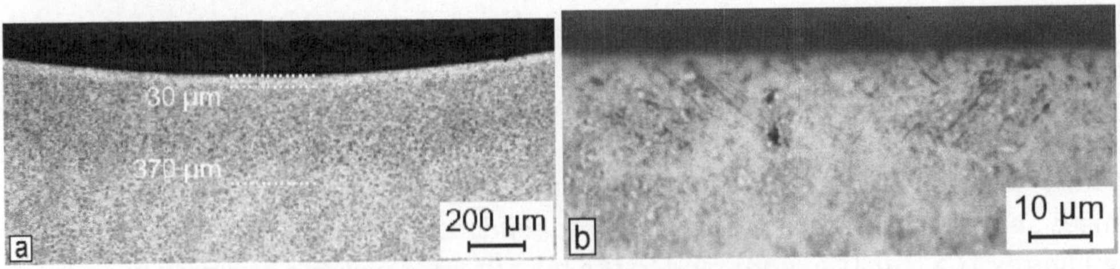

Fig. 17. LOM images of (a) the etched axial microsection of the inner ring of Figure 16a with evaluation of the extended subsurface DER and (b) a detail revealing the near-surface DER

4. Subsurface rolling contact fatigue

Since the historical beginnings with *August Wöhler* in the middle of the 19th century, today's research on material fatigue can draw from extensive experiences. Cyclic stressing in rolling contact, however, even eludes a theoretical description based on advanced multiaxial damage criteria, such as the Dang Van critical plane approach (Ciavarella et al., 2006; Desimone et al., 2006). Although little noticed in the young research field of very high cycle fatigue (VHCF) so far, RCF is the most important type of VHCF in engineering practice. Complex VHCF conditions occur under rapid load changes. The inhomogeneous triaxial stress state exhibits a large fraction of hydrostatic pressure $p_h = -(s_x + s_y + s_z)/3$ (see Figure 1, maximum on the surface) and, in the ideal case of pure radial force transfer, no critical tensile stresses, which is favorable to brittle materials and makes the hardened steel behave ductilely. The number of cycles to failure defining the rolling bearing life is thus by orders of magnitude larger than in comparable push-pull or rotating bending loading (Voskamp, 1996). The RCF performance of hardened steels is difficult to predict. Fatigue damage evolution by gradual accumulation of microplasticity is associated with increasing probability of crack initiation and failure. Microstructural changes during RCF are usually evaluated as a function of the number of ring revolutions (Voskamp, 1996). For the scaled comparison of differently loaded bearings, however, the material inherent RCF progress measure of the minimum XRD peak width ratio, b/B, is more appropriate as it correlates with the statistical parameters of the Weibull life distribution of a fictive lot (see section 3.3).

The influence of hydrogen on rolling contact fatigue is also quantifiable this way, as applied to classical RCF in section 4.3.

4.1 Microstructural changes during subsurface rolling contact fatigue

The characteristic subsurface microstructural alterations in hardened bearing steels occur due to shear induced carbon diffusion mediated phase transformations (Voskamp, 1996), for which a mechanistic metal physics model is introduced in the following. The local material fatigue aging of butterfly formation is already discussed in section 3.3. In Figures 18a to 18c, the XRD material response analysis of a rig tested automobile gearbox ball bearing is evaluated in the region of subsurface RCF. A Hertzian pressure of 3400 MPa is deduced. The joint interdependent profile evaluation is shown in Figure 18b. At the found relative decrease of the X-ray diffraction peak width to b/B»0.71, i.e. still above the XRD L_{10} life equivalent value of roughly 0.64, rolling contact fatigue produces a distinct DER in the microstructure in the depth range predicted by the $FWHM/B$ reduction below 0.84 (cf., Figures 10, 12 and 17a). This exact agreement is emphasized by a comparison of Figures 18a and 18c.

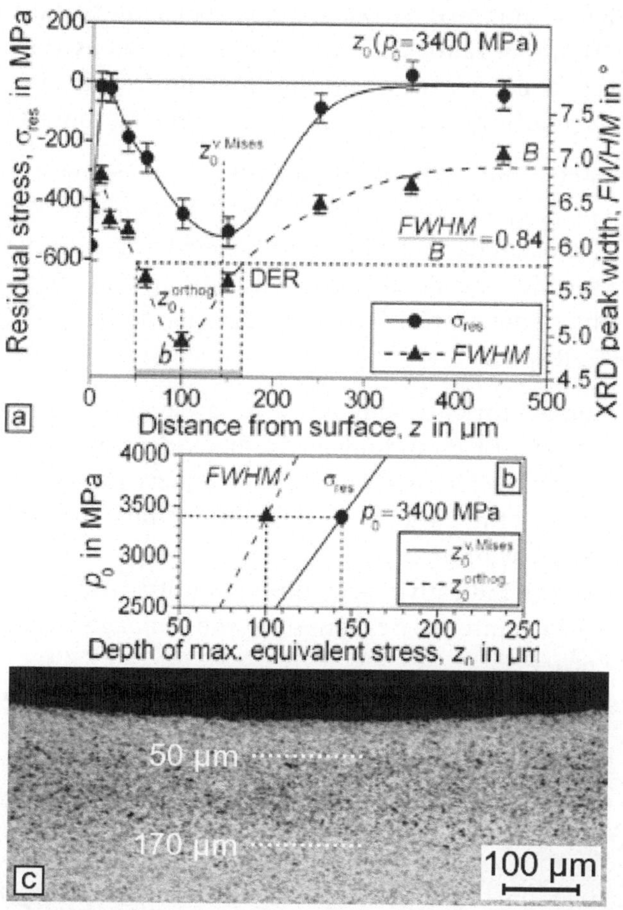

Fig. 18. Subsurface RCF analysis of the IR of a run DGBB including (a) the measured depth distribution of residual stress and XRD peak width (b/B»0.71) with DER prediction, (b) the joint XRD profile evaluation and (c) an etched axial microsection with actual DER extension

Spatial differences in the etching behavior of the bearing steel matrix in metallographic microsections caused by high shear stresses below the raceway surface after a certain stage of material aging by cyclic rolling contact loading are known since 1946 (Jones, 1946). The localized weakening structural changes result from stress induced gradual partial decay of martensite into heavily plasticized ferrite, the development of regular deformation slip bands and alterations in the carbide morphology (Schlicht et al., 1987; Voskamp, 1996). Due to the appearance of the damaged zones after metallographic preparation in an optical microscope, these areas are referred to as dark etching regions (Swahn et al., 1976a). The small decrease in specific volume of less than 1% by martensite decomposition results in a tensile contribution to operational residual stress formation but the effects of opposed austenite decay and local yield strength reduction by phase transformation prevail (Voskamp, 1996). Recent reheating experiments also point to diffusion reallocation of carbon atoms from (partially) dissolving temper as well as globular carbides for dislocational segregation in severely deformed regions (Gegner et al., 2009), which is assumed to be inducible by cyclic material loading in rolling contact (see section 4.2).

The overall quite uniformly appearing DER (see Figures 17a and 18c) is displayed at higher magnification in the LOM micrograph of Figure 19a. On the micrometer scale, affected dark etching material evidently occurs locally preferred in zones of dense secondary cementite. As well as the spatial and size distribution of the precipitation hardening carbides, micro- segregations (e.g., C, Cr) influence the formation of the DER spots.

Subsurface fatigue cracks usually advance in circumferential, i.e. overrolling, direction parallel to the raceway tangent in the early stage of their propagation (Lundberg & Palmgren, 1947), as exemplified in Figure 19b (Voskamp, 1996). The aged matrix material of the dark etching region exhibits embrittlement (see also section 5.5) that is most pronounced around the depth of maximum orthogonal shear stress, where the indicative X-ray diffraction line width is minimal and the microstructure reveals intense response to the damage sensitive preparative chemical etching process.

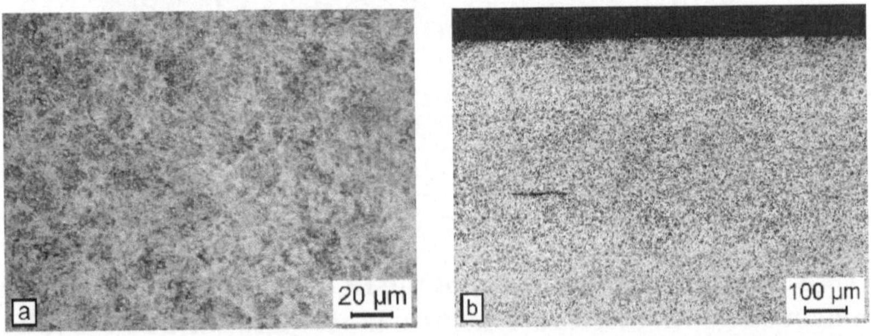

Fig. 19. LOM micrographs of (a) a detail of the DER of Figure 17a and (b) typical subsurface fatigue crack propagation parallel to the raceway around the depth of maximum orthogonal shear stress in the etched radial microsection of the inner ring of a deep groove ball bearing

In the upper subsurface RCF life range of the instability stage above the XRD L_{10} equivalent value, i.e. b/B<0.64 according to Figure 10, shear localization and dynamic recrystallization (DRX) induce (100)[110] and (111)[211] rolling textures that reflect the balance of plastic deformation and DRX (Voskamp, 1996). Regular flat white etching bands (WEB) of elongated parallel carbide-free ferritic stripes of inclination angles b_f of 20° to 32° to the raceway tangent in overrolling direction occur inside the DER (Lindahl & Österlund, 1982; Swahn et al., 1976a, 1976b; Voskamp, 1996). For the automobile alternator and gearbox ball bearing from rig tests, N° 1 and N° 2 in Figure 20a, respectively, *b*/*B* equals about 0.61 and 0.57. Metallography of the investigated inner rings in Figures 20b and 20c confirms the dark etching region predicted by the relative XRD peak width reduction and indicates the discoid flat white bands (FWB) in the axial (N° 1) and radial microsection (N° 2).

Ferrite of the FWB is surrounded by reprecipitated highly carbon-rich carbides and remaining martensite (Lindahl & Österlund, 1982; Swahn et al., 1976a, 1976b). Note that the carbides originally dispersed in the hardened steel are dissolved in the WEB under the influence of the RCF damage mechanism (see section 4.2). The SEM images of Figures 21a and 21b imply that the aged DER microstructure, the embrittlement of which is reflected in

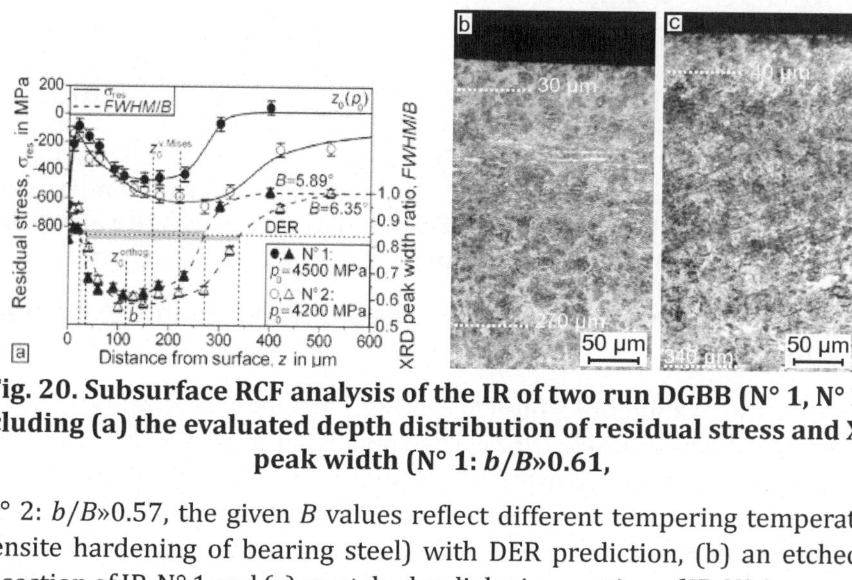

Fig. 20. Subsurface RCF analysis of the IR of two run DGBB (N° 1, N° 2) including (a) the evaluated depth distribution of residual stress and XRD peak width (N° 1: *b*/*B*»0.61,

N° 2: *b*/*B*»0.57, the given *B* values reflect different tempering temperature of martensite hardening of bearing steel) with DER prediction, (b) an etched axial microsection of IR-N° 1 and (c) an etched radial microsection of IR-N° 2, respectively with DER indication and visible FWB the preparatively lacerated material from the chemical attack by the etching process, acts as precursor of WEB formation (dark appearing phase, SEM-SE). The angles b_f are determined to be 29° and 22° (see Figures 20c, 21b) for the inner ring of bearing N° 1 and N° 2, respectively. Texture development as initiating step of WEA evolution is suggested. Steep white bands (SWB) as shown in Figure 21c occur at an advanced RCF state, once a critical FWB density is reached, not until the actual L_{50} life (Voskamp, 1996), which amounts to $5.54'L_{10}$ for ball bearings with a typical Weibull modulus of 1.1. The inclination b_s of 75° to 85° to the raceway in overrolling direction again relates to the stress field. The included angle b_{s-f} between the FWB (30°-WEB) and the SWB (80°-WEB) thus equals about 50°. Note that in Figures 20c, 21b and 21 c, the overrolling direction is respectively from left to right. FWB appear weaker in the etched microstructure. The hardness loss is due to the increasing ferrite content. SWB reveal larger thickness and mutual spacing. The ribbon-like shaped carbide-free ferrite is highly plastically deformed (Gentile et al., 1965; Swahn et al., 1976a, 1976b; Voskamp, 1996).

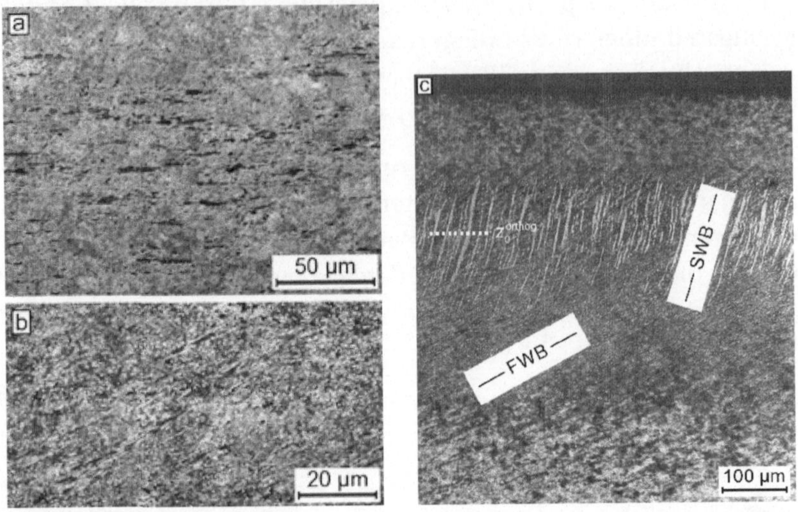

Fig. 21. SEM-SE detail of (a) Figure 20b (preparatively initiated cracks expose the DER) and (b) Figure 20c ($b_f = 22°$) and (c) an etched radial microsection of the IR of a DGBB rig tested at a Hertzian pressure of 3700 MPa with indicated depth of maximum orthogonal shear stress

4.2 Metal physics model of rolling contact fatigue and experimental verification

The classical Lundberg-Palmgren bearing life theory is empirical in nature (Lundberg & Palmgren, 1947, 1952). The application of continuum mechanics to RCF is limited. Material response to cyclic loading in rolling contact involves complex localized microstructure decay and cannot be explained by few macroscopic parameters. Moreover, fracture mechanics does not provide an approach to realistic description of RCF. The stage of crack growth, representing only about 1% of the total running time to incipient spalling (Yoshioka, 1992; Yoshioka & Fujiwara, 1988), is short compared to the phase of damage initiation in the brittle hardened steels. Without a fundamental understanding of the microscopic mechanisms of lattice defect accumulation for the prediction of material aging under rolling contact loading, which is reflected in (visible) changes of the cyclically stressed microstructure that are decisive for the resulting fatigue life, therefore, measures to increase bearing durability, for instance, by tailored alloy design cannot be derived. Physically based RCF models, however, are hardly available in the literature (Fougères et al., 2002). The reason might be that hardened bearing steels reveal complex microstructures of high defect density far from equilibrium. Precipitation strengthening due to temper carbides of typically 10 to 20 nm in diameter governs the fatigue resistance of the material in tempered condition. The mechanism proposed in the following therefore focuses on the interaction between dislocations and carbides or carbon clusters in the steel matrix.

The stress-strain hysteresis from plastic deformation in cyclic loading reflects energy dissipation (Voskamp, 1996). The vast majority of about 99% is generated as heat (Wielke, 1974), which produces a limited temperature increase under the conditions of bearing operation. The remaining 1% is absorbed as internal strain energy. This amount is associated with continuous lattice defect accumulation during metal fatigue and, therefore, damaging changes to the affected microstructure eventually. Gradual decay of retained austenite, martensite and cementite occurs in the instability stage of RCF (see Figure 10), with the dislocation arrangement of a fine sub-grain (cell) structure in the emerging ferrite and white etching band as well as texture development inside the DER in the upper life range (Voskamp, 1996). The phase transformations require diffusive redistribution of carbon on a micro scale, which is assisted by plastification. Strain energy dissipation and microplastic damage accumulation in rolling contact fatigue is described by the mechanistic Dislocation Glide Stability Loss (DGSL) model introduced in Figure 22. The different stages of compressive residual stress formation, XRD peak width reduction and microstructural alteration during advancing RCF are discussed in the framework of this metal physics scheme in the following.

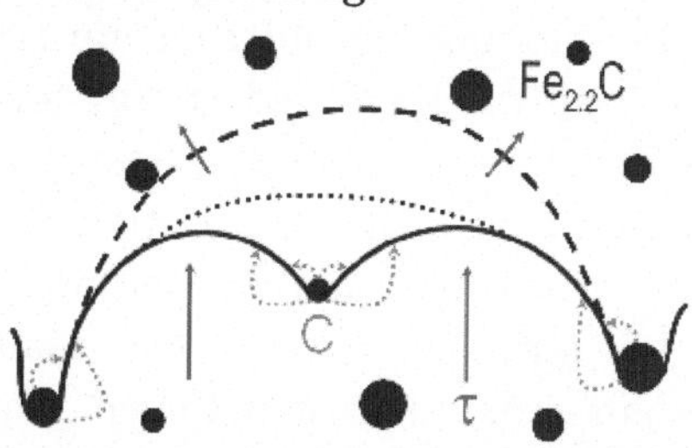

Fig. 22. In the dislocation glide stability loss (DGSL) model of rolling contact fatigue, according to which gradual dissolution of (temper) carbides (spheres) occurs by diffusion (dotted arrows) mediated continuous carbon segregation at pinned dislocations (lines) bowing out under the influence of the cyclic shear stress t (solid arrows), the smallest particles tend to disappear first due to their higher curvature-dependent surface energy so that the obstacles are passed successively and the level of localized microplasticity is increased accordingly

Rolling contact fatigue life is governed by the microcrack nucleation phase. Gradual dissolution of $Fe_{2.2}C$ temper carbides (spheres in Figure 22) driven by carbon segregation at initially pinned dislocations (lines), which bow out under the acting cyclic shear stress t (arrows), causes successive overcoming of the obstacles and local restarting of plastic flow until activation of Frank-Read sources. Fatigue damage incubation in the steady state of apparent elastic material behavior is followed in the instability stage by the microstructural changes of DER formation, decay of globular secondary cementite (in the DGSL model due to dislocation-carbide interaction) and regular ferritic white etching bands developing inside the DER. Strain hardening, which embrittles the aged steel matrix and thus promotes crack initiation, compensates for the diminishing precipitation strengthening in the progress of rolling contact fatigue. This process results in further compressive residual stress build-up from the shakedown level and newly decreasing XRD peak width (see Figure 10). Gradual concentration of local microplasticity and microscopic accumulation of lattice defects characterize proceeding RCF damage. According to the DGSL model, Cottrell segregation of carbon atoms released from dissolving carbides at uncovered cores of dislocations, which are regeneratively generated by the glide movements during yielding, provides an additional contribution to the XRD peak width reduction by cyclic rolling contact loading (Gegner et al., 2009). The experimental proof of this essential prediction is discussed in detail below by means of Figures 23 and 24. The gradually increasing amount of localized dislocation microplasticity represents the fatigue defect accumulation mechanism of the DGSL model of RCF. It is thus associated with a rising probability for bearing failure (cf. Figure 10) due to material aging. The DGSL criterion for local microcracking is based on a critical dislocation density. Orientation and speed of fatigue crack propagation can then also be analyzed.

The proposed dislocation-carbide interaction mechanism explains (partial) fragmentation of uncuttable globular carbides of μm size, which is occasionally observed in microsections, and the increased energy level in the affected region. Localized microplastic deformation is related to energy dissipation. Note that the DGSL fatigue model involves the basic internal friction mechanism of Snoek-Köster dislocation damping under cyclic rolling contact loading. The increasing dislocation density of the aged, highly strained material eventually causes local dynamic recrystallization into the nanoscale microstructure of white etching areas, where carbides are completely dissolved.

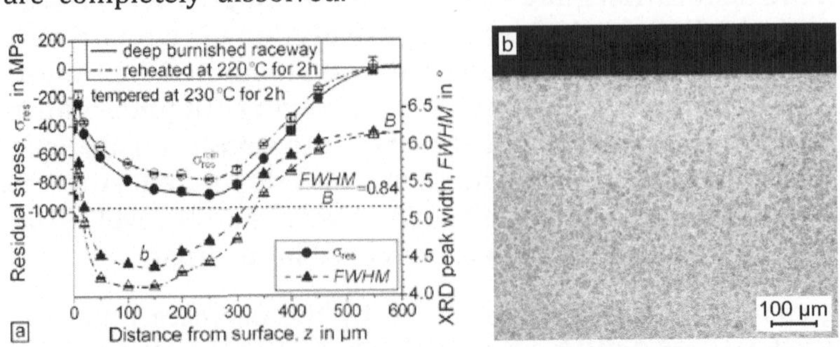

Fig. 23. Investigation of cold working of a martensite hardened OR revealing (a) the residual stress and XRD peak width distributions, respectively after deep ball burnishing (*b*/*B*»0.71) and subsequent reheating below the tempering temperature (unchanged hardness: 61 HRC) and (b) an etched axial microsection after burnishing free of visible microstructural changes

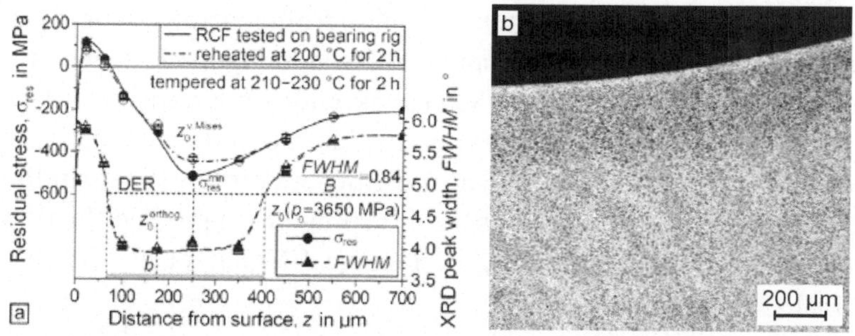

Fig. 24. Experimental investigation of reheating below tempering temperature (unchanged hardness: 60.5 HRC) after RCF loading on the martensite hardened IR of the endurance life tested DGBB of Figures 16 and 17 revealing (a) the initial and final residual stress and XRD peak width distributions (*b*/*B*»0.68) and (b) an etched axial microsection (DER indicated)

This approach also adumbrates an interpretation of the development of (steep) white bands (see Figure 21c) differently from adiabatic shearing (Schlicht, 2008). The DGSL model suggests strain induced reprecipitation of carbon in the form of carbides at a later stage of RCF damage (Lindahl & Österlund, 1982; Shibata et al., 1996). Former austenite or martensite grain boundaries represent sites for heterogeneous nucleation. Reprecipitated carbide films tend to embrittle the material.

Shakedown in Figure 10 can be considered to be a cold working process (Nierlich & Gegner, 2008). As discussed in section 3.3, the XRD line broadening is sensitive to changes of the lattice distortion. The rapid peak width reduction during shakedown occurs due to glide induced rearrangement of dislocations to lower energy configurations, such as multipoles. This dominating influence, which surpasses the opposing effect of the limited dislocation density increase in the defect-rich material of hardened bearing steel, reflects microstructure stabilization. An example of intense shakedown cold working is high plasticity ball burnishing. Figure 23a presents the result of the XRD measurement on the treated outer ring (OR) raceway of a taper roller bearing. The residual stress profile obeys the distribution of the v. Mises equivalent stress below the Hertzian contact (cf. Figure 1). The minimum XRD peak width *b* occurs closer to the surface. The applied Hertzian pressure is in the range of 6000 MPa (6 mm ball diameter). At the same *b*/*B* level of about 0.71 as in Figure 18a, in contrast to rolling contact fatigue, deep ball burnishing does not produce visible changes in the microstructure. The difference is evident from a comparison of the corresponding etched microsections in Figures 18c and 23b. Material alteration owing to mechanical conditioning by the build-up of compressive residual stresses in the shakedown cold working process is restricted to the higher fatigue endurance limit and based on yielding induced stabilization of the dislocation configuration but does not involve carbon diffusion (Nierlich &

Gegner, 2008). Therefore, no dark etching region from martensite decay develops in the microstructure of the burnished ring displayed in Figure 23b, even in the depth zone indicated in Figure 23a by the XRD peak width relationship *FWHM/B*£0.84. Mechanical surface enhancement treatments, like deep burnishing, shot peening, drum deburring and rumbling, as well as finishing operations (e.g. grinding, honing) and manufacturing processes, such as hard turning or (high-speed) cutting, are not associated with microstructural fatigue damage (Gegner et al., 2009; Nierlich & Gegner, 2008).

Figure 23a indicates that an additional stabilization of the plastically deformed steel matrix by dislocational carbon segregation can also be induced thermally by reheating after deep ball burnishing. The associated slight compressive residual stress reduction does not affect a bearing application. The positive effect of this thermal post-treatment on RCF life, in the literature reported for surface finishing (Gegner et al., 2009; Luyckx, 2011), suggests only subcritical partial carbide dissolution. According to the DGSL model, the corresponding amount of *FWHM* decrease should be included in the reduced *b* value in rolling contact fatigue (cf. Figure 22). Therefore, no additional effect by similar reheating below the applied tempering temperature is to be expected. This crucial prediction of the model is confirmed by the experiment. In Figure 24a, the small thermal reduction of the absolute value of the residual stresses is comparable with the alterations for burnishing shown in Figure 23a. However, reheating after RCF loading leaves the XRD peak width unchanged. In Figures 23a and 24a, the plotted s_{res} and *FWHM* values are deduced at separate sites of the raceway (i.e., one individual specimen for each depth) with increased reliability from three and eight repeated measurements, respectively, before and after the thermal treatment. The results of Figure 24a agree well with the XRD data of Figure 16a, determined by successive electrochemical polishing at one position of the racetrack of the same DGBB inner ring. This concordance is also evident for the indicated dark etching regions from a comparison of Figures 24b and 17a. The DGSL model is strongly supported by the discussed findings on the different *FWHM* response to reheating after rolling contact fatigue and cold working.

4.3 Current passage through bearings – The aspect of hydrogen absorption and accelerated rolling contact fatigue

The passage of electric current through a bearing causes damage by arcing across the surfaces of the rings and rolling elements in the contact zone. Fused metal in the arc results in the formation of craters on the racetrack, the appearance of which depends on the frequency. In the literature, the origin of causative shaft voltages in rotating machinery and the sources of current flows, the electrical characteristics of a rolling bearing and the influence of the lubricant properties as well as the development of the typical surface patterns are discussed in detail (Jagenbrein et al., 2005; Prashad, 2006; Zika et al., 2007, 2009, 2010). Complex chemical reactions occur in the electrically stressed oil film (Prashad, 2006). However, the ability of hydrogen released from decomposition products to be absorbed by the steel under the prevailing specific circumstances and subsequently to affect rolling contact fatigue is not yet investigated so far (Gegner & Nierlich, 2011b, 2011c).

Depending on the design of the electric generator, e.g. in diesel engines, alternator bearings may operate under current passage. Possible damage mechanisms become more important today because of the increased use of frequency inverters. Grease lubricated deep groove ball bearings with stationary outer ring, stemming from an automobile alternator rig test, are investigated in the following. The running period is in accordance with the nominal L_{10} life. Rings and balls are made out of martensitically hardened bearing steel. The racetrack in Figure 25a suffers from severe high-frequency electric current passage. Arc discharge in the lubricating gap causes a gray matted surface. The resulting shallow remelting craters of few μm in diameter and depth cover the racetrack densely. The indicated isolated indentation, magnified in Figure 25b, reveals the earlier condition of a less affected area. The tribological properties of the contact surface are still sufficient. The microsection of Figure 25c confirms the small influence zone by a thin white etching layer. However, continuous chemical decomposition of the lubricant and surface remelting promote hydrogen penetration. Thus, a highly increased content of more than 3 ppm by weight is measured for the DGBB outer ring of Figure 25 by carrier gas hot extraction (CGHE). Typical concentrations in the as- delivered state, after through hardening and machining, range from 0.5 to 1.0 ppm H.

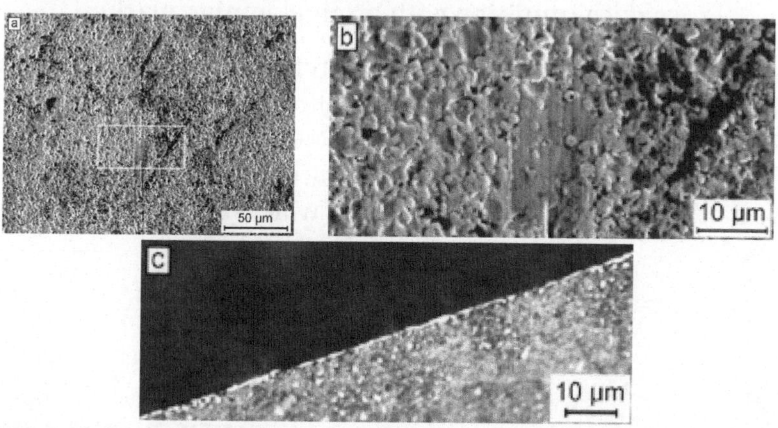

Fig. 25. Characterization of severe high-frequency electric current passage through a DGBB by (a) a SEM-SE overview and (b) the indicated SEM-SE detail of the remelted OR raceway track and (c) a near-surface LOM micrograph of an etched axial microsection

The amount of hydrogen absorbed by the steel depends on the release from the decomposition products of the aging lubricant and the available catalytically active blank metal surface (Kohara et al., 2006). Both affecting factors are enhanced by current passage in service. Fresh blank metal from remelting on the raceway enables the process step from physi- to chemisorption with abstraction of hydrogen atoms, which is otherwise effectively inhibited by the regenerative formation of a passivating protective reaction layer on the surface. The weaker operational high-frequency electric current passage of another bearing from the same rig test series documented in Figure 26a results only in a slightly increased content of 1.3 ppm H. The original honing structure of the raceway is displayed in Figure 26b. For comparison, Figures 25a, 26a and 26b have similar magnification.

An XRD material response analysis is performed in the load zone of the raceway of the hydrogen loaded outer ring of the bearing of Figure 25. According to Figures 27a, a high Hertzian pressure above 5000 MPa is deduced.

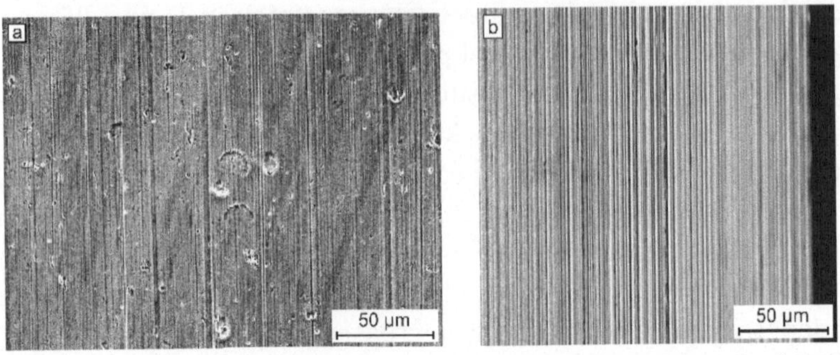

Fig. 26. SEM-SE image of the raceway (a) of the OR of an identical DGBB tested in the same alternator rig as the bearing of Figure 25 after moderate high-frequency electric current passage and (b) in as-delivered (non-overrolled) surface condition with original honing marks

The applied joint evaluation of the depth profiles of the residual stress and XRD peak width in the subsurface zone of classical rolling contact fatigue is shown in Figure 27b. The damage parameter equals $b/B \approx 0.71$. The XRD L_{10} life equivalent is thus not yet exceeded on the outer ring. The microsection in Figure 27c confirms a subsurface dark etching region, the position of which reflects the contact angle.

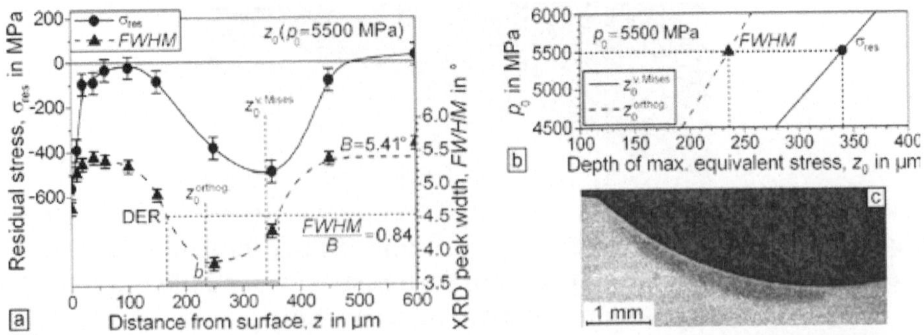

Fig. 27. Material response analysis of the OR of the tested DGBB of Figure 25 including (a) the residual stress and XRD peak width distribution (b/B»0.71, B measured below the shoulder), (b) the joint profile evaluation and (c) an axial microsection with pronounced DER

Inside the wide DER of Figure 27c, extended white etching areas are located (cf. Figure 28a), which evolve from the steel matrix. In the used clean material, butterfly formation is irrelevant and only two early stages are found (see inset of Figure 28a). Etching accentuates the actual RCF damage: the DER identified as brittle by the observed preparative cracking is clearly distinguishable from the chemically less affected material above and below in the indicated SEM-SE detail of Figure 28b. The WEA inside the DER appear smooth black.

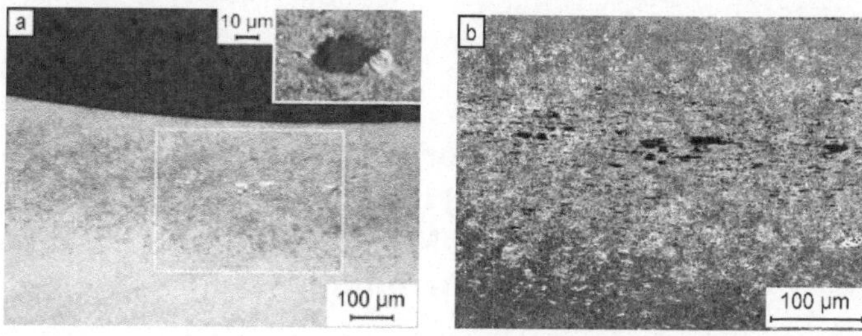

Fig. 28. Etched axial microsection of the DGBB outer ring of Figure 27c revealing (a) a LOM overview (the inset shows an embryo butterfly) and (b) the indicated SEM-SE detail

The LOM micrograph in Figure 29a reveals dense dark etching regions adjacent to the WEA zones. Although reported contrarily in the literature (Martin et al., 1966), the embrittled dark etching region evidently acts as precursor of further phase transformation. The SEM-SE detail of Figure 29b also points to interfacial delamination (see indication) as pre-stage of fatigue crack initiation.

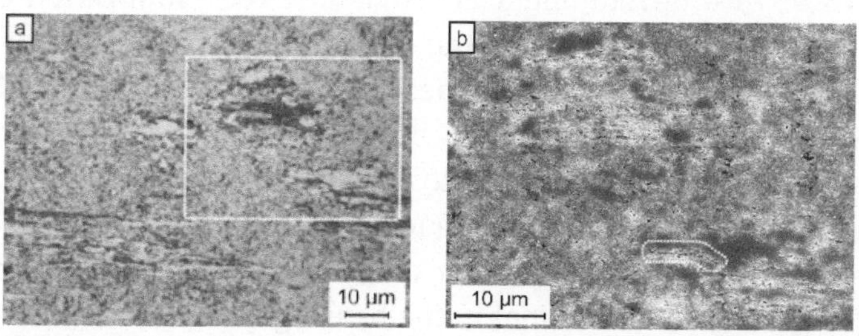

Fig. 29. Etched axial microsection of the DGBB outer ring of Figure 27c revealing (a) a LOM image and (b) the indicated SEM-SE detail

The development of white etching bands is identified in the radial microsection of the investigated outer ring shown in Figure 30a. Dense FWB and distinct SWB of inclinations b_f=25° and b_s=76°, respectively, are visible inside the indicated DER. The central SEM-SE detail of Figure 30b reveals the included angle b_{s-f} of 51° (see section 4.1, Figure 21c). The indication of microcrack initiation on white etching bands by interfacial delamination is confirmed by Figure 30c. It is not observed in pure mechanical rolling contact fatigue (Voskamp, 1996), where actually an influence of WEB (as well as of butterfly) formation on bearing life is not proven (Schlicht, 2008). Therefore, hydrogen induced cracking propensity on WEB suggests higher hardness of the white etching areas and hydrogen embrittlement. Note again the pronounced DER microstructure around the WEA in Figure 30c.

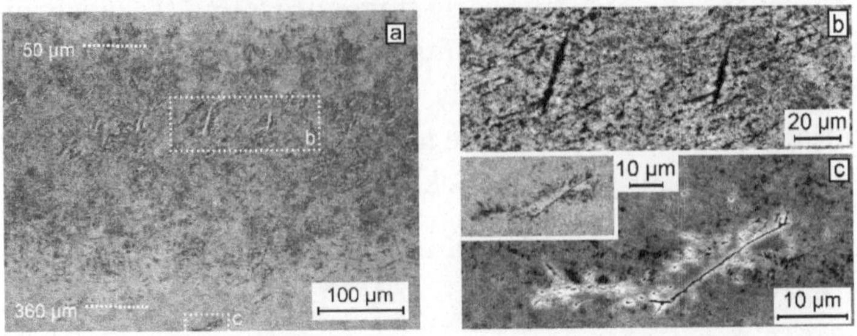

Fig. 30. Etched radial microsection of the OR of Figure 27c revealing (a) a LOM overview with indicated DER, (b) the SEM-SE detail b and (c) the SEM-SE detail c, where the corresponding LOM inset highlights the WEA precursor effect of the surrounding DER

As also emphasized in Figure 31a by grain boundary etching, flat and steep white bands evolve from the distinctive surrounding DER material. The SWB seem to develop in an earlier stage prior to the complete dense formation of FWB (cf. Figure 21c). Particularly the oriented slip bands of FWB exhibit more intense white etching microstructure (cf. Figure 21c). Figure 31b presents the corresponding SEM-SE image of this extended detail of Figure 30b in the center of Figure 30a. The gradual evolution of white etching bands from the DER precursor, as particularly evident from Figure 31a, indicates advancing fatigue processes, e.g. as outlined in section 4.2, presumably correlated with texture development and dynamic recrystallization during rolling contact loading (Voskamp, 1996). On the other hand, this microstructural finding speaks against causative adiabatic shearing (Schlicht, 2008). The preferred occurrence of white etching bands in ball bearings should rather be connected with the higher Hertzian pressure compared to a corresponding roller contact. Note that no WEA of premature rolling contact fatigue damage are formed in the case of Figure 26. This moderate high-frequency electric current passage in operation is connected with only slight hydrogen enrichment in the bearing steel.

Despite the occurrence of white etching bands in the outer ring of the rig tested DGBB of Figure 25, as documented in Figures 28 to 31, the XRD material aging parameter deduced from Figure 27a amounts to just b/B»0.71. The same value is derived from the peak width distribution in Figure 18a, where for pure mechanical subsurface RCF, however, no WEA are formed inside the DER (see Figure 18c). As for the bearing operating under severe high- frequency electric current passage, the XRD L_{10} equivalent of classical rolling contact fatigue without additional chemical loading is not yet exceeded but well developed white etching bands, particularly SWB, already occur, hydrogen charging noticeably accelerates the evolution of microstructural RCF damage (hydrogen accelerated rolling contact fatigue, H-RCF). The dark etching region extends to zones of $FWHM/B$>0.84 near the surface, as evident from a comparison of Figures 27a, 28a and 30a. The calibration relationship between the L_{10} life and the evidently reduced b/B equivalent is modified by the hydrogen embrittled DER.

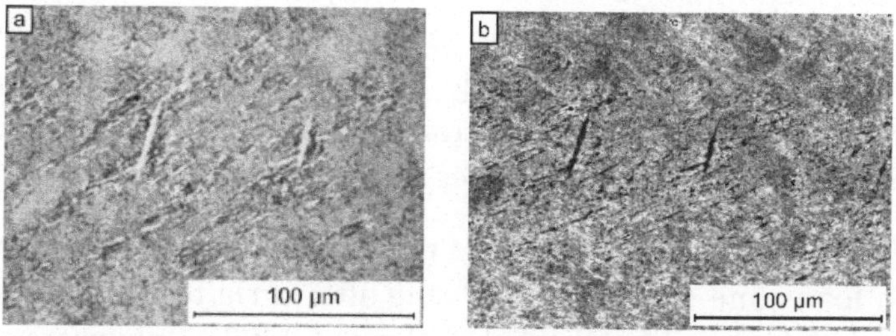

Fig. 31. Detail (approx. b) of Figure 30 comparing (a) a LOM and (b) a SEM-SE micrograph

The metal physics dislocation glide stability loss model, introduced in section 4.2, provides an approach to the mechanistic description of rolling contact fatigue in bearing steels. Hydrogen interacts with lattice defects (Gegner et al., 1996). The response to cyclic loading reflects its high atom mobility even at low temperature. The effect of hydrogen can be illustrated by the DGSL model of Figure 22. The microscopic fatigue processes are considerably promoted by intensifying the increase of dislocation density and glide mobility. Mechanisms of hydrogen enhanced localized plasticity (HELP) are discussed in the literature (Birnbaum & Sofronis, 1994). A comparison of chemically assisted with pure mechanical rolling contact fatigue and shakedown cold working at constant reference level of b/B»0.71 in Figures 18, 23 and 27 to 31, completed by Figures 20, 21 and 24, suggests that material aging is accelerated by enhancing the microplasticity. At the same stage of b/B reduction, microstructural RCF damage is much more advanced. Premature formation of ribbon like or irregularly oriented white etching areas, for instance, might yet occur at lower loads.

5. Surface failure induced by mixed friction in rolling-sliding contact

The practically predominating surface failure mode involves various damage mechanisms. Besides indentations, discussed in detail in section 2.2, mixed friction or boundary lubrication in the rolling contact area occurs frequently in bearing applications. Polishing wear on the raceway, resulting in differently pronounced smoothing of the machining marks, is a characteristic visual indication. The depth of highest material loading is shifted towards the surface by sliding friction in rolling contact. The effect on the distribution and the maximum of the equivalent stress is similar to the scheme shown in Figure 5. The mechanisms of crack initiation on the surface are of utmost technical importance (Olver, 2005). New aspects of rolling contact tribology in bearing failures are presented in the following.

5.1 Vibrational contact loading and tribological model

Near-surface loading is often superimposed by the impact of externally generated three- dimensional mechanical vibrations that represents a common cause of disturbed EHL operating conditions, e.g., in paper making or weaving machines, coal pulverizers, wind turbines, cranes, trains, tractors and fans. Ball bearings in car alternators of four-cylinder diesel engines are another familiar example.

The SEM image of Figure 32a shows the completely smoothed raceway in the rotating main load zone of a CRB inner ring after a rig test time of about 40% of the calculated nominal L_{10} life (Nierlich & Gegner, 2002). Only parts of the deepest original honing grooves are left over on the surface. Causative mixed friction results from inadequate lubrication conditions without sufficient film formation (fuel addition to the oil). Initial micropitting by isolated material delamination of less than 10 μm depth is observed. Figure 32b provides a comparison with the non-overrolled as-finished raceway condition. On the damaged inner ring, a residual stress material response analysis is performed. The result is shown in Figure 33a. No changes of the measured XRD parameters in the depth of the material are found, whereas the XRD peak width on the surface decreases to $b/B^{3}0.79$. The relation symbol accounts for the small *FWHM* reduction of about 0.2° due to the honing process (see section 3.3, Figure 16a). Material aging considerably exceeds the XRD L_{10} equivalent value of 0.86 for the relevant surface failure mode of RCF. The corresponding re-increase of the residual stress on the raceway, discussed in the context of Figures 11 and 12 in section 3.3, reaches –230 MPa.

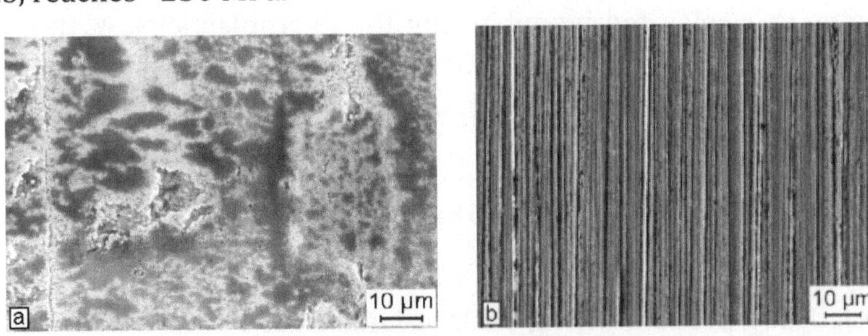

Fig. 32. SEM-SE image of (a) the damaged raceway of the inner ring of a CRB after rig testing under engine vibrations and (b) an original honing structure at the same magnification

The residual stress distribution of Fig. 33a is identified as a type B profile of vibrational loading in rolling-sliding contact (Gegner & Nierlich, 2008). The characteristic compressive residual stress side maximum in a short distance from the surface (here 40 μm), clearly above the depth v.Mises 0 of maximum v. Mises equivalent stress for pure radial load, is reflected in the corresponding reduction of the XRD peak width. The monotonically increasing type A vibration residual stress profile occurs more frequently in practical applications. The result of a material response analysis on a CRB outer ring, the raceway of which does not reveal indentations, represents a prime example in Figure 33b. Bainitic through hardening of the bearing steel results in compressive residual stresses in the core of the material. The XRD life parameter $b/B^{3}0.82$ is taken from the diagram. The running time of 2´108 revolutions indicates low-cycle fatigue under the influence of intermittently acting severe vibrations (Nierlich & Gegner, 2008). The residual stress analysis of the inner ring of a taper roller bearing from a harvester in Figure 34a provides another instructive example. Mixed short-term deeper reaching type A vibrational and near-surface Hertzian micro contact loading of the material are superimposed. Figure34b reveals indentations on the partly smoothed raceway. The applied Hertzian pressure p_0 amounts to 2000 MPa. For comparison, the depth of maximum v. Mises equivalent stress for incipient plastic deformation in pure radial contact loading, i.e. p_0 above 2500 to 3000 MPa, equals about 180 µm.

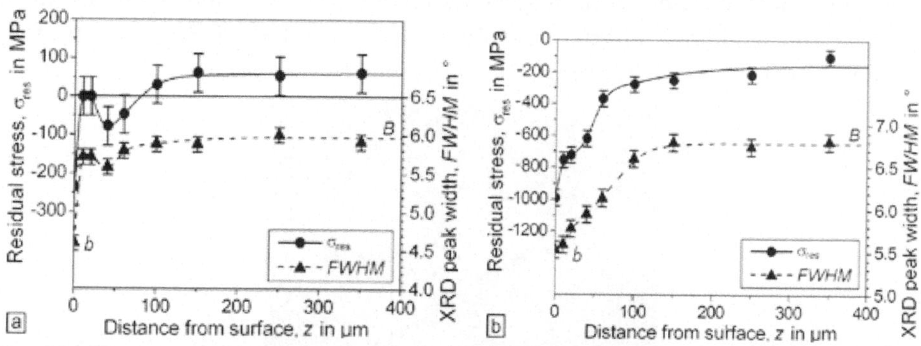

Fig. 33. The two types of vibration residual stress-XRD line width profiles, i.e. (a) type B with near-surface side peaks measured on the IR raceway of a CRB from a motorcycle gearbox test rig and (b) type A with monotonically increasing curves from a field application

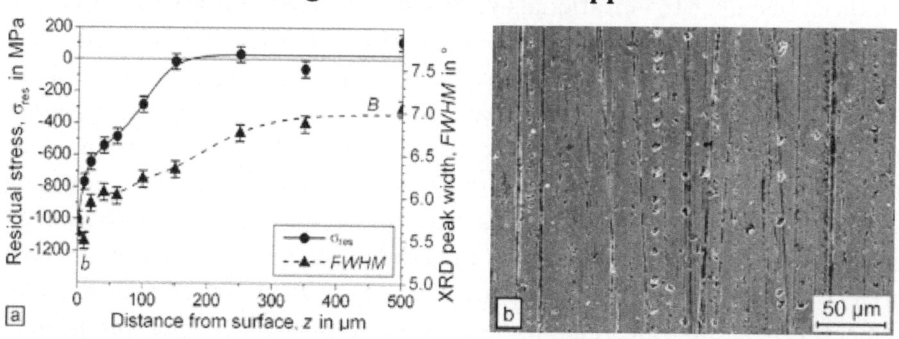

Fig. 34. Investigation of the IR of a vibration-loaded harvester TRB revealing (a) the obtained type A residual stress pattern and (b) a SEM-SE image of the raceway with indentations

Both types of residual stress distributions are simulated experimentally in a specially designed vibration test rig for rolling bearings (Gegner & Nierlich, 2008). A type N CRB is used. The stationary lipless outer ring of the test bearing is displaced and experiences high vibrational loading via the sliding contact to the rollers. It thus becomes the specimen. In addition to the radial load, controlled uni- to triaxial vibrations can be applied in axial, tangential and radial direction. Figure 35 displays a photograph of the rig. It represents a view of the housing of the test bearing and the equipment for the transmission of axial and tangential vibrations (radial excitation from below) with thermocouples and displacement sensors.

A micro friction model of the rolling-sliding contact is introduced by means of Figure 36. It describes the effect of vibrational loading. As shown in Figure 36, tangential forces by sliding friction acting on a rolling contact increase the equivalent stress and shift its maximum toward the surface on indentation-free raceways (Broszeit et al., 1977). A transition, indicated by solid-line curves, occurs between friction coefficients m of 0.2 and 0.3: above and below m=0.25, the increasing maximum of the Tresca equivalent stress is located directly on or near the surface, respectively. If the yield strength of the material is exceeded (cf. Figure 5), therefore, type A or B residual stress depth profiles are generated.

Fig. 35. Housing of the test bearing with devices for vibration generation

Material response to vibrational loading, which causes increased mixed friction, is described in the tribological model by partitioning the nominal contact area *A* into microscopic sections of different friction coefficients (Gegner & Nierlich, 2008). The inset of Figure 36 illustrates the basic idea. In some subdomains, arranged e.g. in the form of dry spots or bands, peak values from $m_{>}$»0.2 (type B) to $m_{>}{}^{3}$0.3 (type A) are supposed to be reached intermittently for short periods. The thixotropy effect supports this concept because shearing of the lubricant by vibrational loading reduces the viscosity, which increases the tendency to mixed friction. In the other subareas of the contact, $m_{<}$ is much lower so that the average friction coefficient $m_{(eff)}$, meeting a mixing rule, remains below 0.1 as typical of running rolling bearings. Besides the verified compressive residual stress buildup, nonuniform cyclic mechanical loading of the contact area by, in general, complex three- dimensional vibrations is also evident from occasionally observed dent-like plastic deformation on the surface, spots of dark etching regions in the microstructure of the outermost material and varying preferred orientation of yielding across the raceway width, reflected in differing tangential and axial components of the residual stresses in the affected edge zone (Gegner & Nierlich, 2008). Friction increase is confirmed by temperature rise in the lubricating gap that correlates with the power loss per contact area. This effect can be exploited to easily assess the vibration resistance of specific oils or greases on the adapted bearing test rig (Gegner & Nierlich, 2008).

Under the influence of vibrations, disturbance of proper contact operating conditions in a way that high shearing stresses are induced in the lubricating film can promote lubricant degradation (Kudish & Covitch, 2010). Reduced lubricity enhances the effect of sliding friction, e.g. described in the tribological model of Figure 36. Further to the discussed mechanical and thermal influence, vibration loading induces chemical aging of the lubricant and its additives (Gegner & Nierlich, 2008). Contaminations, like water or wear debris, increase the effect. The gradual decomposition process and associated acidification of the lubricant promote, for instance, the initiation of surface cracks on the raceway by tribochemical dissolution of nonmetallic MnS inclusion lines, which is discussed in the next section.

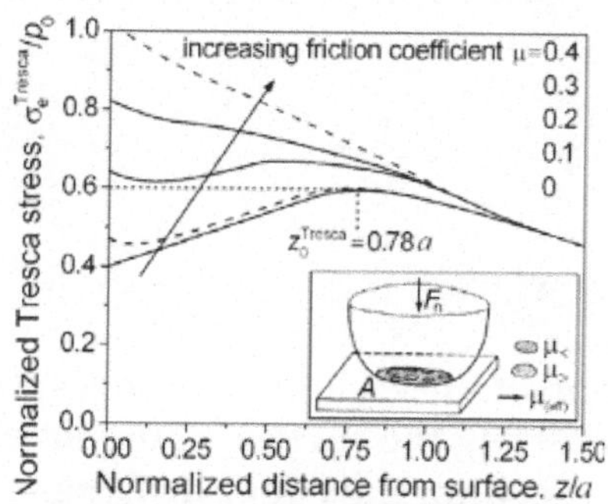

Fig. 36. Distribution of the Tresca equivalent stress below a rolling-sliding contact (z_0 depth indicated for pure radial load, i.e. m=0) and illustration of the tribological model of localized friction coefficient in the inset (F_n is the normal force)

5.2 Tribochemically initiated surface cracks

First, Figure 37 gives a demonstrative example of a corrosive attack by a decomposed lubricant. The etching pattern on the raceway reveals chemical smoothing of the surface.

Copper contamination by abrasion from the graphite-brass ground brush of the diesel electric locomotive gets into the grease of the train wheel bearing and accelerates lubricant aging. As evident from Figure 37, the foreign particles also cause indentations on the raceway. An electrical oil sensor system can be used for online condition monitoring of the lubricant (Gegner et al., 2010). The application in industrial gearboxes, for instance of wind turbines, is of special practical interest. Note that the (e.g., extreme pressure) additives markedly influence the electrical properties of the lubricant (Prashad, 2006).

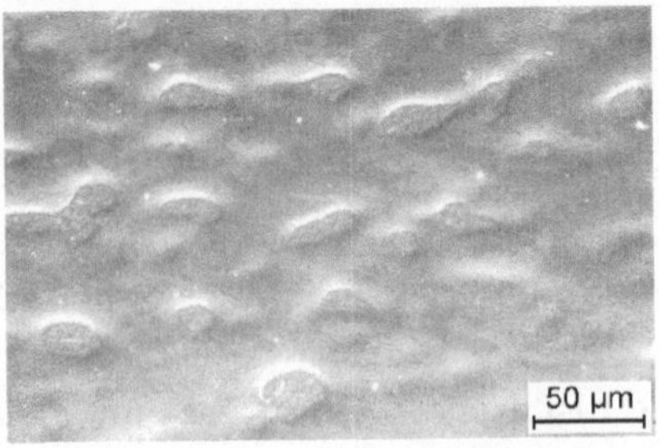

Fig. 37. SEM-SE image of a chemical surface attack on the outer ring raceway of a CRB

As exemplified by Figure 38, some manganese sulfide lines intersect the rolling contact surface. Such inclusions are manufacturing related from the steelmaking process, despite the high level of cleanliness of bearing grades.

Fig. 38. LOM micrograph of the etched metallographic section of a sulfide inclusion line intersecting the surface of the inner ring raceway of a cylindrical roller bearing

On the inner ring raceway of a cylindrical roller bearing of a weaving machine examined in Figure 39, mixed friction is indicated by the mechanically smoothed honing structure. Due to aging of the lubricating oil, as detected under vibration loading, the gradually acidifying fluid attacks the steel surface. Tribochemical dissolution of manufacturing related MnS inclusion lines leaves crack-like defects on the raceway. Sulfur is continuously removed as gaseous H_2S by hydrogen from decomposition products of the lubricant:

$$MnS + H_2 \rightarrow H_2S^{\uparrow} + Mn \qquad (6)$$

The remaining manganese is then preferentially corroded out. This new mechanism of crack formation on tribologically loaded raceway surfaces is verified by chemical characterization using energy dispersive X-ray (EDX) microanalysis on the SEM. The EDX spectra in Figure 39, recorded at an acceleration voltage of 20 kV, confirm residues of manganese and sulfur at four sites (S1 to S4) of an emerging crack, thus excluding accidental intersection. The ring is made of martensitically hardened bearing steel. Reaction layer formation on the raceway is reflected in the signals of phosphorus from lubricant additives and oxygen.

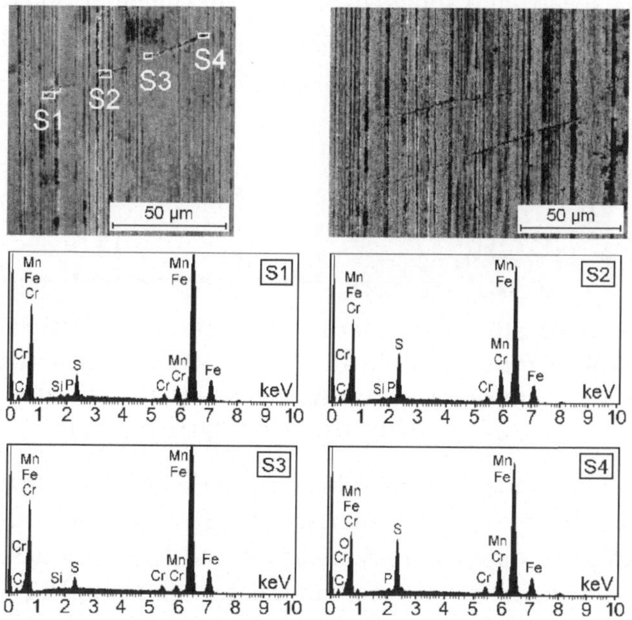

Fig. 39. SEM-SE images of cracks on the IR raceway of a CRB from the gearbox of a weaving machine and EDX spectra S1 to S4 taken at the indicated analysis positions

Crack initiation by tribochemical reaction is also found on lateral surfaces of rollers. In Figure 40, remaining manganese and sulfur are detected by elemental mapping in the insets on the right.

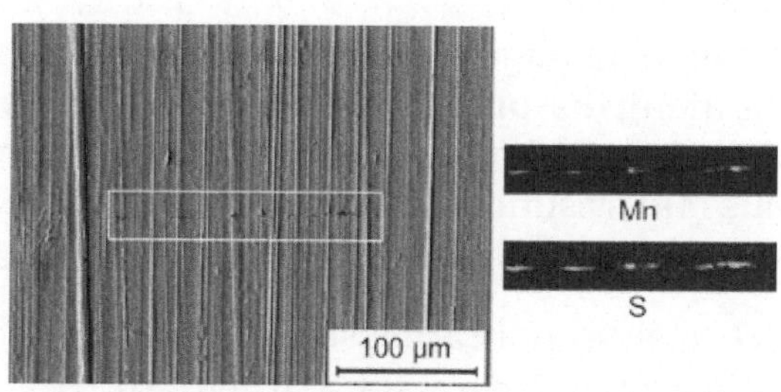

Fig. 40. SEM-SE image of a crack on a CRB roller and elemental mapping (area as indicated)

The tribochemical dissolution of MnS lines on raceway surfaces during the operation of rolling bearings also agrees with the general tendency that inclusions of all types reduce the corrosion resistance of the steel. The chemical attack occurs by the lubricant aged in service. The example of an early stage of defect evolution in Figure 41a points out that continuous dissolution but not fracturing of MnS inclusions gradually initiates a surface crack. Three analysis positions, where residues of manganese and sulfur are found, are indicated in the SEM image. An exemplary EDX spectrum is shown in Figure 41b. The inner ring raceway of the ball bearing from a car alternator reveals high-frequency electric current passage (cf. Figure 26a) that promotes lubricant aging (see section 4.3).

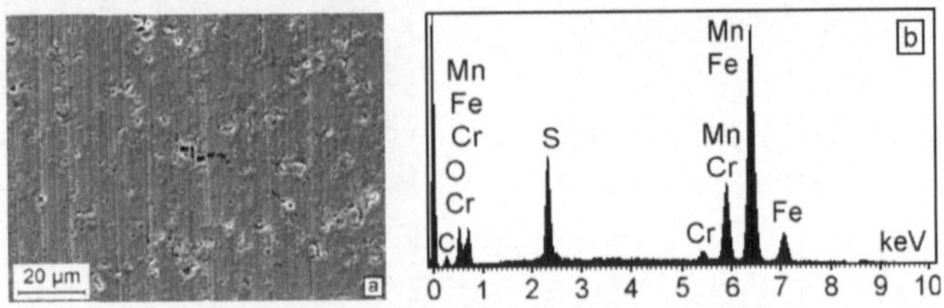

Fig. 41. Tribochemically induced crack evolution on the IR raceway of a DGBB revealing a SEM-SE image with indicated sites where EDX analysis proves the presence of residues of MnS dissolution and (b) a recorded EDX spectrum exemplarily of the analysis results

After defect initiation on MnS inclusions, further damage development involves shallow micropitting (Gegner & Nierlich, 2008). Figure 42a also suggests crack propagation into the depth. Four sites of verified MnS residues are indicated, for which Figure 42b provides a representative detection example. The partly smoothed raceway reflects the effect of mixed friction.

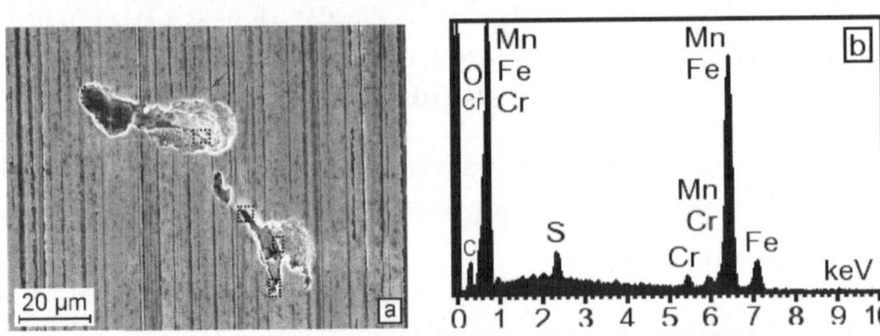

Fig. 42. Documentation of damage evolution by (a) a SEM-SE image of shallow material removals along dissolved MnS inclusions on the IR raceway of a TRB from an industrial gearbox with indication of four positions where EDX analysis reveals MnS residues and (b)EDX spectrum exemplarily of the analysis results recorded at the sites given in Figure 42a

The EDX reference analysis of bearing steel is provided in Figure 43. It allows comparisons with the spectra of Figures 39, 41b and 42b.

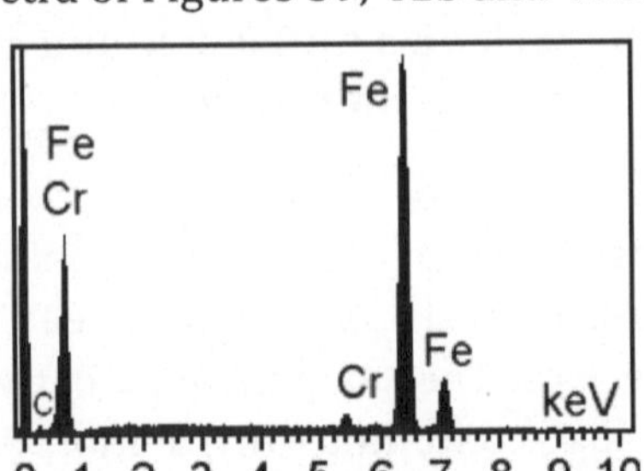

Fig. 43. EDX reference spectrum of bearing steel for comparison of the signals

5.3 Gray staining – Corrosion rolling contact fatigue

Gray staining by dense micropitting, well known as a surface damage on tooth flanks of gears, is also caused by mixed friction in rolling-sliding contact. The flatly expanded shallow material fractures of only few µm depth, which cover at least parts of an affected raceway, are frequently initiated along honing marks. In Figure 44a, propagation of material delamination to the right occurs into sliding direction. Typical features of the influence of corrosion are visible on the open fracture surfaces. The corresponding XRD material response analysis in Figure 44b shows that vibrational loading of the tribological contact can cause gray staining. Note that the shallow micropits do not affect the residual stress state considerably. The smoothed raceway of Fig. 44a, which indicates mixed friction, is virtually free of indentations. A characteristic type A vibration residual stress profile, maybe with some type B contribution in 100 µm depth (cf. Figures 33 and 36, z_0 much larger), is obtained. The XRD rolling contact fatigue damage parameter of $b/B^3$0.83 reaches or slightly exceeds the L_{10} equivalent value of 0.86 for the surface failure mode of roller bearings.

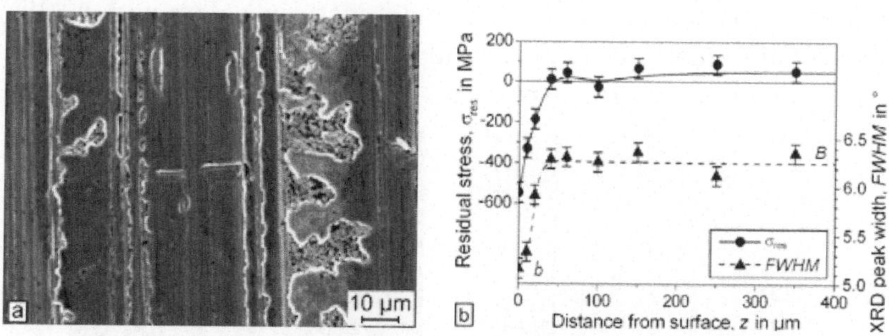

Fig. 44. Investigation of gray staining on the IR raceway of a CRB revealing (a) a SEM-SE image and (b) the measured type A vibration residual stress and XRD peak width distribution

The appearance of the micropits on the raceway is similar to shallow material removals on tribochemically dissolved MnS inclusions, as evident from a comparison of Figures 44a and 42a. Micropitting can occur on small cracks initiated on the loaded surface. The SEM image of Figure 45a indicates such causative shallow cracking induced by shear stresses, slightly inclined to the axial direction. The metallographic microsection in Figure 45b documents crack growth into the material in a flat angle to the raceway up to a small depth of few µm followed by surface return to form a micropit eventually.

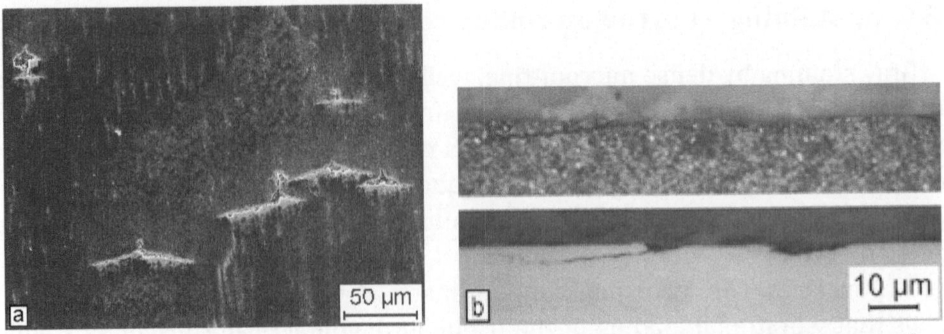

Fig. 45. Investigation of gray staining on the IR raceway of a rig tested automobile gearbox DGBB revealing (a) a SEM-SE image and (b) LOM micrographs of the etched (top) and unetched section of a developing micropit

The SEM overview in Figure 46a illustrates how dense covering of the raceway with micropits results in the characteristic dull matte appearance of the affected surface. On the bottom left hand side of the detail of Figure 46b, damage evolution on axially inclined microcracks results in incipient material delamination. Micropitting on a honing groove illustrates typical band formation. Note that the b/B parameter is reduced on the raceway surface to 0.69.

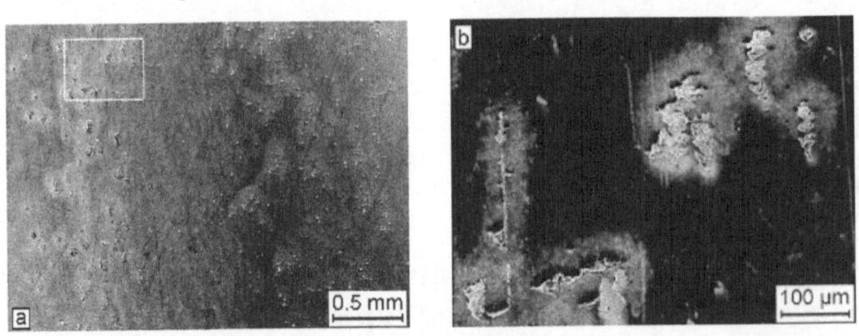

Fig. 46. Investigation of the smoothed damaged inner ring raceway of the deep groove ball bearing of Figure 45a presenting (a) a SEM-SE overview and (b) the indicated detail that reveals near-surface crack propagation in overrolling direction from the bottom to the top

Pronounced striations on the open fracture surfaces of micropits prove a significant contribution of mechanical fatigue to the crack propagation. The SEM details of Figures 47a and 47b confirm this finding. Therefore, it is concluded that a variant of corrosion fatigue is the driving force behind crack growth of micropitting in gray staining.

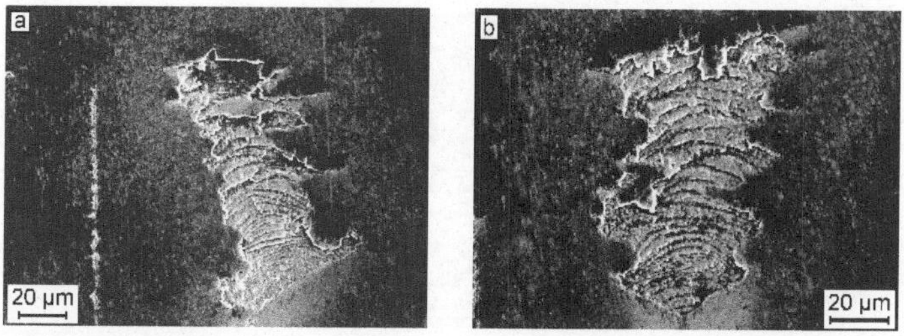

Fig. 47. SEM-SE details of the inner ring raceway of the deep groove ball bearing of Figure 46a revealing (a) distinct striations on a micropit fracture surface and (b) the same microfractographic feature on the open fracture face of another micropit

The additional chemical loading is not considered in fracture mechanics simulations of micropit formation by surface initiation and subsequent propagation of fatigue cracks (Fajdiga & Srami, 2009). The findings discussed above, however, suggest that gray staining can be interpreted as corrosion rolling contact fatigue (C-RCF).

5.4 Surface embrittlement in operation

Although quickly obscured by subsequent overrolling damage in further operation, shallow intercrystalline fractures are sporadically observed on raceway surfaces (Nierlich & Gegner, 2006). Illustrative examples are shown in the SEM images of Figures 48a and 48b.

The microstructure breaks open along former austenite grain boundaries. The affected raceway is heavily smoothed by mixed friction. Figure 48a and 48b characterize the lateral surface of a roller from a rig tested TRB and gray staining on the cam race tracks of a camshaft, respectively. The even appearance of the separated grain boundaries points to intercrystalline cleavage fracture of embrittled surface material by frictional tensile stresses. The micropit on a raceway suffering from gray staining in Figure 49 suggests partly intercrystalline corrosion assisted crack growth. Striation-like crack arrest marks are clearly visible on the fracture surface. Microvoids in the indicated region point to corrosion processes (see section 5.3, C-RCF).

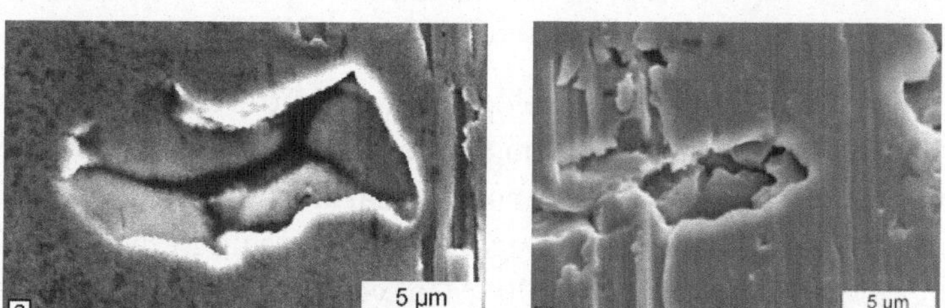

Fig. 48. SEM-SE images of the rolling contact surfaces of (a) a TRB roller and (b) a cam

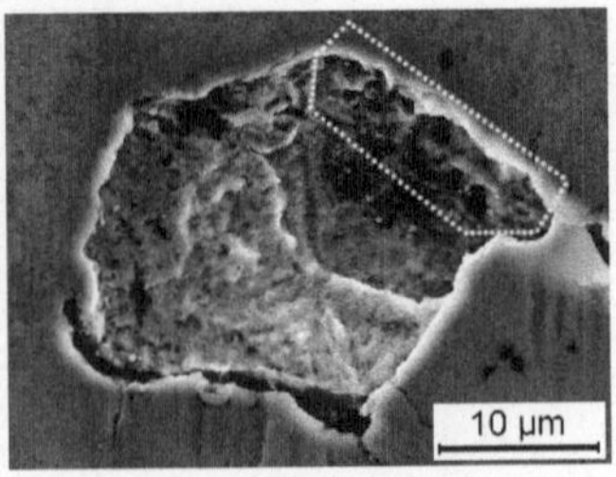

Fig. 49. SEM-SE image of a micropit on the IR raceway of a CRB from a field application

Possible mechanisms of gradual near-surface embrittlement during overrolling are (temper) carbide dissolution by dislocational carbon segregation (see section 4.2, Figure 22), carbide reprecipitation at former austenite or martensite grain boundaries, hydrogen absorption and work hardening by raceway indentations or edge zone plastification in the metal-to-metal contact under mixed friction. The occurrence of plate carbides, for instance, in micropits of gray staining is reported (Nierlich & Gegner, 2006). Due to lower chromium content than the steel matrix, these precipitates are obviously formed during rolling contact operation.

5.5 White etching cracks

Premature bearing failures, characterized by the formation of heavily branching systems of cracks with borders partly decorated by white etching microstructure, occur in specific susceptible applications typically within a considerably reduced running time of 1% to 20% of the nominal L_{10} life. Therefore, ordinary rolling contact fatigue can evidently be excluded as potential root cause, which agrees with the general finding that only limited material response is detected by XRD residual stress analyses. As shown in Figure 50, axial cracks of length ranging from below 1 to more than 20 mm, partly connected with pock-like spallings, are typically found on the raceway in such rare cases. For an affected application, for instance, it is reported in the literature that the actual L_{10} bearing life equals only six months, resulting in 60% failures within 20 months of operation (Luyckx, 2011).

Particularly axial microsections often suggest subsurface damage initiation. An illustrative example is shown in Figure 51.

In the literature, abnormal development of butterflies, material weakening by gradual hydrogen absorption through the working contact and severe plastic deformation in connection with adiabatic shearing are considered the potential root cause of premature bearing damage by white etching crack (WEC) formation (Harada et al., 2005; Hiraoka et al., 2006; Holweger & Loos, 2011; Iso et al., 2005; Kino & Otani, 2003; Kohara et al., 2006; Kotzalas & Doll, 2010; Luyckx, 2011; Shiga et al., 2006). These hypotheses, however, conflict with essential findings from failure analyses (further details are discussed in the following). White etching cracks are observed in affected bearings without and with butterflies (Hertzian pressure higher than about 1400 MPa required, see section 3.3) so that evidently both microstructural changes are mutually independent. Depth resolved concentration determinations on inner rings with differently advanced damage show that hydrogen enrichment occurs as a secondary effect abruptly only after the formation of raceway cracks by aging reactions of the penetrating lubricant, i.e. rapidly during the last weeks to few months of operation but not continuously over a long running time (Nierlich & Gegner, 2011). Hydrogen embrittlement on preparatively opened raceway cracks, reflected in an increased portion of intercrystalline fractures, is restricted to the surrounding area of the original cracks (Nierlich & Gegner, 2011). The undamaged rolling contact surface is protected by a regenerative passivating reaction layer. Adiabatic shear bands (ASB) develop by local flash heating to austenitising temperature due to very rapid large plastic deformation characteristic of, for instance, high speed machining or ballistic impact. Such extreme shock straining conditions obviously do not arise during bearing operation. WEC reveal strikingly branched crack paths, whereas ASB form essentially straight regular ribbons of length in the mm range. Adiabatic shearing represents a localized transformation into white etching microstructure possibly followed by cracking of the brittle new ASB phase. WEC evolve contrary by primary crack growth. Parts of the paths are subsequently decorated with white etching constituents.

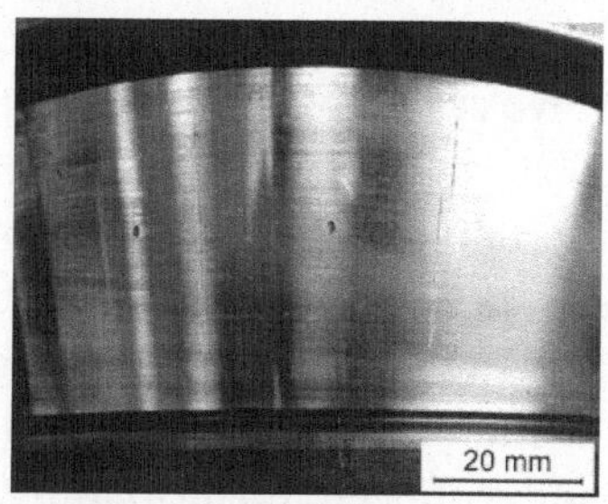

Fig. 50. Macro image of the raceway of a martensitically hardened inner ring out of bearing steel of a taper roller bearing from an industrial gearbox

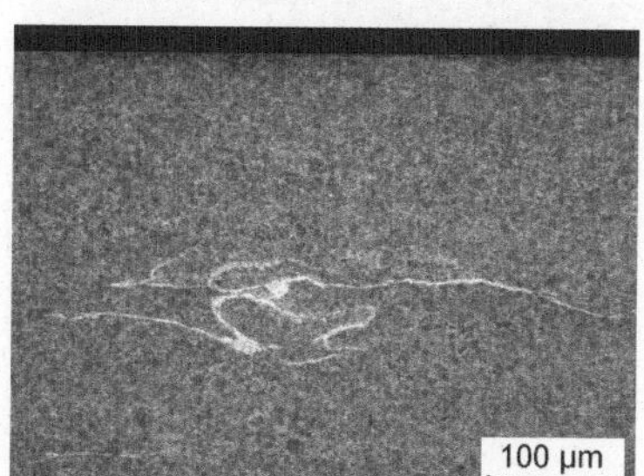

Fig. 51. LOM micrograph of the etched axial microsection of the bainitically hardened inner ring of a spherical roller bearing from a crane lifting unit

The spidery pattern of the white etching areas in Figure 51 indicates irregular crack propagation prior to the microstructural changes on the borders. Equivalent stresses reveal uniform distribution in the subsurface region. The reason for the appearance of Figure 51 is the spreading and branching growth of the cracks in circumferential orientation. Cracks originated subsurface usually do not create axial raceway cracks but emerge at the surface mostly as erratically shaped spalling (cf. Figure 2b). Targeted radial microsections actually reveal the connection to the raceway. Figure 52 points to surface WEC initiation due to the overall orientation and depth extension of the crack propagation in overrolling direction from left to right. One can easily imagine how damage pattern similar to Figure 51 occur in accidentally located etched axial microsections.

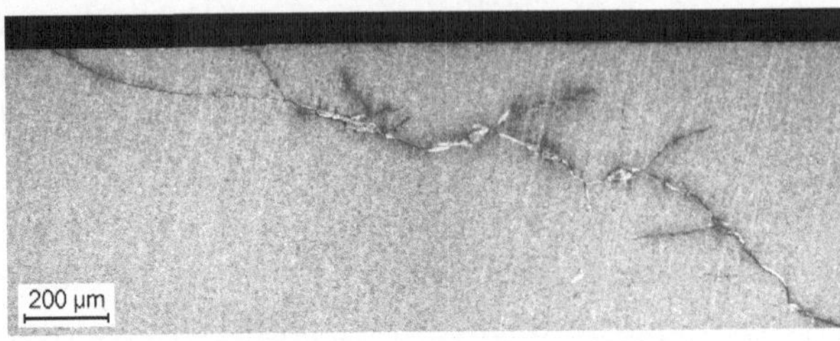

Fig. 52. LOM micrograph of the etched radial microsection of the case hardened inner ring of a CARB bearing from a paper making machine. The overrolling direction is left-to-right

Another example is shown in Figure 53a. The overrolling direction is from left to right so that crack initiation on the surface is evident. Figure 53b reveals the view of the edge of this microsection. No crack is visible at the initiation site on the raceway in the SEM (see section 5.5.1) so that also the detection probability question arises. The intensity of the white microstructure decoration of individual crack segments depends, for instance, on the depth (e.g., magnitude of the orthogonal shear stress) and the orientation to the raceway surface (friction and wear between the flanks). The pronounced tendency of the propagating cracks to branch indicates no pure mechanical fatigue but high additional chemical loading. Together with the regularly observed transcrystalline crack growth, this is typical of corrosion fatigue.

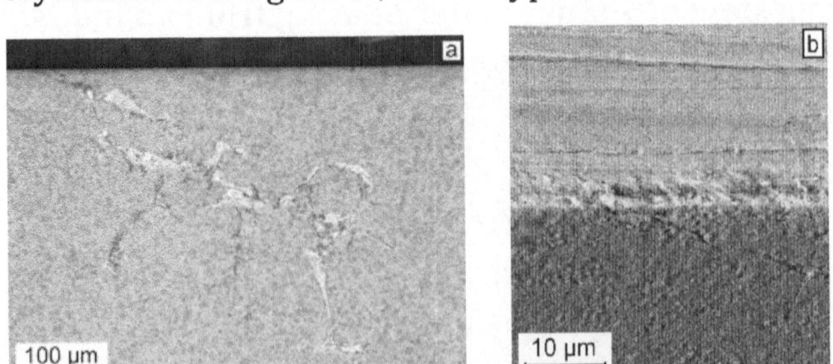

Fig. 53. Investigation of a white etching crack system in the martensitically hardened inner ring of a taper roller bearing from a coal pulverizer revealing (a) a LOM micrograph of the etched radial microsection (overrolling direction from left to right) and (b) a near-surface SEM detail (backscattered electron mode) of the view of the edge of the same microsection

5.5.1 Shear stress induced surface cracking and corrosion fatigue crack growth

Mixed friction in rolling-sliding contact can cause surface cracks on bearing raceways. The shear stress induced initiation mechanism is introduced first. The result of the XRD material response analysis performed on both raceways of a double row spherical roller bearing is depicted in Figures 54a and 54b.

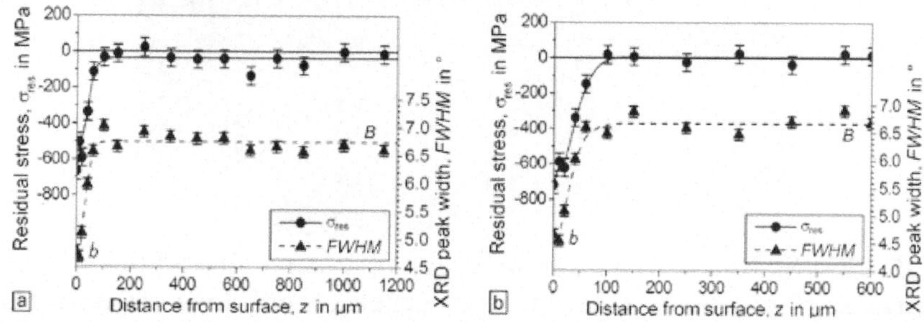

Fig. 54. Material response analysis showing a type A vibration residual stress and XRD peak width distribution below (a) the first and (b) the second raceway surface of the inner ring of a prematurely failed double row spherical roller bearing from a paper making machine

No subsurface changes of the XRD parameters occur. Note that for a Hertzian pressure of p_0=2500 MPa, i.e. incipient plastic deformation in pure radial contact loading, the z_0 depths of maximum v. Mises and orthogonal shear stress equal about 1.15 and 0.85 mm, respectively. Load induced butterfly microstructure transformations on nonmetallic inclusions are not observed in metallographic microsections of this large size roller bearing. Therefore, the maximum applied Hertzian pressure actually does not exceed about 1400 MPa (see section 3.3). Compressive residual stresses are formed near the surface up to a depth of around 60 µm. The original loading conditions relevant to damage initiation are not obscured by overrolling of spalls at a later stage of failure and only isolated indentations are found on the raceway. The characteristic type A residual stress profile in Figures 54a and 54b thus identifies the impact of vibrations. On the surface, advanced material aging of $b/B^3 0.69$ is deduced.

Incipient hairline cracks on the raceway are almost undetectable even in the SEM. The virtually perspective view of the edge of a microsection in Figure 55 provides an example (cf. Figure 53b). A corresponding micrograph of the etched microsection is shown in Figure 56.

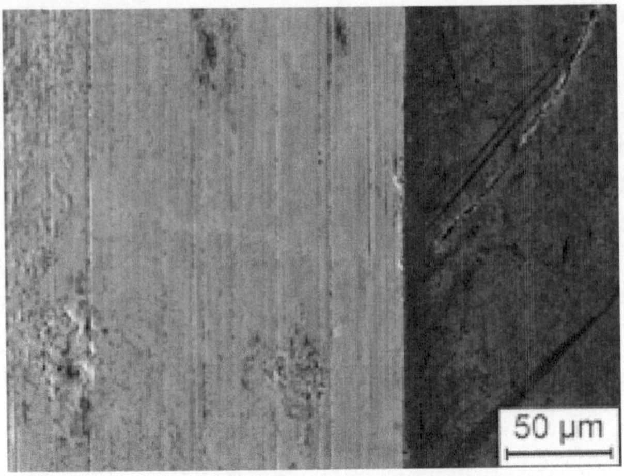

Fig. 55. SEM-SE image of a hairline crack initiation site on the smoothed raceway surface and incipient fatigue crack growth into the material in overrolling direction from bottom to top visible in the cut microsection on the right. The SRB failure of Figure 54 is investigated

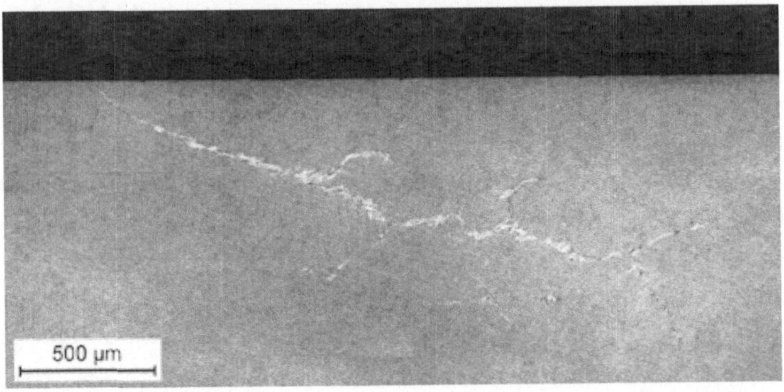

Fig. 56. LOM micrograph of the etched metallographic section on the right of Figure 55. The raceway surface is at the top of the image. The overrolling direction is from left to right

Shear stress control of surface fatigue crack initiation, under varying load and friction- defining slip in the contact area, and subsequent propagation is apparent from crack advance in overrolling direction in a small angle to the raceway tangent. The mechanism is particularly evident from the unbranched crack in Figure 57. The inset zooms in on the edge zone. Compressive residual stresses near the surface (cf. Figure 54) demonstrate the effect of shear stresses required for crack development. According to Figure 58, extended white etching crack systems up to a depth of more than 1 mm are formed, where crack returns to the raceway result in pitting by break-out of the surface eventually. Note that in Figures 56 to 58, the overrolling direction from left to right strikingly indicates top-down WEC propagation.

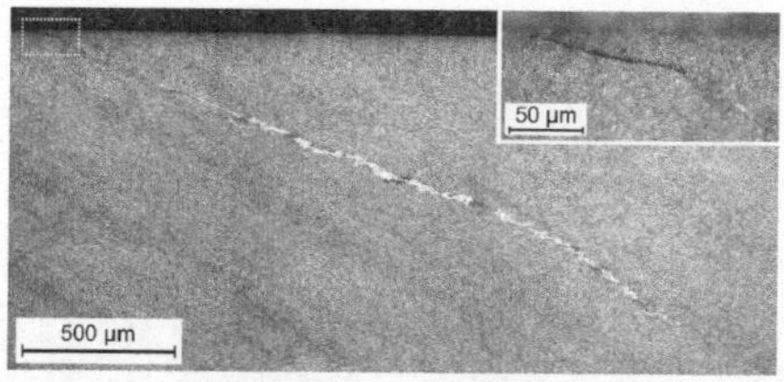

Fig. 57. Same as Figure 56, another crack. The overrolling direction is from left to right

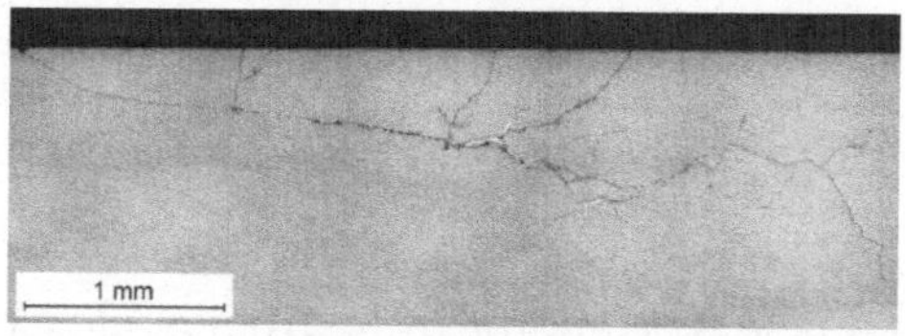

Fig. 58. Same as Figure 56, another WEC system. The overrolling direction from left to right and the orientation of repeated branching proves top-down growth of the CFC crack

Pronounced branching and deep, widely spreading propagation of the transcrystalline cracks essentially under moderate mechanical load of typically p_0»1500 MPa reveals corrosion fatigue in rolling contact as the driving force of crack growth. A comparison of Figure 56 and 57 suggests that also fracture of the new brittle ferritic phase can lead to the initiation of side cracks. Local phase transformation into white etching microstructure along the crack paths is caused by hydrogen (HELP mechanism) released from the highly stressed penetrating lubricant to the adjacent steel matrix. Wear between the crack flanks promotes the degradation reactions on blank metal faces (Kohara et al., 2006). Oil additives can influence the tribochemical release of hydrogen. Accelerated lubricant aging due to vibration loading further supports the chemical assistance of corrosion fatigue cracking (CFC) and microstructure transformation into white etching constituents. Local material aging and embrittlement is manifested in the frequently observed formation of a dark etching region around the cracks. An example is given in the micrograph of Figures 59a. Regular etching induced preparative cracking along the branching CFC path in the corresponding SEM image of Figure 59b reflects plastification in the slip bands of the embrittled DER material.

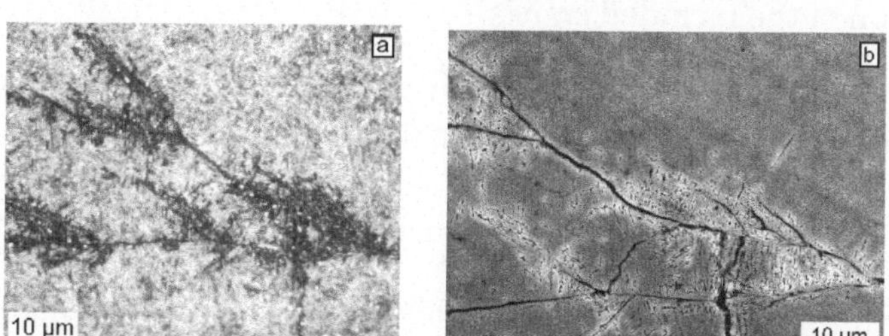

Fig. 59. DER around CFC crack paths indicate localized material aging in (a) a LOM and (b) a SEM micrograph of an etched microsection of the IR of a TRB from an industrial gearbox

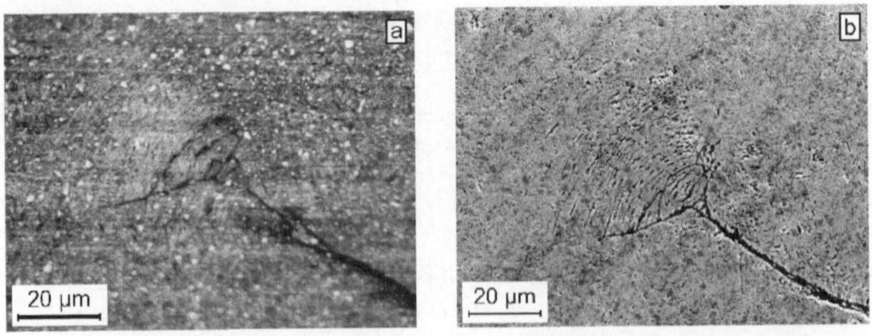

Fig. 60. Carbide dissolution and distinct localized plastification at the multi-branching tip of a CFC crack visible in (a) a LOM and (b) a corresponding SEM micrograph of an etched radial microsection of the inner ring of a cylindrical roller bearing from a weaving machine

Localized fatigue damage is promoted by hydrogen released from decomposition products and possibly contaminations of the lubricant, penetrating through the advancing crack from the raceway surface to the depth. The most intense microstructural changes thus occur on multi-branching sites of CFC cracks (cf. Figure 59). Particularly at these most effective hydrogen sources, pronounced carbide dissolution (see DGSL model, section 4.2) in the proceeding phase transformation is visible in the microsection. The region of the heavily branching tip of a CFC crack in the LOM micrograph of Figure 60a provides an illustration. Localized plasticity in the area of carbide dissolution is evident from the corresponding SEM image of Figure 60b. Weaker material aging and incipient phase transformation (DER) also occurs along unbranched crack paths. The etching process emphasizes the actual microstructure damage. The secondary hydrogen embrittlement around CFC cracks, linked to DER formation, is reflected in the increased susceptibility of the locally aged steel matrix to preparative stress corrosion cracking, which from its first detection is referred to as *Zang* structure. The example of Figures 61a and 61b documents that the local dark etching region around corrosion fatigue cracks can be perceived as precursor of WEA (see also section 4.3). The developed banana-shaped WEA, surrounded by the preliminary DER structure, nestles to the CFC crack at a multi-branching site. Its harder material (more than 1000 HV) appears smoothed and darker in the SEM detail of Figure 61b, where texturing is indicated by reorientation of the included cracks.

The observation of enhanced, evidently hydrogen induced phase transformation at (multi-) branching sites agrees with regular finding of pronounced white etching area decoration at these positions of WEC systems. Note that in Figure 61, the match of the curved shape of the WEA with the crack path excludes primary WEA evolution.

In the outer zone of the overrolled material, the shear stresses for dislocation glide in the described dynamic (nano-) recrystallization process of white etching microstructure formation around CFC cracks, which offer the hydrogen source for accelerated local fatigue aging, increase with depth. This is one reason why the decorating constituents in a WEC are often found less intense near the raceway surface (see, e.g., Figures 52, 56 and 57). The overall hydrogen content of 0.9 ppm measured at the inner ring of Figures 54 to 58 is consistent with the typical delivery condition. This finding reflects the limited damage of the investigated bearing. Depending on the density of the raceway cracks, gradual secondary hydrogen absorption from the surface to the bore is verified at the final stage of service life (Nierlich & Gegner, 2011).

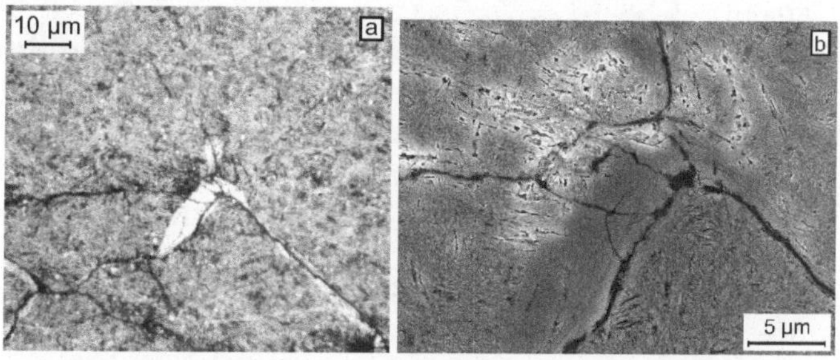

Fig. 61. Curved white etching area along a multi-branching site of a WEC with surrounding embrittled DER material, identified as WEA precursor, in (a) a LOM and (b) a SEM micrograph of an etched microsection of the inner ring of a TRB from an industrial gearbox

Figure 62 completes the investigation of the SRB failure of Figures 54 to 58. The fracture face of a preparatively opened crack at the initiation site on the surface is shown. The inner ring raceway is visible at the top. Following the brittle incipient crack of about 5 µm depth, dense striations indicate the fatigue nature of crack propagation almost from the surface.

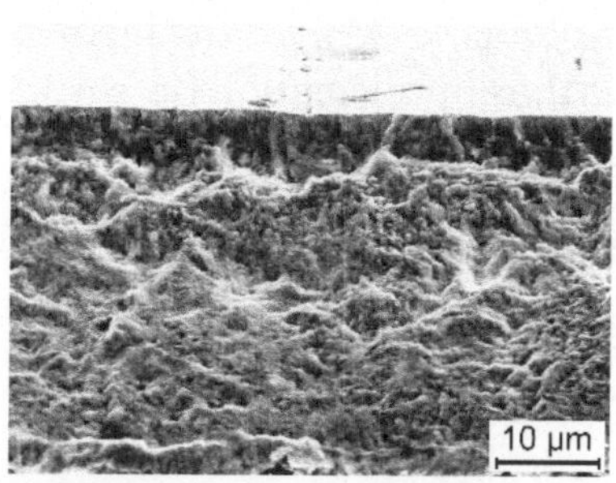

Fig. 62. SEM-SE fractograph of the original fracture surface of a subsequently opened raceway crack on the inner ring of the spherical roller bearing of Figures 54 to 58

5.5.2 Frictional tensile stress induced surface cracking and normal stress hypothesis

Figure 63 reveals a micropit on the smoothed inner ring raceway of a CARB bearing from a paper making machine. Material removal is caused by a brittle Mg-Al-O spinel inclusion that breaks off from the surface under tribomechanical loading of the rolling-sliding contact.

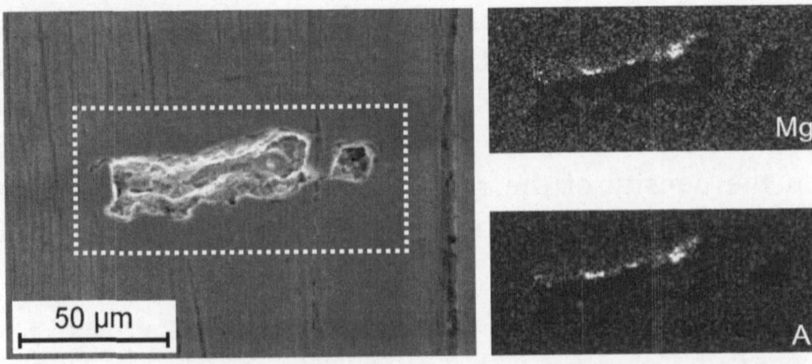

Fig. 63. SEM-SE image of the IR raceway of a CARB bearing and indicated elemental mapping (on the right) revealing an oxide inclusion (Al and Mg detected) that breaks off from the surface under frictional rolling contact loading to cause a micropit eventually

In Figure 63, the demonstrative elemental distribution images of magnesium and aluminum are mapped over the damaged region. Sharp-edged axial surface cracks on tribochemically dissolved MnS inclusions (see section 5.2, e.g. Figure 42a), which advance vertically downwards into the material (Nierlich & Gegner, 2006), as well as grain boundary cleavage (cf. Figure 48) further indicate the action of frictional tensile stresses. Another type of failure causing loading by differently disturbed bearing kinetics is thus reflected in brittle spontaneous crack initiation on raceway surfaces.

Application of the tribological model introduced in section 5.1 in the inset of Figure 36 allows the estimation of the development of the frictional tangential normal stressesyy with depth z. The classical analytical solution of a uniform infinite rolling-sliding line contact (Karas, 1941), for the highest tension level evaluated at the runout y=-a, is used for the approximation (m=$m_>$):

$$\frac{\sigma_{yy}}{p_0} = \sinh\alpha\sin\beta\left(1 - \frac{\sinh 2\alpha + \mu\sin 2\beta}{\cosh 2\alpha - \cos 2\beta}\right) - (\sin\beta - 2\mu\cos\beta)\exp(-\alpha) \tag{7}$$

$$\sinh\alpha = \frac{1}{a}\sqrt{\frac{1}{2}\left[y^2 + z^2 - a^2 + \sqrt{\left(y^2 + z^2 - a^2\right)^2 + 4a^2z^2}\right]},\ \text{for } -a \le y \le 0 \tag{8}$$

$$y = a\cosh\alpha\cos\beta,\ \alpha \ge 0 \tag{9}$$

$$z = a\sinh\alpha\sin\beta,\ \beta \ge 0 \tag{10}$$

The relationships of Eqs. (9) and (10) hold for the elliptic coordinates. Figure 64 shows a graphical representation of calculated depth distributions for increased friction coefficients m of 0.2, 0.3 and 0.4. On the raceway surface at z=0, maximum tension of 2mp_0 is reached.

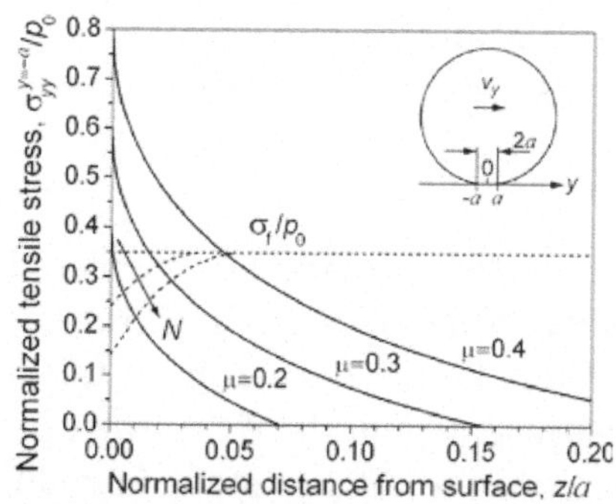

Fig. 64. Normalized distribution of the equivalent normal stress below a rolling-sliding contact (rolling occurs in *y* direction at velocity v_y, see inset) and indication of the level of the critical fracture strength s_f»R_e for typical peak loading with illustration of the expanding failure range by gradual in-service surface embrittlement (cf. section 5.4) with running time

Note that Figure 1 represents the stress field in the center of the Hertzian contact area. At the runout (y=-a, see inset of Figure 64), where the maximum sliding friction induced circumferential tensile stresses of Eq. (7) occur in the surface zone of the material, the hydrostatic pressure reduces to zero. A graphical illustration is provided in Figure 65.

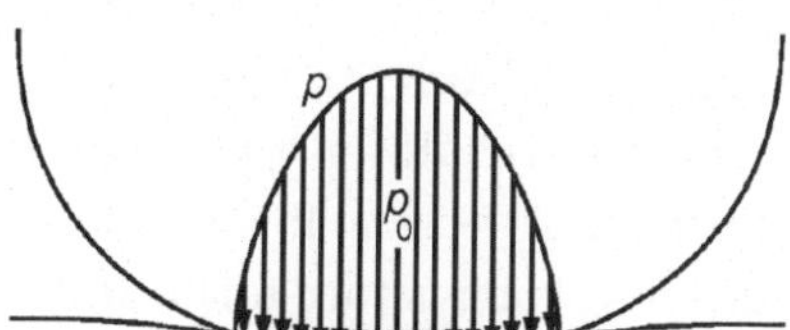

Fig. 65. Schematic representation of the macro contact area with elliptical Hertzian distribution of the pressure *p* (maximum p_0 in the center is indicated)

Preferential surface cracking occurring vertically in axial direction on raceways of larger roller bearings (providing high *a* values) that run under (intermittently) increased mixed friction points to the validity of a normal stress fracture criterion (Nierlich & Gegner, 2011):

$$\sigma_e^{nsh} = \sigma_{yy}^{y=-a} \tag{11}$$

Modification of the equivalent normal stress s^{nsh}, for instance by residual stresses (e.g. from surface finishing or cold working, cf. Figures 16a and 23a) or stress raising nonmetallic inclusions, is neglected in Figure 64 for the sake of simplicity. In a rough approximation, the relevant critical fracture strength s_f of brittle spontaneous crack initiation is, due to almost deformationless material separation (see Figures 66 and 67 later in the text), estimated as the elastic limit R_e»800 MPa, which falls significantly below the yield strength for hardened bearing steel. In cyclic tension-compression tests, for instance, the material changes its response from elastic to microplastic at a stress level around 500 MPa (Voskamp, 1996). The failure range of the introduced normal stress hypothesis can then be determined as follows:

$$\sigma_{yy}^{y=-a} \geq \sigma_f \qquad (12)$$

As spontaneous incipient crack formation is considered, the illustration of Figure 64 realistically refers to short-term loading of high Hertzian pressure $p_0{}^{3}$2000 MPa and friction coefficient m^{3}0.2. Rough indication of the relative s$_f$/p_0 level occurs accordingly. Note that the exact magnitude of the fracture strength s$_f$£$R_{p0.2}$ does not make an essential difference to the validity of the introduced normal stress failure hypothesis but only influences the frequency of the rare events of raceway cracking as critical peak load operating conditions can cause tensile stresses 2mp_0»2000 MPa on the surface. The length of the brittle mode I propagation of a frictionally initiated cleavage-like raceway crack depends on the stress intensity factor K_I and the fracture toughness K_{Ic} according to K_I>K_{Ic}. The depth effect of operational material embrittlement (see section 5.4) on the critical fracture strength s$_f$ (also valid for K_{Ic}), which increases with the number *N* of ring revolutions, is schematically included in Figure 64, where larger size bearings with *a* in the range of 0.5 mm are considered. A concrete calculation example is given in the literature (Nierlich & Gegner, 2011). The semiminor axis *a* of the contact ellipse influences the extension of the failure range according to Eq. (12) and Figure 64. The micro friction model of Figure 36 is regarded. As deduced in section 5.1 from the effect of the induced equivalent shear stresses on plastification and the resulting type A or B residual stress patterns, vibrational loading can intermittently cause locally increased mixed friction. Under peak load operating conditions, such short-term states generally coincide with the impact of high Hertzian pressures. As the detection of type A residual stress distributions (see Figure 54) indicates, friction coefficients above 0.3 can occur temporarily in subareas of the rolling contact. Larger size roller bearings are most sensitive to brittle cracking.

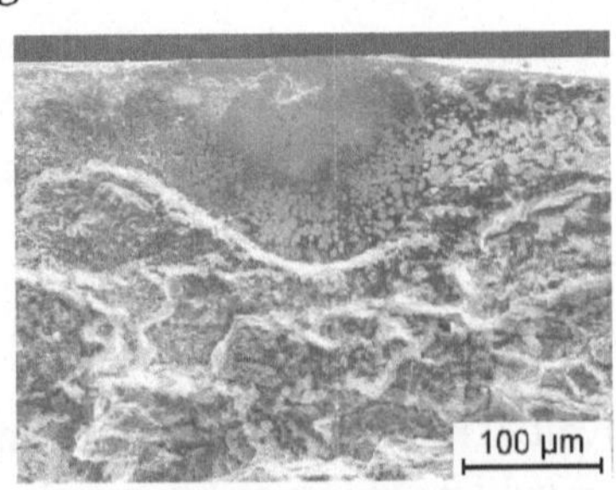

Fig. 66. SEM-SE fractograph of the original fracture surface of a preparatively opened axial crack on the inner ring raceway of a failed taper roller bearing from an industrial gearbox

Further fractographic verification of normal stress failures is provided in the following. The steep gradient of the causative frictional tensile stress in Figure 64 indicates limited advance and rapid stop of an initiated brittle spontaneous mode I surface crack. The fracture faces of two preparatively opened vertical axial raceway cracks in the SEM images of Figures 66 and 67 confirm this prediction. The development of the (semi-) circular shape of the spontaneous cracks may be described by the depth dependence of the stress intensity factor and an energy balance criterion to minimize the interface energy.

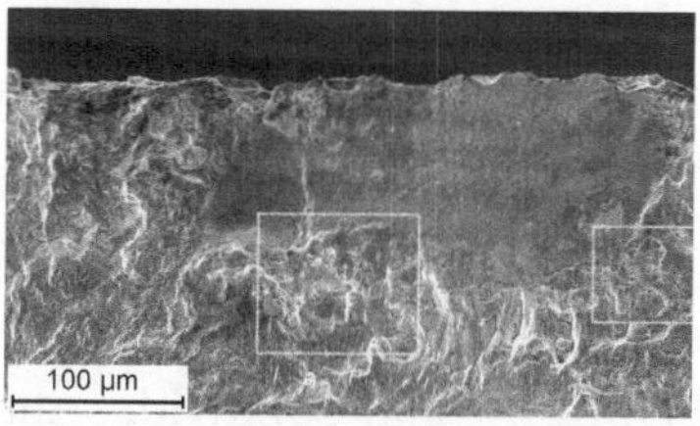

Fig. 67. SEM-SE fractograph of the original fracture surface of a preparatively opened axial crack on the inner ring raceway of a failed taper roller bearing from an industrial gearbox

The low-deformation transcrystalline lenticular cracks of about 150 µm depth act as incipient cracks of subsequent corrosion fatigue cracking into the depth, to the sides and on the surface. The distinct change of the fracture pattern in Figures 66 and 67, respectively with a demarcating bulge or crack network, on the latter of which Figures 68a and 68b zoom in, is evident. The crack arrest indicating numerous side cracks in Figure 68 reflect local material embrittlement as observed in the affected DER microstructure around CFC cracks in the SEM micrographs of etched microsections in Figures 59b and 60b.

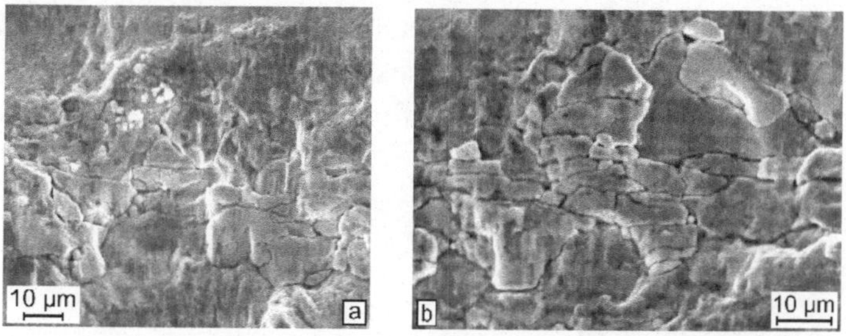

Fig. 68. SEM-SE details of Figure 67 as indicated (a) in the lower middle and (b) on the right

In the area of the spontaneous crack of Figure 66, a mixed TiCN-MnS nonmetallic inclusion near the raceway surface in a depth of about 25 µm acts as stress raising crack nuclei. The appearance of the microsections, revealing white etching crack systems, is similar to Figures 56 to 61.

Hydrogen releasing aging reactions of the lubricant during corrosion fatigue crack growth are proven by EDX microanalysis on preparatively opened fracture surfaces. As an example, Figure 69a shows an overview of the deep CFC region below the brittle lenticular crack of Figure 67. The area of the performed EDX analysis is marked in the SEM fractograph. Sulfur, phosphorus and zinc in the recorded spectrum of Figure 69b indicate reacted residues of oil additives near the crack tip in a depth of about 1 mm in higher concentration than on the low-deformation spontaneous incipient crack visible at top left of Figure 69a, where chemical attack is restricted to subsequent surface corrosion. Furthermore, numerous side cracks characterize corrosion fatigue fracture faces (see also, for instance, the microsections of Figures 56, 58 and 59). The forced rupture from preparative crack opening stands out clearly at the bottom and bottom left of Figure 69a against the dark original CFC fracture structure.

The bearing applications of Figures 66 and 67 operate under vibrations. The observed local crack initiation on the raceway agrees with the approach of the tribological model in Figure 36 that subdivides the contact area into regions of different loading levels. Brittle spontaneous cracking occurs in subdomains of increased friction coefficient. Compared with the competing fatigue crack initiation mechanism discussed in section 5.5.1, lower slip in the moment of surface cracking is suggested.

It is worth noting that post-machining thermal treatment (PMTT) of ground and honed rings and rollers, previously proposed in the literature for material reinforcement in the mechanically influenced edge zone (Gegner, 2006b; Gegner et al., 2009), is recently reported to be an effective countermeasure against premature bearing failures by white etching crack formation (Luyckx, 2011). The short reheating process of, e.g., 0.5 to 1 h after the finishing operation occurs below the tempering or transformation temperature to avoid undesired hardness loss (cf. section 4.2, Figure 23). As only the plastically deformed material in the outermost layer up to a depth of about 10 μm is microstructurally stabilized, a success of this simple treatment would provide further indication of surface WEC failure initiation.

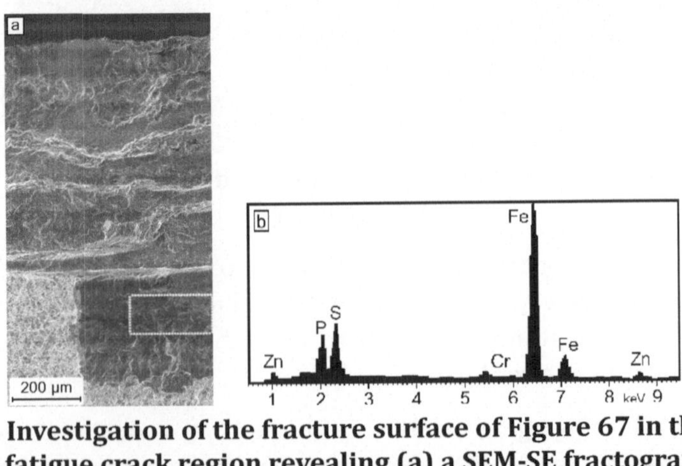

Fig. 69. Investigation of the fracture surface of Figure 67 in the deep corrosion fatigue crack region revealing (a) a SEM-SE fractograph and (b) the EDX spectrum taken at the indicated position where the presence of the tracer elements S, P and Zn of the oil additives proves the assistance of fatigue crack growth by chemical reactions, i.e. CFC, in a depth of 1 mm

6. Conclusion

The present chapter deals with important aspects of rolling contact tribology in bearing failures. Following the introduction, the fundamentals are presented in sections 2 and 3. The subsurface and (near-) surface failure modes of rolling bearings are outlined. X-ray diffraction (XRD) based residual stress analysis identifies the depth of highest loading and provides information about the material response and the stage of damage. The measurement technique, evaluation methodology and application procedure are discussed in detail. The loading induced reduction of the XRD peak width ratio b/B of minimum to initial value is used as a life calibrated measure of material aging to correlate the successive microstructural changes during rolling contact fatigue (RCF) with the Weibull bearing failure distribution. Therefore, it also permits the prediction of gradual alterations of the hardened steel matrix, which are roughly assigned to the corresponding bearing life in the final phase of the three stage model of RCF (shakedown, steady state, instability). Strong indication is given that dark etching regions (DER) from martensite decay act as the precursor of subsequently occurring ferritic white etching areas (WEA). The WEA are formed in regular parallel flat (30°) and steep (80°) bands within the aged matrix or along propagating corrosion fatigue cracks. Rolling contact fatigue in the subsurface region can also occur on nonmetallic inclusions. The generation and growth of butterflies are briefly discussed, based on recent findings.

Section 4 focuses on classical subsurface RCF, which may lead to fatigue wear. Raceway spalling is initiated by cracks from the depth of the material eventually. The microstructural changes that characterize the progression of subsurface rolling contact fatigue in the steel matrix are metallographically examined, including scanning electron microscopy. A distinction is made from the shakedown stage during the short running-in period, which is identified as a cold working process of local (micro-) plastic deformation. Rapid compressive residual stress formation in this phase, in response to the exceedance of the yield strength by the v. Mises equivalent stress at Hertzian pressures above 2500 to 3000 MPa, occurs without visible microstructural alterations, the development of which requires carbon diffusion. The mechanistic metal physics dislocation glide stability loss (DGSL) model of rolling contact fatigue is introduced and examined by reheating experiments. As a new aspect of material damage by severe in-service high-frequency electric current passage through bearings, continuous absorption of hydrogen is found to accelerate subsurface RCF. Steep white bands that occur not until the L_{50} life (50% failure probability) in pure mechanical loading appear earlier at much lower b/B reduction in chemically promoted rolling contact fatigue. The accelerating effect of dissolved hydrogen is demonstrated by a comparison of the microstructures at b/B»0.71, also considering cold working. The chosen reference level is yet above the XRD equivalent value of b/B»0.64 of the rating L_{10} bearing life in pure mechanical subsurface rolling contact fatigue. The additional chemical loading accelerates material aging by enhancing the dislocation mobility and microplasticity, as evident from the DGSL model. Hydrogen absorption also causes crack initiation in the pre-embrittled microstructure by interfacial delamination at white etching bands that is not observed in pure mechanical RCF.

In section 5 of the present chapter, the effect of mixed friction in the rolling contact area, which occurs frequently in bearing applications, is discussed in detail. Smoothing of the machining marks by polishing wear on the raceway is a characteristic visual indication. Several mechanisms of mixed friction induced failure initiation are introduced. The impact of externally generated three-dimensional mechanical vibrations represents a common cause of disturbed elastohydrodynamic lubrication conditions. Larger size roller bearings operating typically at low to moderate Hertzian pressure are most susceptible to frictional surface loading. Tangential forces by sliding friction acting on a rolling contact increase the v. Mises equivalent stress and shift its maximum, i.e. the position of incipient plastic deformation, toward the surface. The resulting build-up of compressive residual stresses in the edge zone at Hertzian pressures below 2500 MPa is observed for indentation-free raceways under the action of, e.g. engine, vibrations in operation. Material response is described by a tribological model that partitions the contact area into microscopic subdomains of intermittently different friction coefficients up to peak values above 0.3. The distinguishable type A and B vibrational residual stress distributions are explained. Vibrations can reduce the shear-sensitive viscosity of the lubricant. The generated temperature increase is associated with the contact area related power loss.

Also, mixed friction or lubricant contamination, e.g. by water or wear debris, promotes chemical aging of the oil and its additives. As a consequence, the gradually acidifying fluid attacks the steel surface. Tribochemical dissolution of manufacturing related manganese sulfide inclusion lines leaves crack-like defects on the raceway. Further damage evolution by shallow micropitting occurs similar to gray staining that is also caused by, e.g. vibration induced, mixed friction. Reasons are given for the hypothesis that the crack propagation mechanism is a variant of corrosion fatigue in rolling contact. The material shows indication of in-service (near-) surface embrittlement.

White etching cracks can cause premature bearing failures in specific susceptible applications. The development of heavily branching and widely spreading transcrystalline crack systems at essentially low to moderate mechanical load indicate chemically assisted crack growth by corrosion fatigue under the influence of the penetrating aging lubricant. Released hydrogen locally induces collateral microstructural changes (HELP, DGSL) resulting in the decorating white etching constituents around parts of the crack paths eventually. Surface failure initiation by mixed friction is detected. Shear and tensile stress controlled damage mechanisms are identified. The formation of fatigue microcracks on the surface, comparable with gray staining, and initial crack extension in overrolling direction at a small angle to the raceway tangent are caused by the variation of load and friction- defining slip in the contact area. The characteristic orientation of crack propagation reveals failure promoting shear stresses. The established tribological model also explains competing frictional tensile stress induced failure initiation in rolling-sliding contact. Vertical brittle spontaneous hairline cracking of limited depth and surface length of respectively about 0.1 to 0.2 mm occurs mainly in axial direction on the raceway. The normal stress hypothesis is thus proposed. Illustrative case examples are discussed. Failure metallography, fractography and residual stress analysis are applied. Whereas the circumferential tensile stress in the affected subdomains, referring to the introduced tribological model, must be high (maximum on the surface, $\mu m p_0$) to initiate cleavage-like raceway cracks, the contact area related frictional power loss ($\mu m p_0 v_s$) is limited so that no smearing (adhesive wear) occurs. This interrelation leads to the conclusion that the rare events of brittle spontaneous raceway cracking in premature bearing failures can be considered as a consequence of specific (three-dimensional) vibration conditions of high Hertzian pressure p_0 and local friction coefficient m at low sliding speed v_s (*gluing* effect). The shear stress induced inclined flat fatigue-like incipient microcracks, in contrast, are characterized by lower frictional tensile stresses, i.e. smaller mp_0 value (v_s less important). From both of these crack initiation mechanisms, smearing is clearly differentiated by the much higher contact area related power loss.

References

1. Balzar, D. (1999). Voigt Function Model in Diffraction-line Broadening Analysis. In: *Defect and Microstructure Analysis by Diffraction*, R. Snyder, J. Fiala, H.J. Bunge (Eds.), Oxford University Press, Oxford, Great Britain, pp. 94-126
2. Birnbaum, H.K. & Sofronis, P. (1994). Hydrogen-Enhanced Localized Plasticity – a Mechanism for Hydrogen Related Fracture. *Materials Science and Engineering*, Vol. A176, No. 1/2, pp. 191-202
3. Böhm, K.; Schlicht, H.; Zwirlein, O. & Eberhard, R. (1975). Nonmetallic Inclusions and Rolling Contact Fatigue. In: *Bearing Steels: The Rating of Nonmetallic Inclusions*, ASTM STP 575, J.J.C. Hoo, P.T. Kilhefner, J.J. Donze (Eds.), American Society for Testing and Materials (ASTM), West Conshohocken, Pennsylvania, USA, pp. 96-113
4. Böhmer, H.-J.; Hirsch, T. & Streit, E. (1999). Rolling Contact Fatigue Behaviour of Heat Resistant Bearing Steels at High Operational Temperatures. *Material Science and Engineering Technology*, Vol. 30, No. 9, pp. 533-541
5. Bragg, W.H. & Bragg, W.L. (1913). The Reflection of X-rays by Crystals. *Proceedings of the Royal Society of London*, Vol. 88A, No. 605, pp. 428-438

6. Broszeit, E.; Heß, F.J. & Kloos, K.H. (1977). Werkstoffanstrengung bei oszillierender Gleitbewegung. *Zeitschrift für Werkstofftechnik*, Vol. 8, No. 12, pp. 425-432

7. Broszeit E.; Preussler, T.; Wagner, M. & Zwirlein, O. (1986). Stress Hypotheses and Material Stresses in Hertzian Contacts. *Zeitschrift für Werkstofftechnik*, Vol. 17, No. 7, pp. 238- 246

8. Brückner, M.; Gegner, J.; Grabulov, A.; Nierlich, W. & Slycke, J. (2011). Butterfly Formation Mechanisms in Rolling Contact Fatigue. *Proceedings of the 5th International Conference on Very High Cycle Fatigue*, pp. 101-106, C. Berger, H.-J. Christ (Eds.), DVM German Association for Materials Research and Testing, Berlin, Germany, June 28-30, 2011

9. Ciavarella, M.; Monno, F. & Demelio, G. (2006). On the Dang Van Fatigue Limit in Rolling Contact Fatigue. *International Journal of Fatigue*, Vol. 28, No. 8, pp. 852-863

10. Dally, J.W. & Riley, W.F. (2005). *Experimental Stress Analysis*, College House, Knoxville, Tennessee, USA

11. Desimone, H.; Bernasconi, A. & Beretta, S. (2006). On the Application of Dang Van Criterion to Rolling Contact Fatigue. *Wear*, Vol. 260, No. 4-5, pp. 567-572

12. Fajdiga, G. & Srami, M. (2009). Fatigue Crack Initiation and Propagation under Cyclic Contact Loading. *Engineering Fracture Mechanics*, Vol. 76, No. 9, pp. 1320-1335

13. Faninger, G. & Wolfstieg, U. (1976). Aufnahmeverfahren, Auswertung der Interferenzlinien und d_{jy}/e_{jy}, sin2y-Zusammenhang. *Härterei-Technische Mitteilungen*, Vol. 31, No. 1- 2, pp. 13-32

14. Fougères, R.; Lormand, G.; Vincent, A.; Nelias, D.; Dudragne, G.; Girodin, D.; Baudry, G. & Daguier, P. (2002). A New Physically Based Model for Predicting the Fatigue Life Distribution of Rolling Bearings. In: *Bearing Steel Technology*, ASTM STP 1419, J.M. Beswick (Ed.), American Society for Testing and Materials (ASTM), West Conshohocken, Pennsylvania, USA, pp. 197-212

15. Furumura, K.; Murakami, Y. & Abe, T. (1993). The Development of Bearing Steels for Long Life Rolling Bearings under Clean Lubrication and Contaminated Lubrication. In: *Creative Use of Bearing Steels*, ASTM STP 1195, J.J.C. Hoo (Ed.), American Society for Testing and Materials (ASTM), West Conshohocken, Pennsylvania, USA, pp. 199- 210

16. Gegner, J. (2006a). Materialbeanspruchungsanalyse und ihre Anwendung auf Prüfstandsversuche zum Oberflächenausfall (Nierlich-Schadensmodus) von Wälzlagern. *Materialwissenschaft und Werkstofftechnik*, Vol. 37, No. 3, pp. 249-259

17. Gegner, J. (2006b). Post-Machining Thermal Treatment (PMTT) of Hardened Rolling Bearing Steel. *Proceedings of the 4th International Conference on Mathematical Modeling and Computer Simulation of Material Technologies*, Vol. 1, Chap. 2, pp. 66-75, College of Judea and Samaria, Ariel, Israel, September 11-15, 2006

18. Gegner, J.; Hörz, G. & Kirchheim, R. (1996). Hydrogen Interaction with 0-, 1-, and 2- Dimensional Defects. In: *Hydrogen Effects in Materials*, A.W. Thompson, N.R. Moody (Eds.), TMS The Minerals, Metals and Materials Society, Warrendale, Pennsylvania, USA, pp. 35-46
19. Gegner, J.; Kuipers, U.; Mauntz, M. (2010). Ölsensorsystem zur Echtzeit-Zustandsüberwachung von technischen Anlagen und Maschinen. *TM Technisches Messen*, Vol. 77, No. 5, pp. 283-292
20. Gegner, J.; Nierlich, W. & Brückner, M. (2007). Possibilities and Extension of XRD Material Response Analysis in Failure Research for the Advanced Evaluation of the Damage Level of Hertzian Loaded Components. *Material Science and Engineering Technology*, Vol. 38, No. 8, pp. 613-623
21. Gegner, J. & Nierlich, W. (2008). Operational Residual Stress Formation in Vibration-Loaded Rolling Contact. *Advances in X-ray Analysis*, Vol. 52, pp. 722-731
22. Gegner, J. & Nierlich, W. (2011a). Mechanical and Tribochemical Mechanisms of Mixed Friction Induced Surface Failures of Rolling Bearings and Modeling of Competing Shear and Tensile Stress Controlled Damage Initiation. *Tribologie und Schmierungstechnik*, Vol. 58, No. 1, pp. 10-21
23. Gegner, J. & Nierlich, W. (2011b). Hydrogen Accelerated Classical Rolling Contact Fatigue and Evaluation of the Residual Stress Response. *Materials Science Forum*, Vol. 681, pp. 249-254
24. Gegner, J. & Nierlich, W. (2011c). Sequence of Microstructural Changes during Rolling Contact Fatigue and the Influence of Hydrogen. *Proceedings of the 5th International Conference on Very High Cycle Fatigue*, pp. 557-562, C. Berger, H.-J. Christ (Eds.), DVM German Association for Materials Research and Testing, Berlin, Germany, June 28-30, 2011
25. Gegner, J.; Schlier, L. & Nierlich, W. (2009). Evidence and Analysis of Thermal Static Strain Aging in the Deformed Surface Zone of Finish-Machined Hardened Steel. *Powder Diffraction*, Vol. 24, No. 2-supplement, pp. 45-50
26. Gentile, A.J.; Jordan, E.F. & Martin, A.D. (1965). Phase Transformations in High-Carbon, High-Hardness Steels under Contact Loads. *Transactions AIME*, Vol. *233*, No. 6, pp. 1085-1093
27. Gohar, R. (2001). *Elastohydrodynamics*, Imperial College Press, London, Great Britain
28. Harada, H.; Mikami, T.; Shibata, M.; Sokai, D.; Yamamoto, A. & Tsubakino, H. (2005). Microstructural Changes and Crack Initiation with White Etching Area Formation under Rolling/Sliding Contact in Bearing Steel. *ISIJ International*, Vol. 45, No. 12, pp. 1897-1902
29. Harris, T.A. (2001). *Rolling Bearing Analysis*, Wiley, New York, USA
30. Hauk, V. & Wolfstieg, U. (1976). Röntgenographische Elastizitätskonstanten. *Härterei- Technische Mitteilungen*, Vol. 31, No. 1-2, pp. 38-42

31. Hauk, V.M. & Macherauch, E. (1984). A Useful Guide for X-ray Stress Evaluation (XSE). *Advances in X-ray Analysis*, Vol. 27, pp. 81-99

32. Hauk, V.M., (Ed.), (1997). *Structural and Residual Stress Analysis by Nondestructive Methods – Evaluation, Application, Assessment*, Elsevier, Amsterdam, The Netherlands

33. Hiraoka, K.; Nagao, M. & Isomoto, T. (2006). Study on Flaking Process in Bearings by White Etching Area Generation. *Journal of ASTM International*, Vol. 3, No. 5, Paper ID JAI14059

34. Holweger, W. & Loos, J. (2011). Interaction of Rolling Bearing Fatigue Life with New Material Phenomenons of Special Applications. *Proceedings of the 14th Heavy Drive Train Conference*, pp. 223-238, G. Jacobs, K. Hameyer, C. Brecher (Eds.), Apprimus, Aachen, Germany, RWTH Aachen University, Aachen, Germany, March 29-30, 2011

35. Hooke, C.J. (2003). Dynamic Effects in EHL Contacts. In: *Tribological Research and Design for Engineering Systems*, D. Dowson, A.A. Lubrecht, M. Priest (Eds.), Tribology Series, Vol. 41, Elsevier, Amsterdam, The Netherlands, pp. 69-78, Proceedings of the 29th Leeds-Lyon Symposium on Tribology, Leeds, Great Britain, September 03-06, 2002

36. Ioannides, E. & Harris, T.A. (1985). A New Fatigue Life Model for Rolling Bearings. *ASME Journal of Tribology*, Vol. 107, No. 3, pp. 367-378

37. Iso, K.; Yokouchi, A. & Takemura, H. (2005). Research Work for Clarifying the Mechanism of White Structure Flaking and Extending the Life of Bearings. In: *Engine Lubrication and Bearing Systems*, SAE SP-1967, Paper No. 2005-01-1868, Society of Automotive Engineers (SAE), SAE International, Warrendale, Pennsylvania, USA

38. Jagenbrein, A.; Buschbeck, F.; Gröschl, M. & Preisinger, G. (2005). Investigation of the Physical Mechanisms in Rolling Bearings during the Passage of Electric Current. *Tribotest*, Vol. 11, No. 4, pp. 295-306

39. Jatckak, C.F. & Larson, J.A. (1980). *Retained Austenite and Its Measurement by X-ray Diffraction*, SAE SP-453, Society of Automotive Engineers (SAE), Warrendale, Pennsylvania, USA

40. Johnson, K.L. & Jefferis, J.A. (1963). Plastic Flow and Residual Stresses in Rolling and Sliding Contact. *Proceedings of the Institute of Mechanical Engineers*, Vol. 177, No. 25, pp. 54- 65

41. Johnson, K.L. (1985). *Contact Mechanics*, Cambridge University Press, Cambridge, Great Britain

42. Jones, A.B. (1946). Effect of Structural Changes in Steel on Fatigue Life of Bearings. *Steel*, Vol. 119, No. 9, pp. 68-70 and 97-100

43. Karas, F. (1941). Die äußere Reibung beim Walzendruck. *Forschung auf dem Gebiet des Ingenieurwesens*, Vol. 12, No. 6, pp. 266-274

44. Kino, N. & Otani, K. (2003). The Influence of Hydrogen on Rolling Contact Fatigue Life and its Improvement. *JSAE Review*, Vol. 24, No. 3, pp. 289-294

45. Kohara, M.; Kawamura, T. & Egami, M. (2006). Study on Mechanism of Hydrogen Generation from Lubricants. *Tribology Transactions*, Vol. 49, No. 1, pp. 53-60
46. Kotzalas, M.N. & Doll, G.L. (2010). Tribological Advancements for Reliable Wind Turbine Performance. *Philosophical Transactions of the Royal Society A*, Vol. 368, No. 1929, pp. 4829-4850
47. Kudish, I.I. & Covitch, M.J. (2010). *Modeling and Analytical Methods in Tribology*, Chapman and Hall, CRC Press, Boca Raton, Florida, USA
48. Lindahl, E. & Österlund, R. (1982). 21/2D Transmission Electron Microscopy Applied to Phase Transformations in Ball Bearings. *Ultramicroscopy*, Vol. 9, No. 4, pp. 355-364
49. Lundberg, G. & Palmgren, A. (1947). *Dynamic Capacity of Roller Bearings*, Acta Polytechnica, Mechanical Engineering Series, Vol. 1, No. 3, Royal Swedish Academy of Engineering Sciences, Stockholm, Sweden
50. Lundberg, G. & Palmgren, A. (1952). *Dynamic Capacity of Roller Bearings*, Acta Polytechnica, Mechanical Engineering Series, Vol. 2, No. 4, Royal Swedish Academy of Engineering Sciences, Stockholm, Sweden
51. Luyckx, J. (2011). WEC Failure Mode on Roller Bearings: From Observation via Analysis to Understanding and a Solution. *Proceedings of the VDI Symposium Gleit- und Wälzlagerungen: Gestaltung, Berechnung, Einsatz*, VDI-Berichte 2147, VDI Wissensforum, Düsseldorf, Germany, pp. 31-42, Schweinfurt, Germany, May 24-25, 2011
52. Macherauch, E. & Müller, P. (1961). Das sin2y-Verfahren der röntgenographischen Spannungsmessung. *Zeitschrift für angewandte Physik*, Vol. 13, No. 7, pp. 305-312
53. Macherauch, E. (1966). X-ray Stress Analysis. *Experimental Mechanics*, Vol. 6, No. 3, pp. 140- 153
54. Martin, J.A.; Borgese, S.F. & Eberhardt, A.D. (1966). Microstructural Alterations in Rolling Bearing Steel Undergoing Cyclic Stressing. *Transactions ASME Journal of Basic Engineering*, Vol. 88, No. 3, pp. 555-567
55. Marx, K.-W. (1966). *Röntgenographische Eigenspannungsmessungen an einem gehärteten und angelassenen Wälzlagerstahl 100 Cr Mn 6*, Thesis, Aachen University of Technology, Aachen, Germany
56. Muro, H. & Tsushima, N. (1970). Microstructural, Microhardness and Residual Stress Changes due to Rolling Contact. *Wear*, Vol. 15, No. 5, pp. 309-330
57. Muro, H.; Tsushima, N.; Nunome, K. (1973). Failure Analysis of Rolling Bearings by X-ray Measurement of Residual Stress. *Wear*, Vol. 25, No. 3, 1973, pp. 345-356
58. Nierlich, W.; Brockmüller, U. & Hengerer, F. (1992). Vergleich von Prüfstand- und Praxisergebnissen an Wälzlagern mit Hilfe von Werkstoffbeanspruchungsanalysen. *Härterei-Technische Mitteilungen*, Vol. 47, No. 4, pp. 209-215
59. Nierlich, W. & Gegner, J. (2002). Material Response Analysis of Rolling Bearings Using X-ray Diffraction Measurements. *Proceedings of the Materials Week 2001, International Congress on Advanced Materials, their Processes and Applications,*

CD-ROM, Paper No. 413, Werkstoffwoche-Partnerschaft, Frankfurt, ISBN 3-88355-302-6, Munich, Germany, October 1-4, 2001

60. Nierlich, W. & Gegner, J. (2006). Material Response Models for Sub-Surface and Surface Rolling Contact Fatigue. *Proceedings of the 4th International Conference on Mathematical Modeling and Computer Simulation of Material Technologies*, Vol. 1, Chap. 1, pp. 182-192, College of Judea and Samaria, Ariel, Israel, September 11-15, 2006
61. Nierlich, W. & Gegner, J. (2008). X-ray Diffraction Residual Stress Analysis: One of the Few Advanced Physical Measuring Techniques that have Established Themselves for Routine Application in Industry. *Advances in Solid State Physics*, Vol. 47, pp. 301-314
62. Nierlich, W. & Gegner, J. (2011). Einführung der Normalspannungshypothese für Mischreibung im Wälz-Gleitkontakt. *Proceedings of the VDI Symposium Gleit- und Wälzlagerungen: Gestaltung, Berechnung, Einsatz*, VDI-Berichte 2147, VDI Wissensforum, Düsseldorf, Germany, pp. 277-290, Schweinfurt, Germany, May 24- 25, 2011
63. Noyan, I.C. & Cohen, J.B. (1987). *Residual Stress – Measurement by Diffraction and Interpretation*, Springer, New York, New York, USA
64. Olver, A.V. (2005). The Mechanism of Rolling Contact Fatigue: An Update. *Proceedings of the Institution of Mechanical Engineers, Part J: Journal of Engineering Tribology*, Vol. 219, No. 5, pp. 313-330
65. Österlund, R.; Vingsbo, O.; Vincent, L. & Guiraldenq, P. (1982). Butterflies in Fatigued Ball Bearings - Formation Mechanisms and Structure. *Scandinavian Journal of Metallurgy*, Vol. 11, No. 1, pp. 23-32
66. Palmgren, A. (1964). *Grundlagen der Wälzlagertechnik*, Franckh, Stuttgart, Germany, 1964 Prashad, H. (2006). *Tribology in Electrical Environments*, Elsevier, Amsterdam, The Netherlands
67. Rollmann, J. (2000). *Wälzfestigkeit von induktiv randschichtgehärteten bauteilähnlichen Proben*, Thesis, Darmstadt University of Technology, Shaker, Aachen, Germany
68. Schlicht, H. (2008). Über adiabatic shearbands und die Entstehung der „Steilen Weißen Bänder“ in Wälzlagern. *Materialwissenschaft und Werkstofftechnik*, Vol. 39, No. 3, pp. 217-226
69. Schlicht, H.; Schreiber, E. & Zwirlein, O. (1987). Ermüdung bei Wälzlagern und deren Beeinflussung durch Werkstoffeigenschaften. *Wälzlagertechnik*, No. 1, pp. 14–22
70. Schlicht, H.; Schreiber, E. & Zwirlein, O. (1988). Effects of Material Properties on Bearing Steel Fatigue Strength. In: *Effect of Steel Manufacturing Processes on the Quality of Bearing Steels*, ASTM STP 987, J.J.C. Hoo (Ed.), American Society for Testing and Materials (ASTM), West Conshohocken, Pennsylvania, USA, pp. 81-101

71. Schreiber, E. (1992). Analyse realer Beanspruchungsverhältnisse im Wälzkontakt. In: *Randschichtermüdung im Wälzkontakt*, F. Hengerer (Ed.), Association for Heat Treatment and Materials Technology (AWT), Wiesbaden, Germany, pp. 35-51
72. Shibata, M.; Gotoh, M.; Oguma, N. & Mikami, T. (1996). A New Type of Micro-Structural Change due to Rolling Contact Fatigue on Bearings for the Engine Auxiliary Devices. *Proceedings of the International Tribology Conference*, Vol. 3, pp. 1351-1356, Japanese Society of Tribologists, Tokyo, Japan, Yokohama, Japan, October 29- November 2, 1995
73. Shiga, T.; Umeda, A. & Ihata, K. (2006). *Method and Apparatus for Designing Rolling Bearing to Address Brittle Flaking*, United States Patent, Assignee: Denso Corporation, Publication No.: US 2006/0064197 A1, Publication Date: March 23, 2006
74. Swahn, H.; Becker, P.C. & Vingsbo, O. (1976a). Martensite Decay during Rolling Contact Fatigue in Ball Bearings. *Metallurgical Transactions A*, Vol. 7A, No. 8, pp. 1099-1110
75. Swahn, H.; Becker, P.C. & Vingsbo, O. (1976b). Electron Microscope Studies of Carbide Decay during Contact Fatigue in Ball Bearings. *Metal Science*, Vol. 10, No. 1, pp. 35- 39
76. Takemura, H. & Murakami, Y. (1995). Rolling Contact Fatigue Mechanism (Elasto-plastic Analysis around Inclusion). In: *Fatigue Design 1995*, G. Marquis, J. Solin (Eds.), VTT Manufacturing Technology, Espoo, Finland, pp. 345-356
77. Vincent, A.; Lormand, G.; Lamagnère, P.; Gosset, L.; Girodin, D.; Dudragne, G. & Fougères, R. (1998). From White Etching Areas Formed around Inclusions to Crack Nucleation in Bearing Steels under Rolling Contact Fatigue. In: *Bearing Steels: Into the 21st Century*, ASTM STP 1327, J.J.C. Hoo, W.B. Green (Eds.), American Society for Testing and Materials (ASTM), West Conshohocken, Pennsylvania, USA, pp. 109-123
78. Voskamp, A.P. (1985). Material Response to Rolling Contact Loading. *ASME Journal of Tribology*, Vol. 107, No. 3, pp. 359-366
79. Voskamp, A.P. (1987). Rolling Contact Fatigue and the Significance of Residual Stresses. In: *Residual Stresses in Science and Technology*, Vol. 2, E. Macherauch, V.M. Hauk (Eds.), Deutsche Gesellschaft für Metallkunde (DGM) Informationsgesellschaft, Oberursel, Germany, pp. 713-720
80. Voskamp, A.P. (1996). *Microstructural Changes during Rolling Contact Fatigue – Metal Fatigue in the Subsurface Region of Deep Groove Ball Bearing Inner Rings*, Thesis, Delft University of Technology, Delft, The Netherlands
81. Voskamp, A.P. (1998). Fatigue and Material Response in Rolling Contact. In: *Bearing Steels: Into the 21st Century*, ASTM STP 1327, J.J.C. Hoo, W.B. Green (Eds.), American Society for Testing and Materials (ASTM), West Conshohocken, Pennsylvania, USA, pp. 152-166
82. Wielke, B. (1974). Hysteresis Loop of an Elastic-Plastic l/2 Oscillator. Physica Status Solidi, Vol. 23, No. 1, pp. 237-244

83. Yhland, E. (1983). Statische Tragfähigkeit – Shakedown. Kugellager-Zeitschrift, Vol. 56, No. 211, pp. 1-8

84. Yoshioka, T. (1992). Acoustic Emission and Vibration in the Process of Rolling Contact Fatigue (4th Report): Measurement of Propagation Initiation and Propagation Time of Rolling Contact Fatigue Crack. Japanese Journal of Tribology, Vol. 37, No. 2, pp. 205-217

85. Yoshioka, T. & Fujiwara, T. (1988). Measurement of Propagation Initiation and Propagation Time of Rolling Contact Fatigue Cracks by Observation of Acoustic Emission and Vibration. In: Interface Dynamics, D. Dowson, C.M. Taylor, M. Godet, D. Berthe (Eds.), Tribology Series, Vol. 12, Elsevier, Amsterdam, The Netherlands, pp. 29-33, Proceedings of the 14th Leeds-Lyon Symposium on Tribology, Lyon, France, September 08-11, 1987

86. Zika, T.; Buschbeck, F.; Preisinger, G. & Gröschl, M. (2007). Electric Erosion – Current Passage through Bearings in Wind Turbine Generators. *Proceedings of the 6th Chinese Electrical Machinery Development Forum*, pp. 85-99, Shanghai, China, October 10, 2007

87. Zika, T.; Gebeshuber, I.C.; Buschbeck, F.; Preisinger, G. & Gröschl, M. (2009). Surface Analysis on Rolling Bearings after Exposure to Defined Electric Stress. *Proceedings of the Institution of Mechanical Engineers, Part J: Journal of Engineering Tribology*, Vol. 223, No. 5, pp. 778-787

88. Zika, T.; Buschbeck, F.; Preisinger, G.; Gebeshuber, I.C. & Gröschl, M. (2010). Surface Damage of Rolling Contacts Caused by Discrete Current Flow. *Tribologie und Schmierungstechnik*, Vol. 57, No. 3, pp. 11-14

3

Methodology of Calculation of Dynamics and Hydromechanical Characteristics of Heavy-Loaded Tribounits, Lubricated with Structurally-Non-Uniform and Non-Newtonian Fluids

1. Introduction

Friction units, in which the sliding surfaces are separated by a film of liquid lubricant, generally, consist of three elements: a journal, a lubricating film and a bearing. Such tribounits are often referred to as journal bearings. Tribounits with the hydrodynamic lubrication regime and the time-varying magnitude and direction of load character are hydrodynamic, heavy-loaded (unsteady loaded). Such tribounits include connecting-rod and main bearings of crankshafts, a "piston-cylinder" coupling of internal combustion engines (ICE); sliding supports of shafts of reciprocating compressors and pumps, bearings of rotors of turbo machines and generators; support rolls of rolling mills, etc. The presence of lubricant in the friction units must provide predominantly liquid friction, in which the losses are small enough, and the wear is minimal.

The behavior of the lubricant film, which is concluded between the friction surfaces, is described by the system of equations of the hydrodynamic theory of lubrication, a heat transfer and friction surfaces are the boundaries of the lubricant film, which really have elastoplastic properties. During the simulation and calculation of heavy-loaded bearings researchers tend to take into account as many geometric, force and regime parameters as possible and they provide adequacy of the working capacity forecast of the hydrodynamic tribounits on the early stages of the design.

2. The system of equations

In the classical hydrodynamic lubrication theory of fluid the motion in a thin lubricating film of friction units is described by three fundamental laws: conservation of a momentum, mass and energy. The equations of motion of movable elements of tribounits are added to the equations which are made on the basis of conservation laws for heavy-loaded bearings.

The problem of theory of hydrodynamic tribounits is characterized by the totality of methods for solving the three interrelated tasks:

1. The hydrodynamic pressures in a thin lubricating film, which separates the friction surfaces of a journal and a bearing with an arbitrary law of their relative motion, are calculated.
2. The parameters of nonlinear oscillations of a journal on a lubricating film are detected and the trajectories of the journal center are calculated.
3. The temperature of the lubricating film is calculated.

The field of hydrodynamic pressures in a thin lubricating film depends on:

- the relative motion of the friction surfaces;
- the temperature parameters of the tribounit lubricant film during the period of loading, sources of lubricant on these surfaces are taken into account;
- the elastic deformation of friction surfaces under the influence of hydrodynamic pressure in the lubricating film and the external forces;
- the parameters of the nonlinear oscillation of a journal on the lubricating film with a nonstationary law of variation of influencing powers;
- the supplies-drop performance of a lubrication system;
- the characteristics of a lubricant, including its rheological properties.

Complex solution of these problems is an important step in increasing the reliability of tribounits, development of friction units, which satisfy the modern requirements. However, this solution presents great difficulties, since it requires the development of accurate and highly efficient numerical methods and algorithms.

The simulation result of heavy-loaded tribounits is accepted to assess by the hydromechanical characteristics. These are extreme and average per cycle of loading values for the minimum lubricant film thickness and maximum hydrodynamic pressure, the mean- flow rate through the ends of the bearing, the power losses due to friction in the conjugation, the temperature of the lubricating film. The criterions for a performance of tribounits are the smallest allowable film thickness and maximum allowable hydrodynamic pressure.

2.1 Determination of pressure in a thin lubricating film

The following assumptions are usually used to describe the flow of viscous fluid between bearing surfaces: bulk forces are excluded from the consideration; the density of the lubricant is taken constant, it is independent of the coordinates of the film, temperature and pressure; film thickness is smaller than its length; the pressure is constant across a film thickness; the speed of boundary lubrication films, which are adjacent to friction surfaces, is taken equal to the speed of these surfaces; a lubricant is considered as a Newtonian fluid, in which the shear stresses are proportional to the shear rate; the flow is laminar; the friction surfaces microgeometry is neglected.

The hydrodynamic pressure field is determined most accurately by employment of the universal equation by Elrod (Elrod, 1981) for the degree of filling of the clearance q by lubricant:

$$\frac{1}{r^2}\frac{\partial}{\partial\varphi}\left[\frac{h^3}{12\mu}\beta g\frac{\partial\theta}{\partial\varphi}\right]+\frac{\partial}{\partial z}\left[\frac{h^3}{12\mu}\beta g\frac{\partial\theta}{\partial z}\right]=\frac{(\omega_2-\omega_1)}{2}\frac{\partial}{\partial\varphi}(h\theta)+\frac{(w_2-w_1)}{2r}\frac{\partial}{\partial z}(h\theta)+\frac{\partial}{\partial t}(h\theta)\ . \quad (1)$$

Where r is the radius of the journal; j , z are the angular and axial coordinates, accordingly (Fig. 1); h (j , z,t) is film thickness; m is lubricant viscosity; b is lubricant compressibility factor; w_1 ,w_2 are the angular velocity of rotation of the bearing and the journal in the inertial coordinate system; w_1 , w_2 are forward speed of bearing and journal, accordingly; t is time; g is switching function, $g=\begin{cases}1, & if\ \theta\geq 1;\\ 0, & if\ \theta<1.\end{cases}$

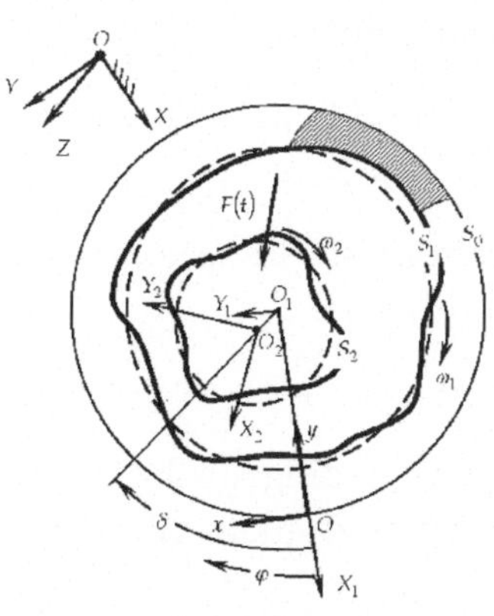

Fig. 1. Cross section bearing

If (w_2 - w_1) = 0 , then we get an equation for the tribounit with the forward movement of the journal (piston unit). If (w_2 - w_1) = 0 , we get the equation for the bearing with a rotational movement of the shaft (radial bearing).

The degree of filling q has the double meaning. In the load region q = r / rc , where r is homogeneous lubricant density; rc is the lubricant density if a pressure is equal to the pressure of cavitation pc . In the area of cavitation $p = pc$, r = r_c and q determines the mass content of the liquid phase (oil) per a unit of space volume between a journal and a bearing. The relation between hydrodynamic pressure *p* (j , *z*) and q (j , *z*) can be written as

$$p = p_c + g \cdot \beta \ln \theta . \tag{2}$$

The equation (1) allows us to implement the boundary conditions by Jacobson-Floberga- Olsen (JFO), which reflect the conservation law of mass in the lubricating film

$$\begin{aligned} &p(\varphi_g, z) = \partial p / \partial \varphi(\varphi_g, z) = p(\varphi_r, z) = p_a; \\ &p(\varphi, z = \pm B/2) = p_a; \, p(\varphi, z) = p(\varphi + 2\pi, z), \end{aligned} \tag{3}$$

where j_g , j_r are the corners of the gap and restore of the lubricating film; *B* is bearing width; *pa* is atmospheric pressure.

The conditions of JFO can quite accurately determine the position of the load region of the film. The algorithms of the solution of equation (1), which implement them, are called "a mass conserving cavitation algorithm".

On the other hand the field of hydrodynamic pressures in a thin lubricating film is determined from the generalized Reynolds equation (Prokopiev et al., 2010):

$$\frac{1}{r^2}\frac{\partial}{\partial \varphi}\left[\frac{h^3}{12\mu}\frac{\partial p}{\partial \varphi}\right] + \frac{\partial}{\partial z}\left[\frac{h^3}{12\mu}\frac{\partial p}{\partial z}\right] = \frac{(\omega_2 - \omega_1)}{2}\frac{\partial h}{\partial \varphi} + \frac{(w_2 - w_1)}{2r}\frac{\partial h}{\partial z} + \frac{\partial h}{\partial t}. \tag{4}$$

The equation (4) was sufficiently widespread in solving problems of dynamics and lubrication of different tribounits.

When integrating the equation (4) in the area W = (j Î 0, 2p ; z Î -*B* / 2, *B* / 2) mostly often

Stieber-Swift boundary conditions are used, which are written as the following restrictions on the function *p* (j , *z*) :

$$p(\varphi, z = \pm B/2) = p_a; \, p(\varphi, z) = p(\varphi + 2\pi, z); p(\varphi, z) \geq p_a, \tag{5}$$

If the sources of the lubricant feeding for the film locate on the friction surfaces, then equations (3) and (5) must be supplemented by

$$p(\varphi, z) = p_S \quad на \, (\varphi, z) \in \Omega_S, S = 1,2 \ldots S^*, \tag{6}$$

where W*S* is the region of lubricant source, where pressure is constant and equal to the supply pressure *pS* ; S^* is the number of sources.

To solve the equations (1) and (3) taking into account relations (3), (5), (6) we use numerical methods, among which variational-difference methods with finite element (FE) models and methods for approximating the finite differences (FDM) are most widely used. These methods are based on finite-difference approximation of differential operators of the boundary task with free boundaries. They can most easily and quickly obtain solutions with sufficient accuracy for bearings with non-ideal geometry. These methods also can take into account the presence of sources of lubricant on the friction surface.

One of the most effective methods of integrating the Reynolds equation are multi-level algorithms, which allows to reduce significantly the calculation time. Equations (1) and (4) are reduced to a system of algebraic equations, which are solved, for example, with the help of Seidel iterative method or by using a modification of the sweep method.

2.2 Geometry of a heavy-loaded tribounit

The geometry of the lubricant film influences on hydromechanical characteristics the greatest. Changing the cross-section of a journal and a bearing leads to a change in the lubrication of friction pairs. Thus technological deviations from the desired geometry of friction surfaces or strain can lead to loss of bearing capacity of a tribounit. At the same time in recent years, the interest to profiled tribounits had increased. Such designs can substantially improve the technical characteristics of journal bearings: to increase the carrying capacity while reducing the requirements for materials; to reduce friction losses; to increase the vibration resistance. Therefore, the description of the geometry of the lubricant film is a crucial step in the hydrodynamic calculation.

Film thickness in the tribounit depends on the position of the journal center, the angle between the direct axis of a journal and a bearing, as well as on the macrogeometrical deviations of the surfaces of tribounits and their possible elastic displacements.

We term the tribounit with a circular cylindrical journal and a bearing as a tribounit with a perfect geometry. In such a tribounit the clearance (film thickness) in any section is equal constant for the central shaft position in the bearing ($h^*(j, Z_1) = const$). Where j, Z_1 are circumferential and axial coordinates.

For a tribounit with non-ideal geometry the function of the clearance isn't equal constant ($h^*(j, Z_1)\ ^1$ const). This function takes into account profiles deviations of the journal and the bearing from circular cylindrical forms as a result of wear, manufacturing errors or constructive profiling.

If the tribounit geometry is distorted only in the axial direction, that is $h^*(Z_1)\ ^1$ const , we term it as a tribounit with non-ideal geometry in the axial direction, or a non-cylindrical tribounit. If the tribounit geometry is distorted only in the radial direction, that is $h^*(j)\ ^1$ const , we term it as a tribounit with a non-ideal geometry in the radial direction or a non- radial tribounit (Prokopiev et al., 2010).

For a non- radial tribounit the macro deviations of polar radiuses of the bearing and the journal from the radiuses r_{i0} of base circles (shown dashed) are denoted by

D_1 (j), D_2 (j,t). Values Di don't depend on the position z and are considered positive (negative) if radiuses $ri0$ are increased (decreased). In this case, the geometry of the journal friction surfaces is arbitrary, the film thickness is defined as

$$h(\varphi,t)=h^{*}(\varphi,t)-e\cos(\varphi-\delta)\,. \tag{7}$$

Where h^* (j,t) is the film thickness for the central position of the journal, when the displacement of mass centers of the journal in relation to the bearing equals zero ($e(t)=0$). It is given by

$$h^{*}(\varphi,t)=\Delta_0+\Delta_1(\varphi)-\Delta_2(\varphi,t),\ \Delta_0=(r_{10}-r_{20})\,. \tag{8}$$

The function h^* (j,t) can be defined by a table of deviations D_i (j,t), analytically (functions of the second order) or approximated by series.

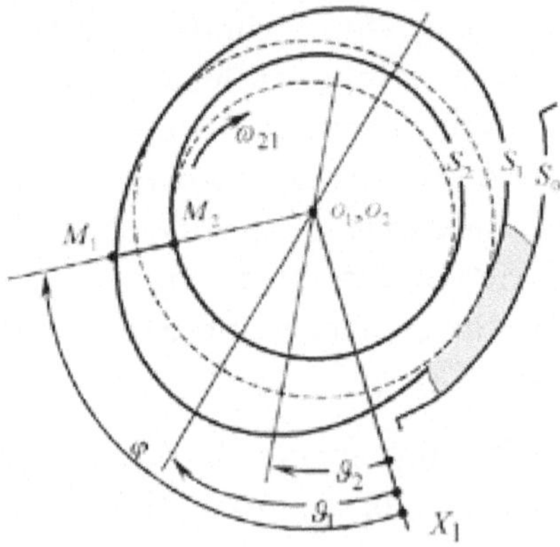

Fig. 2. Scheme of a bearing with the central position of a journal

If a journal and a bearing have the elementary species of non-roundness (oval), their geometry is conveniently described by ellipses. For example, the oval bearing surface is represented as an ellipse (Fig. 2) and the journal surface is represented as a one-sided oval – a half-ellipse.

Using the known formulas of analytic geometry, we represent the surfaces deflection Di of a bearing and a journal from the radiuses of base surfaces $r_0i = bi$ in the following form

$$\Delta_i=b_i\left\{\nu_i\left[\nu_i^2-\left(\nu_i^2-1\right)\cos^2(\varphi-\vartheta_i)\right]^{-0,5}-1\right\}, \tag{9}$$

where the parameter ni is the ratio of high ai to low bi axis of the ellipse, JI are angles which determine the initial positions of the ovals. Due to fixing of the polar axis O_1X_1 on the bearing, the angle J_1 doesn't depend on the time, and the angle J20 , which determines the location of the major axis of the journal elliptic surface with $t=t_0$, is associated with a relative angular velocity w21 by the following relation

$$\vartheta_2(t) = \vartheta_{20} + \int_{t_0}^{t} \omega_{21} dt\,. \tag{10}$$

In an one-sided oval of a journal equation (9) is applied in the field $(\pi/2+\vartheta_2)\leq\phi\leq(3/2\pi+\vartheta_2$ but off it D2 = 0

If the macro deviations D_1 (j), D_2 (g_2) of journal and bearing radiuses r_i (j) from the base circles radiuses *ri*0 are approximated by truncated Fourier series, then they can be represented as (Prokopiev et al., 2010):

$$\Delta_i(\psi) = \tau_{i0} + \tau_i \sin(k_i\psi + \alpha_i)\,, \tag{11}$$

where $i = 1$ for a bearing, $i = 2$ for a journal; $\psi = \phi$ if $i = 1$, $\psi = \gamma_2 = \phi + \vartheta_1 - \vartheta_2$ if $i = 2$; $\vartheta_2 = \int_0^t \omega_{21}(t)dt$; k_i is a harmonic; t*i*0 is a permanent member of the Fourier series, which is defined by

$$\tau_{i0} = \frac{1}{2\pi}\int_0^{2\pi} \Delta_i(\psi)d\varphi\,. \tag{12}$$

For elementary types of non-roundness (oval ($k = 2$); a cut with three ($k = 3$) or four ($k = 4$) vertices of the profile) t*i*0 = 0 .

The thickness of the lubricant film, which is limited by a bearing and a journal having elementary types of non-roundness, after substituting (12) in (7), is given by

$$h(\varphi,t) = \Delta_0 + \tau_1 \sin(k_1\varphi + \alpha_1) - \tau_2 \sin(k_2\gamma_2 + \alpha_2) - e\cos(\varphi - \delta)\,. \tag{13}$$

For tribounits with geometry deviations from the basic cylindrical surfaces in the axial direction the film thickness at the central position of the journal in an arbitrary cross-section Z_1 is written by the expression

$$h^*(Z_1) = \Delta_0 + \Delta_1(Z_1) - \Delta_2(Z_1)\,. \tag{14}$$

Where D_i (Z_1), $i = 1, 2$ are the deviations of generating lines of bearing surfaces and the journal surfaces from the line (positive deviation is in the direction of increasing radius). Then, taking into account the expressions (8) and (14) we can write the general formula for a lubricant film thickness with the central position of the journal in the bearings with non- ideal geometry as

A barreling, a saddle and a taper are the typical macro deviations of a journal and a bearing from a cylindrical shape (Fig. 3).

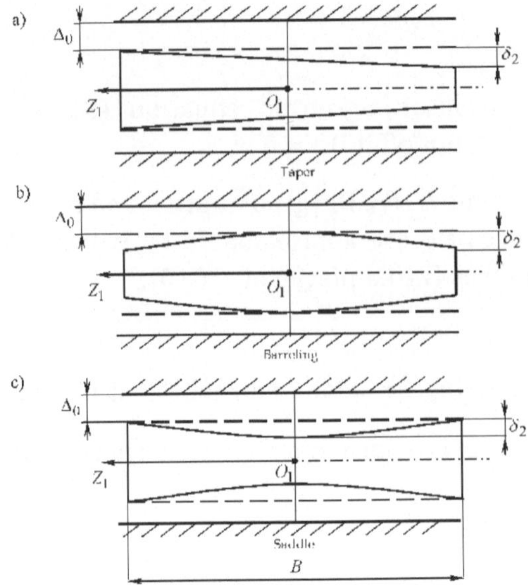

Fig. 3. Types of non-cylindrical journals

The non-cylindrical shapes of the bearing and the journal in the axial direction are defined by the maximum deviations d_1 and d_2 of a profile from the ideal cylindrical profile and are described by the corresponding approximating curve. Then the film thickness at the central position of the journal (Prokopiev et al., 2010) is given by

$$h^*(Z_1) = \Delta_0 + k_1 Z_1^{l_1} + k_2 Z_1^{l_2}\,, \tag{16}$$

where *ki* defines the deviation of the approximating curve per unit of the width of the bearing, the degree of the parabola is accepted: barrel and saddle journals. *li* = 1 for the conical journals; *li* = 2 for

For the circular cylindrical bearing for known formula: D*i* = 0 the film thickness is determined by the well-

$$\bar{h}(\varphi, \bar{t}) = 1 - \chi\cos(\varphi - \delta)\,. \tag{17}$$

For the circular cylindrical journal its rotation axis is parallel to the axis $O_1 Z_1$. In practice, the axis of the journal may be not parallel to the axis of the bearing, so there is a so-called "skewness". These deviations may be as due to technological factors (the inaccuracy of manufacturing during the production and repair) as to working conditions (wear, bending of shafts, etc.).

Position of the journal, which is regarded as a rigid body, in this case you can specify by two coordinates *e*,d of the journal center O_2 and by three angles (g , e , q_2). Angle g is skewness of journal axis; e is the deviation angle of skewness plane from the base coordinate plane; q_2 is the rotation angle of the journal on its own axis $O_2 Z_2$.

When journal axis is skewed the film thickness at a random cross-section $Z_1 i$ of the bearing depends on the eccentricity *ei* and the angle d*i* for this cross-section

$$h(\varphi, Z_{1i}, t) = h^*(\varphi, Z_{1i}) - e_i \cos(\varphi - \delta_i)\,, \tag{18}$$

where h^*(j ,$Z_1 i$) is the film thickness with the central journal position in *i*-th cross -section. We term the tgg = 2*s* / *B* , where *s* is the distance between the geometric centers of the journal and the bearing at the ends of the tribounit; *B* is the width of the tribounit. The expression for the lubricant film thickness, taking into account the skewness, is written in the form

$$h(\varphi, Z_1, t) = h^*(\varphi, Z_1) - e\cos(\varphi - \delta) - Z_1 \cdot \frac{2s}{B}\cos(\varphi - \varepsilon)\,. \tag{19}$$

It should be also taken into account that the bearing surfaces are deformed under the action of hydrodynamic pressures. The value D(*p*) is the radial elastic displacement of the bearing sliding surface under the action of hydrodynamic pressure *p* in the lubricant film. Function D(*p*) is defined in the process of calculating of the bearing strain (for a "hard" bearing D(*p*) = 0) and is written in the form of a component in the equation for the lubricant film thickness.

Thus, the film thickness, taking into account the arbitrary geometry of friction surfaces of a journal and a bearing, the skewness of the journal and elastic displacements of the bearing, is determined by the equation:

$$h(\varphi, Z_1, t) = h^*(\varphi, Z_1) - e\cos(\varphi - \delta) - Z_1 \cdot 2s/B \cdot \cos(\varphi - \varepsilon) + \Delta(p) \tag{20}$$

where h^*(j ,Z_1) is the film thickness with the central position of the journal in the bearing with non-ideal geometry; *e*(*t*) is displacement of journal mass centers in relation to the bearing; e (*t*) - an angle that takes into account the skewness of axes of a bearing and a journal . The values *e*(*t*) ,d (*t*) ,e (*t*) are determined by solving the equations of motion.

2.3 The calculation of thermal processes

The theory of thermal processes in the heavy-loaded tribounit of fluid friction is based on a generalized equation of energy (heat transmission) for a thin film of viscous incompressible fluid, which is between two moving surfaces S_1 and S_2 . If we assume a low thermal conductivity in the direction of the coordinate axes *Oxz* (the axis *Oy* is normal to the surface *S*1) (Fig. 1), the temperature distribution *T*(*x*, *y*, *z*,*t*) in the lubricating film will be described by the equation (Prokopiev&Karavayev, 2003)

$$\rho c_0 \frac{\partial T}{\partial t} + \rho c_0 \left(V_x \frac{\partial T}{\partial x} + V_y \frac{\partial T}{\partial y} + V_z \frac{\partial T}{\partial z} \right) - \lambda_0 \frac{\partial^2 T}{\partial y^2} = Д\,. \tag{21}$$

Where r is density; c_0 , l_0 are specific heat capacity and thermal conductivity of lubricant (usually taken as constant); *t* is the time; *Д* is the dissipation function, which is defined for non-Newtonian fluid by the approximate expression

$$Д \approx \mu^* I_2 . \quad (22)$$

Thethreeapproachestotheintegrationoftheequation(21)(thermohydrodynamic (nonisothermal), adiabatic, isothermal) can be used, depending on the assumptions which are used about the temperature distribution in a thin lubricating film.

When thermohydrodynamic approach is applied the temperature will change in all directions, including across the oil film. In this case, the boundary conditions are stated quite simply and are the most adequate to the real thermal processes. With this approach, we get information about the local properties of the temperature field of lubricating film: a temperature distribution $T(x, y, z,t)$; maximum temperature Tmax , instantaneous average temperature Tav (t) ; zones of elevated temperatures.

If adiabatic approach is applied the change of the temperature across the oil film (along the axis Oy) is ignored, the journal and the bearing are assumed ideal thermal insulators. We introduce a computational averaged over the width of the bearing temperature $T^* = T^*(x,t)$. We substitute it into the equation (21) and receive a differential equation for the temperature distribution along the coordinate x . Since in this case the heat transfer to the journal and the bearing is not taken into account, the calculated temperatures are too high. It reduces the accuracy of the results.

The isothermal approach assumes that the calculated current temperature $Tc = Tc\ (t)$ is the same at all points of the lubricant film. This temperature is a highly inertial parameter and it is determined by solving the heat balance equation

$$A_N^*(t) = A_Q^*(t) . \quad (23)$$

This equation reflects the equality of the average values of the heat A* , which is dissipated in the lubricating film, and the average values of the heat A^* , which is drained by lubricant into the ends of the tribounit during the loading cycle.

The accurate definition of the current temperature can be performed: at each time step of the calculation; once per a cycle of loading the tribounit, at each time step of the calculation taking into account the thermal interaction between the lubricant film with a journal, with a bearing and a lubrication groove.

2.4 The equations of heavy-loaded bearing dynamics

To study the dynamics of bearings of liquid friction the motion of the journal on the lubricant film in the bearing is usually considered (Fig. 4). In the coordinates space OXYZ the movement of the journal, which rotates with the relative angular velocity and the angular acceleration, taking into account the axle skewness of a journal and a bearing, is described by approximate differential equations

$$\tilde{m}\ddot{U}(t) = \tilde{F}(t) + \tilde{R}(U, \dot{U}) . \quad (24)$$

Where $m^{\sim}$ is the matrix of inertia of the journal: $\{m^{\sim}ii\} = \{m,m,m,J_X,J_Y,J_Z\}$, m, J_X, J_Y, J_Z are mass and moments of inertia of the journal, $m^{\sim}ij^1i = 0$, $i = 1,...,6$, $j = 1,...,6$;

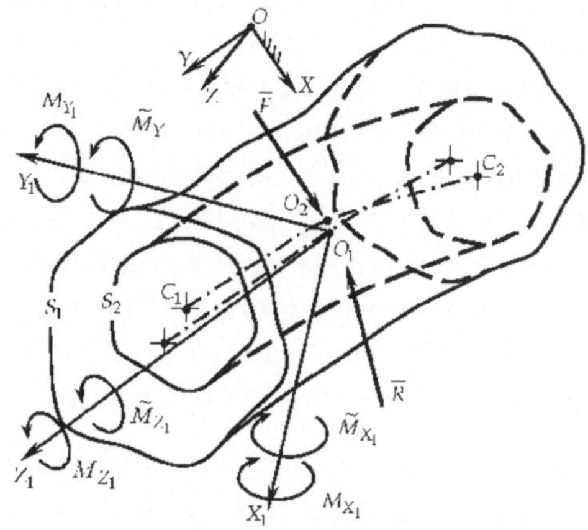

Fig. 4. Scheme of a heavy-loaded bearing with arbitrary geometry of the lubricant film

$U(t) = \{X, Y, Z, \gamma_X, \gamma_Y, \gamma_Z\}$ is the vector of generalized coordinates of the journal centre;

$F^{\sim}(t) = \{F_X, F_Y, F_Z, M_X, M_Y, M_Z\}$ is the vector of known loads on the journal, presented by the power F with its projections FX, FY, FZ on the axes of the coordinate system $OXYZ$ and a moment of forces M with its projections MX, MY, MZ; $R^{\sim}(U, U^{\cdot}) = \{RX, RY, RZ, MX, MY, MZ\}$ is the vector of loads due to the hydrodynamic pressure in the lubricant film. The time derivatives are denoted by points. The forces of friction and weight, as well as gyroscopic moments of the rotating journal are considerably less than other loads, so they aren't taken into account in the equations of motion.

For the dynamics of radial bearings of ICE the level of loads F acting on the journal is higher than its own inertial forces. The system of equations of motion (24) in this case is rewritten as

$$\tilde{F}(t) + \tilde{R}(U, \dot{U}) = 0 . \quad (25)$$

Projections of linear and angular positions and velocities and loads F, M, R,M onto the axis OZ are excluded from the employed vectors.

In the case of planar motion of a journal on the lubricant film the solution to the problem of the dynamics of the radial bearings can be obtained more easily. The skewness of axes of a journal and a bearing are neglected:

$$\gamma_X = \gamma_Y = \gamma_Z = M_X = M_Y = M_Z = 0, \quad \tilde{M}_X = \tilde{M}_Y = \tilde{M}_Z = 0 . \quad (26)$$

The vectors of coordinates, velocities and loads include only their projections on the axes OX, OY.

The solution of the systems of equations of motion (24) or (25) can be found only numerically, because the loads, which are caused by the hydrodynamic pressure, are determined by the repeated numerical solution of the differential equations by Elrod (1) or by Reynolds (4). If we discretize the system of equations of motion over time, then the decision when passing to the next time step can be obtained by using the explicit or implicit method of calculation. In an explicit scheme the unknowns are the pressure and the coordinates of the journal center, in an implicit scheme the unknowns are the pressure and the rate of position change of the journal. However, the implementation of explicit schemes of integrating the motion equations is sensitive to the accumulation of rounding errors. Therefore, implicit schemes for integrating the equations of motion over time are realized in several studies, which are dedicated to the dynamics of heavy-loaded tribounits.

The most common methods for solving equations of motion of type (24) are: Newton's method, Runge-Kutta's method with Merson's modification, the modified method of linear acceleration (Wilson's method), the method of non-central third-order differences (method by Habolt). To solve the system of the form (25) it is expedient to use special techniques, which are adapted to the systems of "stiff" differential equations (method based on the use of differentiation backward formulas (DBF) of the first- and second-order, method by Fowler, Wharton and others). The standard procedure for solving differential equations (25), which are unsolved relatively to derivatives, consists in the formal integration of the equations $U^{\cdot} = f(U,t)$ and determining derivatives with the help of Newton method.

When the character of applied loads is periodical the initial values of variables U and their derivatives $U^{\cdot}$ can be set arbitrarily. With that the integration continues until the time when the values U and $U^{\cdot}$, which are separated by a period tc of load changes, will not be repeated.

Ability to use a particular method of integration depends on the type of a tribounit, the character of acting loads and the possibility to set an initial approximation for the successful solution of (24) or (25). Currently, universal methods for solving the dynamics of heavy- loaded tribounits are not designed. The result of calculating the dynamics of heavy-loaded bearings is a trajectory of mass center of the journal, as well as hydro-mechanical characteristics of tribounits.

The construction of the heavy-loaded tribounit is evaluated by parameters of the calculated trajectory and interconnected hydro-mechanical characteristics (HMCh). There is the lowest and average per cycle of loading values of: the lubricant film thickness inf $h\min$, * $h\min$, μm; the hydrodynamic pressure in the lubricant film max sup$p$, * mp ax , MPa; the unit load $f\max$, *m f ax , MPa; the relative total length of the regions $h\partial on$ α , where the values of $h\min$ less than allowable values $h\partial on$, %; the relative total length of the regions $p\partial on$ α , where the values of $p\max$ greater than allowable values $p\partial on$, %; mean-value losses due to friction N^* ,W, the leakage of lubrication in the bearing ends Q^* ,$м3$ s and temperature of the lubricant film T,C .

3. Lubrication with non-Newtonian and multiphase fluids

The development of technology is inextricably linked with the improvement of lubricants, which today remain an important factor that ensures the reliability of machines. Currently, for lubrication of tribounits of ICE multigrade oils are widely used, rheological behavior of which does not comply with the law of Newton-Stokes equations on a linear relationship between shear stress and shear rate (Whilkinson, 1964):

$$\tau = \mu \cdot \dot{\gamma} \ , \qquad (27)$$

where t – shear stress; m – dynamic viscosity, which is a function of temperature *T* and pressure *p* (Newtonian viscosity); g˙ – shear rate, $\dot{\gamma} = \sqrt{I_2}$; I_2 second invariant of shear rate *I* $I_2 \approx (\partial \dot{V}_x / \partial y)^2 + (\partial \dot{V}_z / \partial y)^2$, $\dot{V}_x, \dot{V}_y, \dot{V}_z$ velocity component of the elementary volume lubrication, which is located between the two surfaces.

Particularly, the viscosity depends not only on the temperature and pressure, but also on the shear rate in a thin lubricating film separating the surfaces of friction pairs. These oils are called non-Newtonian.

Theoretical studies of the dynamics of friction pairs, which take into account non- Newtonian behavior of lubricant, are based on the modification of the equations for determining the field of hydrodynamic pressures by using different rheological models. One classification of a rheological model is shown in Fig. 5.

In general, non-Newtonian behavior includes any anomalies observed in the flow of fluid. In particular, the presence of viscous polymer additives in oils leads to a change in their properties. Oils with additives can be characterized as structurally viscous and viscoelastic Viscoelastic fluids are those exhibiting both elastic recovery of form and viscous flow. There are various models of viscoelastic fluids, among which the best known model is the

Maxwell $\tau + \lambda \dfrac{\partial \tau}{\partial t} = \mu^{*} \dot{\gamma}$. Here l – relaxation time, characterizing the delay of shear stress changes in respect to changes of shear rates; m *(T , *p*,g˙) – dynamic viscosity (non-Newtonian viscosity). In this case, the liquid is called the Maxwell (Maxwell viscoelastic liquid).

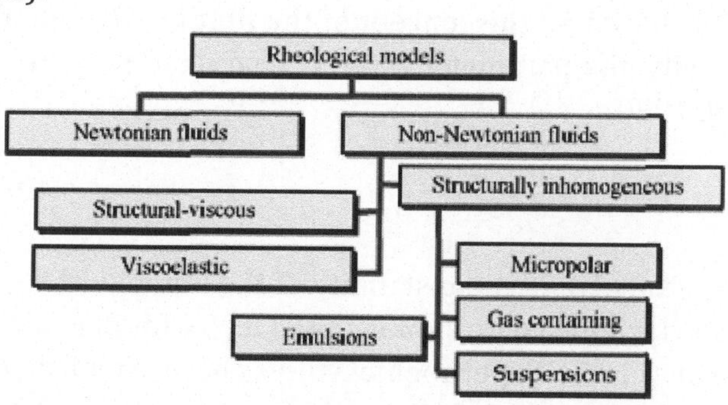

Fig. 5. Classification of rheological models of lubricating fluids

It is assumed that the viscoelastic properties of thickened oils have a positive impact on the operation of sliding bearings, help to increase the thickness of the lubricant film. Qualitative influence of viscoelastic properties (relaxation time) of the lubricant is reflected in Fig. 6.

With the increase of the relaxation time of lubrication, the mean-value of the minimum lubricating film thickness and power loss due to friction increase. It is seen that the character of the dependence is the same, but the values are shifted back to the rotation angle of the crankshaft.

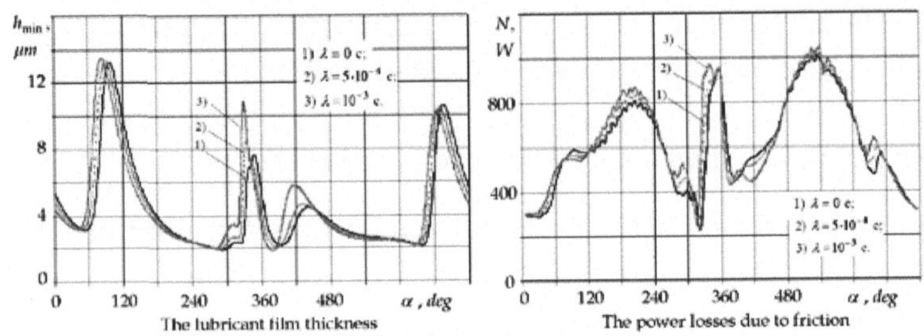

Fig. 6. The dependence of the characteristics from the angle of rotation of crankshaft

Structural-viscous oils have the ability to temporarily reduce the viscosity during the shear, so they are called "energy saving", because they help to reduce power losses due to friction in internal combustion engines and, consequently, fuel consumption (according to various estimates by 2-5%).

The most well-known mathematical model describing the behavior of the structural-viscous oils, is a power law of Ostwald-Weyl, according to which the dependence of viscosity versus shear rate is defined as (Whilkinson, 1964)

$$\mu^* = k\dot{\gamma}^{n-1}. \tag{28}$$

Where *k* – measure the fluid consistency; *n* – index characterizing the degree of non- Newtonian behavior.

Gecim suggested the dependence of viscosity on the second invariant of shear rate, which is based on the concept of the first m_1 (T) and the second m_2 (T) Newtonian viscosity, the parameter K_c (T), characterizing the shear stability of lubricants (Gecim, 1990):

$$\mu^*(\dot{\gamma}) = \mu_1 \frac{K_c + \mu_2 \cdot \dot{\gamma}}{K_c + \mu_1 \cdot \dot{\gamma}}. \tag{29}$$

The higher *Kc*, the higher is the stability of the liquid with respect to the shift. At low shear viscosity value corresponds to the m_1, with increasing shear rate the viscosity tends to m_2 (Fig. 7). Experimental studies have established that multigrade oils of the same viscosity grade of SAE may have different shear stability.

The application of structural-viscous oils, along with a reduction of power losses to friction leads to a decrease in the lubricating film thickness, temperature and to the increase of lubrication flow rate.

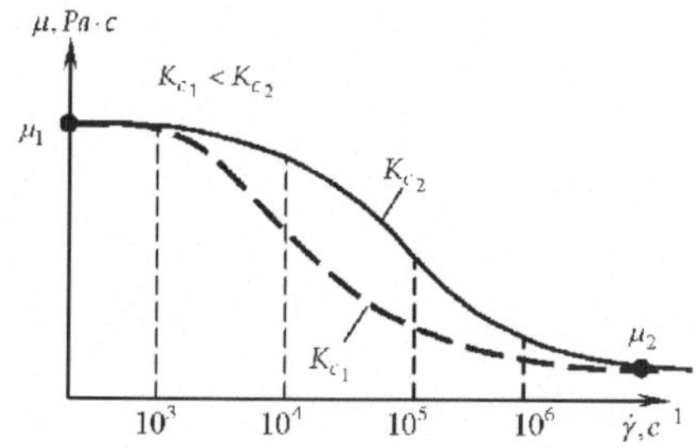

Fig. 7. Fundamental character of the non-Newtonian oils viscosity

Comparative results of the calculation of hydro-mechanical characteristics of the connecting rod bearing for the dependence of oil viscosity versus shear rate and without it are presented in Table. 1 and Fig. 8.

All results were founded for connecting-rod bearing of engine type ЧН 13/15 (Co ltd. "ChTZ-URALTRAC") with follow parameters: rotating speed 219.91 c-1; length 0.033 m; journal radius 0.0475 m; radial clearance 51.5 mm.

The results indicate that the application of structural-viscous oils leads to a reduction of power losses due to friction in the range 15-20%. Consumption of lubricant through the bearing increases, the mean-value of the temperature decreases by 2-3° C. However, there is a decrease in the minimum lubricating film thickness by an average of 14-20%.

This fact confirms the view that the use of low-viscosity oil at high temperature and shear rate is justified only if it is allowed by the engine design, in particular, of crankshaft bearings.

Table 1. The results of the calculation of HMCh of the connecting rod bearing

Hydromechanical characteristics	N^*, W	T, ºC	Q^*_B, l/s	h^*_{min}, μm	sup p_{max}, MPa	inf h_{min}, μm	a*, %
Newtonian fluid	610,5	105,9	0,02345	4,416	280,3	1,93	0
Structural-viscous liquid (28)	518,4	102,6	0,02512	3,75	309,8	1,52	16,9
Structural-viscous liquid (29)	539,0	103,4	0,0246	3,789	307,8	1,66	11,9

In recent years, the oil, which has in its composition the so-called friction modifiers, for example, particles of molybdenum, is widespread. These additives are introduced into the base oil to improve its antiwear and extreme pressure properties to reduce friction and wear under semifluid and boundary lubrication regimes.

Oils with such additives are called "micropolar". They represent a mixture of randomly oriented micro-particles (molecules), suspended in a viscous fluid and having its own rotary motion.

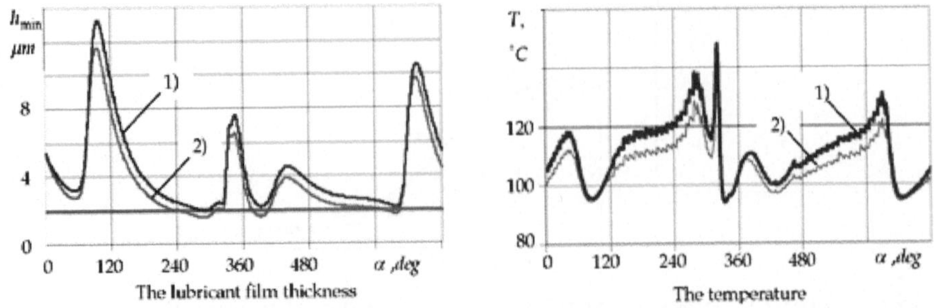

Fig. 8. The dependence of the hydromechanical characteristics from the rotation angle of crankshaft: 1) Newtonian fluid, and 2) the structural-viscous liquid (28)

Micropolar fluid along with the viscosity m additionally characterized by two physical constants m_1 ,ℓ . Parameter m_1 , called the coefficient of eddy viscosity, takes into account the resistance **to** micro-rotation of particles. Length parameter ℓ characterizes the size of microparticles or molecular lubricant. With the help of the coefficient m_1 and the parameter ℓ **you** can calculate the so-called micropolar parameters

$$N = \left(\frac{\mu_1}{2\mu + \mu_1} \right)^{1/2}, \quad L = \frac{h_0}{\ell}, \qquad (30)$$

where h_0 – characteristic film thickness.

The presence of micro-particles in the lubricant leads to an increase in the resultant shear stress in the lubricating film. The calculations of heavy-loaded bearings using micropolar fluid theory suggest that this phenomenon significantly affects the **HMCh** of a bearing, in particular, leads to an increase of lubricating film thickness. The results of the calculation of the connecting rod bearing, taking into account the structural heterogeneity of lubricants (based on the model of micropolar fluids with the parameters $L = 10$, $N^2 = 0,5$) are reflected in Fig. 9 and Table. 2.

Table 2. The results of the calculation of hydro-mechanical characteristics of the connecting rod bearing, taking into account the structural heterogeneity of lubrication

Hydromechanical characteristics	N^*, W	T, ºC	Q^*_B, l/s	h^*_{min}, μm	**sup** p_{max}, MPa	**inf** h_{min}, μm	a^*, %
Newtonian fluid	610,5	105,9	0,02355	4,416	280,3	1,93	0
Structurally heterogeneous fluid (30)	727,4	110,6	0,0215	5,84	237,9	2,9	0

It is obvious, that the results will prove valuable for practice, only in case of experimental determination of the value of the micropolarity parameters *N* and *L*. Further studies of the authors are focused on the experimental basis of these values for modern thickened oils.

The calculation of the structural heterogeneity of the lubricant is a very complicated mathematical problem, since it is necessary to take into account many factors: the speed and shape of particles, their distribution, elasticity, etc.

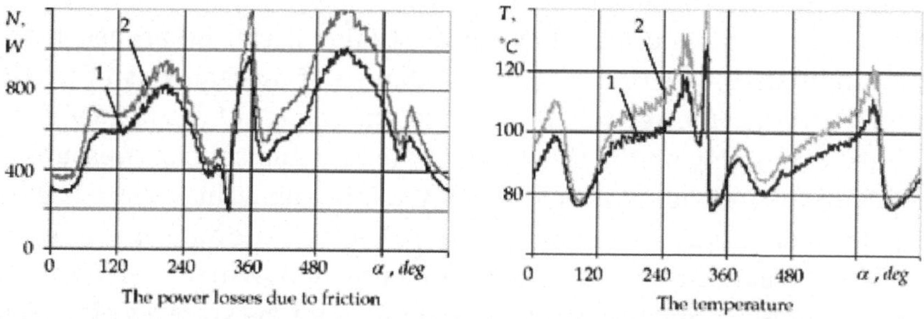

Fig. 9. Dependence of the hydromechanical characteristics from the rotation angle of the crankshaft: 1 - Newtonian fluid; 2 - structurally heterogeneous fluid (30)

Sometimes simplified dependence is used. For example it is assumed that the viscosity of suspensions depends on the concentration volume of solid particles, which may be the wear products, external contaminants or finely divided special additives. In this case, the viscosity of the lubricant is sufficiently well described by the Einstein formula:

$$\mu = \mu^{*}(1 + \xi \cdot \varphi)\,. \tag{31}$$

Where x – shape factor of particles, for asymmetric particles x ³ 2, 5 .

Separate scientific problem is the availability of records in the lubricant gas component. Experimental studies have shown that the engine lubrication system always contains air dissolved in the form of gas bubbles. The proportion of bubbles in the total amount of oil may reach 30%.

The viscosity of gassy oils can be calculated with a sufficient degree of accuracy with the help of the formula:

$$\mu = \mu^{*}(1 - \delta)\,. \tag{32}$$

Where the coefficient d = *VT* / *VM* is equal to the ratio of the volume fraction of gas *VT* in the bubble mixture to the volume fraction of pure oil *VM* at temperature *T*.

When you select computer models you must take into account not only the working conditions, regime and geometric characteristics of tribounits under consideration, but also features of rheological behavior of used lubricants.

At present, as a result of parallel and interdependent modifications of ICE and production technologies of motor oils, the most loaded sliding bearings of an engine work at the minimum design film thickness of about 1 micron in the steady state and less - at low frequencies of crankshaft rotation, that is with film thicknesses comparable to twice the height of surface roughness of tribounits. In this case the life of one and the same friction unit can vary in 3 ... 5 times when using different motor oils, and be by orders of magnitude greater than the resource when using other grease lubricants at the same bulk rheological properties.

Based on experimental and theoretical studies it can be argued that under changing conditions of friction a repeated change of mechanisms of friction and wear occurs, in which the key role is played by the change of rheological properties of lubricants, depending on the thickness of the film, the contact pressure, surface roughness and the individual properties of the lubricant. Thus, there is a need for the computational models depending on the rheological properties of lubricating oil on the factors related to the availability, quantity and structure of the antifriction and antiwear additives and lubricants interaction with the surfaces of the friction.

One model describing the dependence of viscosity of lubricant on thickness is proved in (Mukhortov et al., 2010) and has the following form:

$$\mu_i = \mu_0 + \mu_S \exp\left(-\frac{h_i}{l_h}\right), \qquad (33)$$

where l_h – characteristic parameter having the dimension of length, which value is specific for each combination of lubricant and the solid surface; μ_S – parameter having the meaning of the conditional values of the viscosity at infinitely small distance from the bounding surface; μ0 - viscosity in entirety.

The impact of the availability of a highly viscous boundary film on the friction surfaces on the HMCh of the rod bearing is illustrated in Fig.10 and Table. 3.

In the hydrodynamic friction regime the presence of adsorption films leads to an increase in the minimum lubricating film thickness by 40-45%, the temperature at 6-7%, the maximum hydrodynamic pressure by 4-5%.

Table 3. The results of the calculation of hydro-mechanical characteristics of the connecting rod bearing in the light of high-viscosity boundary film lubrication

Hydromechanical characteristics	N^*, W	T, ºC	h^*_{min}, μm	sup p_{max}, MPa	inf h_{min}, μm
numerical value	610,5 [1)]	105,9	4,416	280,3	1,93
	681,2 [2)]	113,3	5,665	294,9	3,59

- Newtonian fluid; 2 - taking into account the highly viscous boundary film.

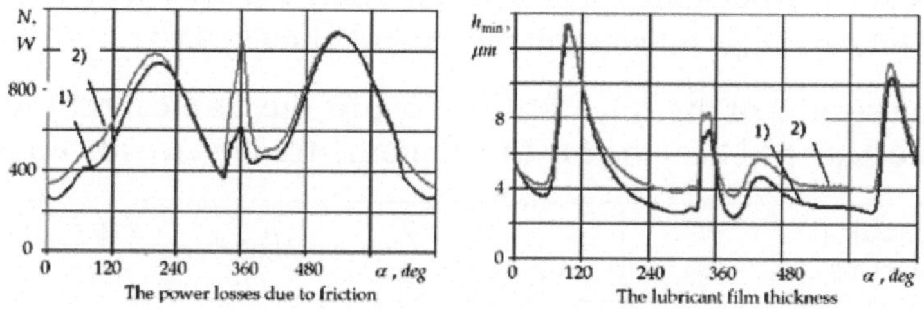

Fig. 10. The dependence of the hydromechanical characteristics from the rotation angle of crankshaft: 1 - Newtonian fluid, 2 - with the boundary layer (33)

These models should be used in accordance with the terms of the friction pairs. In this case, the use of non-Newtonian models of lubricants does not exclude taking into account the dependence of oil viscosity on temperature and pressure in the lubricating film of friction pairs (Prokopiev V. et al., 2010):

$$\mu(T) = C_1 \cdot \exp\left(C_2/(T + C_3)\right), \qquad (34)$$

where C_1, C_2, C_3 - constants, which are the empirical characteristics of the lubricant.

The coefficients *Ci* are calculated using the formula following from the dependence (34):

$$C_3 = \frac{-\left[T_1(T_3 - T_2)\ln\left(\frac{\mu_1}{\mu_2}\right) - T_3(T_2 - T_1)\ln\left(\frac{\mu_2}{\mu_3}\right)\right]}{\left[(T_3 - T_2)\ln\left(\frac{\mu_1}{\mu_2}\right) - (T_2 - T_1)\ln\left(\frac{\mu_2}{\mu_3}\right)\right]}; \qquad ; \qquad (35)$$

$$C_2 = \frac{\ln\left(\frac{\mu_1}{\mu_2}\right) \cdot (T_1 + C_3) \cdot (T_2 + C_3)}{(T_2 - T_1)}; \quad C_1 = \frac{\mu_1}{\exp(C_2/T_1)}$$

To account for the dependence of viscosity on the hydrodynamic pressure the Barus formula is acceptable:

$$\mu_p = \mu_0 e^{\alpha \cdot p}, \qquad (36)$$

where m_0 – viscosity of the lubricant at atmospheric pressure; *p* – hydrodynamic pressure in the lubricating film; a – piezoelectric coefficient of viscosity, which depends on temperature and chemical composition of lubricants.

On the base of a combination of models (28), (34) and (36) the authors propose to use a combined dependence of viscosity versus shear rate, pressure and temperature:

$$\mu^* = k \cdot \dot{\gamma}^{n-1} \cdot e^{\alpha(T) \cdot p} \cdot C_1 \cdot e^{(C_2/(T + C_3))}. \qquad (37)$$

The effect of hydrodynamic pressure in the film of lubricant on the HMCh of the connecting rod bearing is reflected in the Table 4 and Fig. 11.

Table 4. The results of the calculation of hydro-mechanical characteristics of the connecting rod bearing for the dependence of viscosity on pressure

Hydromechanical characteristics	N^*, W	T, ºC	Q^*_B, l/s	h^*_{min}, μm	**sup** p_{max}, MPa	**inf** h_{min}, μm
numerical value	610,5 [1)]	105,9	0,02345	4,416	280,3	1,930
	670,4 [2)]	106,9	0,02420	5,712	588,6	2,560

1) - oil viscosity is independent of pressure, 2) - viscosity depends on pressure.

As seen from Table 4 and Figure 11, in the case of taking into account the effect of hydrodynamic pressure on the viscosity of the lubricant, all the values of HMCh of the bearing increase. In particular, the mean-power losses increase by 8-9%, the minimum film thickness by 20-25%, the temperature by 1-2%. It is important to note that the instantaneous maximum hydrodynamic pressure is increased by 50-52%.

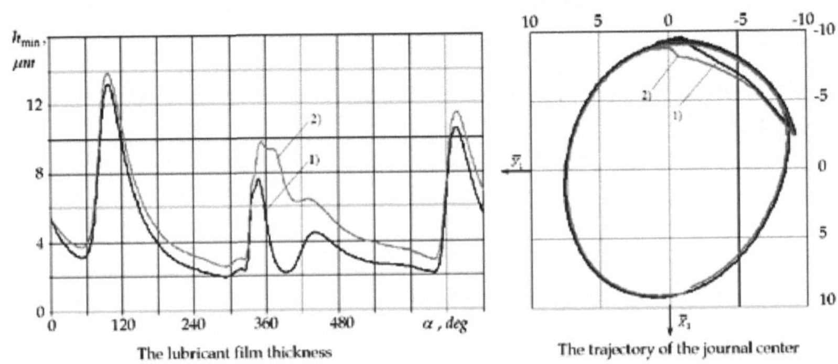

Fig. 11. The hydromechanical characteristics: 1 - Newtonian fluid, 2 - viscosity depends on pressure

Thus, accounting for one of the properties of the lubricant does not reflect the real process occurring in a thin lubricating film. Each of these properties of the lubricant and the dependence of viscosity on one of the parameters (p,T ,g˙,j , etc.) either improves or worsens the hydro-mechanical characteristics of tribounits. Therefore, the choice of rheological models used to calculate heavy-loaded tribounits, depends on the type of working conditions of lubricant and tribounits, as well as on the objectives pursued by the design engineer.

Further research should be focused on experimental substantiation of the parameters of rheological models, as well as the creation of calculation methods for assessing the simultaneous influence of various non-Newtonian properties of the lubricant on the dynamics of heavy loaded tribounits. This will provide simulation of real processes occurring in the lubricant film, and ultimately, will improve accuracy.

4. Effect of elastic properties of the construction

Elastohydrodynamic (EHD) regime of lubrication of bearings is characterized by a significant effect of dynamically changing strain of a bearing and (or) a journal on the clearance in the tribounit. Under unsteady loading the dynamic change in the geometry of the elements of tribounit caused by the finite stiffness of the bearing and the journal, leads to a change in the nature of the lubricant, hydromechanical parameters and supporting forces of tribounits and must be taken into account in the methods of its calculation.

The effect of finite stiffness of a bearing and a journal on the change of the profile of the clearance depends on the geometry of the bearing, the ratio of properties which are in contact through the lubricating film surfaces and other factors. In massive bearings local contact deformation of the surface film of the bearing and the journal prevail over the general changes of form of the bearing and the latter are usually neglected. These tribounits are usually referred to as contact-hydrodynamic (elasto-hydrodynamic). Examples of such units can be gears, frictionless (rolling) bearings, journal bearings with an elastic liner and rigid housing. For their calculations it is reasonable to use the methods of contact hydrodynamics (elasto-hydrodynamic lubrication theory).

However, there is also a large group of hydrodynamic friction pairs in which the general corps deformations make a significant contribution to changing the profile of the clearance. They are characterized by the presence of a continuous gradient of the deformation field, which is independent of the load location, the significant (compared with the contact and hydrodynamic tribounits) values of lubricating film thickness and values of the displacement of the friction surfaces, caused by the bending deformation of the housing, which are commensurate with them. These tribounits are called elasto-yielding (EY TU) or elastohydro- dynamic. The most typical representative of the EY TU is a connecting rod bearing of a crank mechanism (crank) of engine vehicles. The desire of engine designers to maximally reduce the weight of movable elements of a crank reduces the stiffness of the bearing (crank crosshead), which makes the mode of EHD lubrication working for the connecting rod bearings. The above features - comparable with the clearance of dynamically changing elastic displacements and a continuous gradient of deformations - prevent the direct application of methods of contact-hydrodynamic lubrication theory to the calculation of EY TU.

A mathematical model of EY TU differs from the "absolutely rigid" units model by the dependence of the instantaneous value of the lubricating film thickness h (j , z,t, p) on the elastic displacements of the friction surface of a bearing W (j , z,t, p) , which, in their turn, are determined by structural rigidity of the bearing and by the hydrodynamic pressure in the lubricating film p: h (f , $z,t, p) = h_{rig}$ (f , z, t) + W (f , z, t, p) . Where h_{rig} (j , z, t) - film thickness in the "absolutely rigid" bearing.

To determine it the expression (6) is used. Thus, the determination of pressures in the lubricating film and HMCh of EY TU is the related objective of the hydrodynamic lubrication theory and the theory of elasticity.

Modeling of EY TU, compared with "absolutely rigid" bearings is supplemented by an elastic subproblem the purpose of which is to determine the strain state of the friction surface of a crank crosshead under the influence of complex loads. The method of solving the elastic subproblem is chosen according to the accepted approximating model of an EY bearing. In today's solutions for EY TU the compliance and stiffness matrix of the bearing is usually constructed using the FE method.

The other side of modeling the elastic subsystem is adequate description of the entire complex of loads, causing the elastic deformation of the bearing housing and the conditions of fixing of FE model. One must consider not only the hydrodynamic pressure, but also the volume forces of rod inertia.

The known methods of solving the elastohydrodynamic lubrication problem can be classified as follows: direct methods or methods of successive approximations, in which the solutions of the hydrodynamic and elastic subtasks are performed separately, with the subsequent jointing of the results in the direct iterative process; and system, oriented for the joint solution of equations of fluid flow and elastic deformation.

In solving the problem of elastohydrodynamic lubrication of a bearing with the help of a direct iterative method, the hydrodynamic and elastic subproblems at each step of time discretization are solved sequentially in an iterative cycle. The main disadvantage of direct methods for the calculation of EHD is their slow convergence and the associated time- consumption. These difficulties are partially overcome by carefully selected prediction scheme and a number of techniques that accelerate the convergence of the iterative process in the form of restrictions on movement, load and move calculation.

Among the systemic methods the Newton-Raphson method is considered one of the most sustainable and effective solutions for elastohydrodynamic problems. In the literature it is known as the Newton-Kantorovich method or Newton (MN). The algorithm for system solutions of elastohydrodynamic problem consists of three nested iteration loops: the inner - loop of implementation by the Newton method of simultaneous solution of hydrodynamic and elastic subproblems; the average - the cycle of calculation of the cavitation zone and the boundary conditions; external - the cycle of calculation of the trajectory of the journal center. Algorithm for the numerical realization of MN is based on the finite-difference or finite element discretization of the linearized system of equations of EHD problem (Oh&Genka 1985; Bonneau 1995).

The application of the theory of elastohydrodynamic lubrication allows to predict lower mean-value as of the minimum lubricating film thickness as of the maximum hydrodynamic pressure. Thus, for the rod bearing of an engine, these changes may reach 35 ... 40%. The values of the maximum hydrodynamic pressure generated in the lubricating film of a EY bearing, are also smaller than for the "absolutely rigid" one. Reduction of the maximum hydrodynamic pressure is accompanied by an increase in the size of the bearing area. This fact, together with some increase in the clearance caused by the elastic deformation of the bearing, increases the flow of lubricating fluid through the ends of the bearing. Although the pressure gradient, on which the end consumption directly depends, is reduced. The difference in the instantaneous values of the mechanical flow between "absolutely rigid" and EY TU reaches 30%.

Calculation of the bearing, taking into account the elastohydrodynamic lubrication regime, not only improves the quality of design of friction units, but also clarifies the dynamic loading of mating parts such as the engine crank.

Thermoelastichydrodynamic (TEHD) regime of lubrication of journal bearings – is the mode of journal bearings, which are characterized by the influence on the magnitude of the clearance in tribounit thermoelastic deformations of a bearing and a journal, commensurate with the contribution of the displacement of the force nature.

Accounting for changes in the shape of thermoelastic friction surfaces of the journal and the bearing is possible in the case of inclusion in the resolution system of equations for the EY TU of energy equations and the relations of elasticity theory with the effects of temperature to determine the temperature fields and thermoelastic displacements caused by them. The sources of thermal fields can be either external to the tribounit, for example, a combustion chamber of an internal combustion engine for a connecting rod bearing, and internal - lubricating film, in which heat generating is essential for the calculation of TEHD lubrication regime. Thus, the most complete version to solve the problem of TEHD lubrication of tribounits requires the joint consideration of problems of heat distribution in the journal, the bearing and lubricating film. The task is complicated by the fact that journal and the bearing are some idealized concepts. In reality they are rather complex shape parts (crankshaft, connecting rod, crankcase, etc.). Therefore, the methods for solving problems of TEHD lubricants are usually based on the method of FE, allowing to solve problems for bodies of complex geometric shapes the easiest.

Transient thermal fields are typical for a lubricating film of heavy-loaded bearings, which requires the simultaneous solution of equations of fluid dynamics, energy, and elasticity at each step of the calculation of trajectories. In this case, to solve all the subproblems the method of FE and schemes, similar to the systemic methods of solving the elastohydrodynamic lubrication problems, are used.

However, practically important solution of TEHD lubrication of a tribounits is obtained for the steady thermal state, which is justified by the high inertia of the thermal fields in comparison with the rapidly changing power impacts. For calculations of tribounits with TEHD lubrication regime, this approach allows the use of a simple iterative scheme to implement solutions using the method of FE once on the preliminary stage of calculating the EY TU with regard to thermal deformations.

Nonautonomous journal bearings include bearings of multisupporting shafts of piston and rotary engines. Their distinctive feature consists in the interconnectedness of the processes occurring in various tribounits. A typical representative of nonautonomous heavy-loaded bearings is the root supports of the crankshaft of ICE. Main bearings of ICE are a part of a complex tribomechanical system, which also typically includes a crankshaft and a crankcase. Crankshaft journals are interconnected through a resilient connection - crankshaft ICE. Bushings of main bearings are installed in the holes of crankcase walls, and thus, main bearings are interconnected via a flexible design of the crankcase. Therefore, in the most general formulation to calculate the bearings it is necessary to solve a related EHD (or TEHD) problem for a system of "crankshaft - lubricating films - the crankcase". An additional feature of this system is the dependence of the loads acting on the indigenous support on the elastic properties of the crankshaft and crankcase.

In the first methods of calculation of non-autonomous main bearings of ICE a crankshaft model was used as the core spatial frame mounted on a linear-elastic mounts. The use of approximate methods for calculating the trajectories of main crankshaft journal bearings allowed to evaluate the influence of nonlinear properties of the lubricant film on the dynamics and the loading of the crankshaft. Simultaneously, we took into account the effect of necks (journals) deviation, as well as linear and bending stiffness of tribounits on the HMCh of main bearings (Zakharov, 1996a). However, to obtain the results significant estimates and approximate models of lubricating films of main bearings are used.

Currently, for the calculation of bearings the development of computer technology allows to use the exact solution of the Reynolds (4) and Elrod (1) equation for the bearing of finite length, to carry out the assessments of the impact of the crankcase and crankshaft construction supports, and of misalignments of bearings and shaft journals on HMCh. To solve such problems it is advisable to use an iterative algorithm that requires consistent calculation of loads acting on each of the bearings of the shaft and the calculation of the trajectories of its journals in the bearings.

Such calculations of the system of engine bearings allow to determine not only the optimal geometric parameters and position of sources for supplying lubricant to the main bearings, but also limiting in terms of supports performance tolerances concerning the position and shape of the friction surfaces of bearings and of the crankshaft journals.

The application of modern methods of calculation of main bearings allows to specify the values of loads acting on the crankcase and crankshaft, and their strength characteristics. But the task of elastohydrodynamic lubrication for a system "crankshaft - lubricating films - the crankcase" in the most general setting is still not solved.

5. Performance criteria

The development of criteria for evaluating the performance of heavy-loaded hydrodynamic bearings of piston and rotary engines is an integral part of modern methods of calculation and design. At the same time, the reliability of methods themselves is largely dependent on the applied criteria (Zakharov, 1996b). Reliability of criteria is usually assessed on the base of comparison of calculated results with those obtained experimentally or during operation. On the basis of experimental studies and modern methods of calculation the criteria of performance of hydrodynamic tribounits are developed: the smallest allowable film thickness *hper* , maximum allowable hydrodynamic pressure *pper* , minimum film thickness reduced to the diameter of the journal, the maximum unit load *f*max . According to calculations of crankshaft engine bearings of several dimensions the maximum permissible loading parameters listed in the table 5 are obtained.

Assessment of performance of bearings is also done according to the calculated value of the relative total lengths of areas per the cycle of loading a_h and a_p*per*, where the values of inf *h*min are less, and sup *p*max are bigger than acceptable values. Experience has shown that these parameters should not exceed 20% (Fig. 12).

Table 5. Maximum permissible loading parameters of sliding bearings of a crankshaft of automotive internal combustion engines

<table>
<tr><th rowspan="5">Engine group</th><th colspan="12">Loading parameters</th></tr>
<tr><th colspan="4">Maximum specific load fmax , MPa</th><th colspan="4">Reduced to the diameter of the journal minimum film thickness, mm/100 mm</th><th colspan="4">The largest hydrodynamic pressure in the lubricating film sup pmax , MPa</th></tr>
<tr><th colspan="12">Bearing Type</th></tr>
<tr><th colspan="2">Crank</th><th colspan="2">Main</th><th colspan="2">Crank</th><th colspan="2">Main</th><th colspan="2">Crank</th><th colspan="2">Main</th></tr>
<tr><th colspan="12">Antifriction material:
SB - stalebronzovye inserts coated with lead bronze,
SA - staleallyuminievye inserts coated aluminum alloy AMO 1-20</th></tr>
<tr><th></th><th>SB</th><th>SA</th><th>SB</th><th>SA</th><th>SB</th><th>SA</th><th>SB</th><th>SA</th><th>SB</th><th>SA</th><th>SB</th><th>SA</th></tr>
<tr><td>Highly accelerated</td><td>55</td><td>49</td><td>41</td><td>37</td><td colspan="2">2,0</td><td colspan="2">1,2</td><td>448</td><td>397</td><td>336</td><td>305</td></tr>
<tr><td>Medium accelerated</td><td>52</td><td>46</td><td>34</td><td>30,5</td><td colspan="2">2,3</td><td colspan="2">1,5</td><td>420</td><td>377</td><td>275</td><td>245</td></tr>
<tr><td>Low accelerated</td><td>45</td><td>39,5</td><td>31</td><td>27</td><td colspan="2">2,5</td><td colspan="2">1,9</td><td>367</td><td>326</td><td>255</td><td>225</td></tr>
</table>

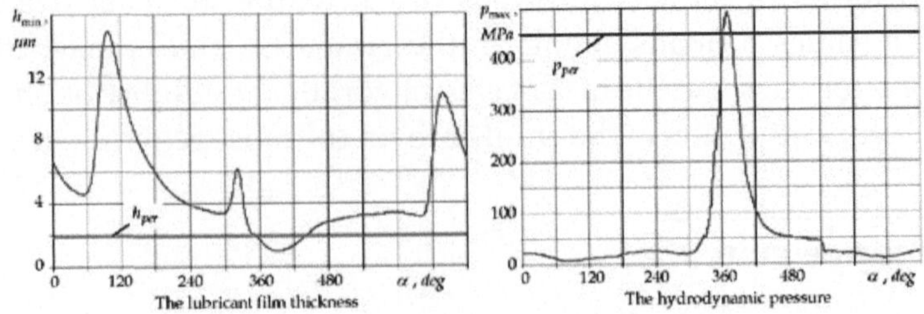

Fig. 12. The dependence of the hydromechanical characteristics on the rotation angle of crankshaft

6. Conclusion

Thus, the methodology of calculating the dynamics and HMCh of heavy-loaded tribounits lubricated by structurally heterogeneous and non-Newtonian fluids, consists of three interrelated tasks: defining the field of hydrodynamic pressures in a thin lubricating film that separates the friction surfaces of a journal and a bearing with an arbitrary law of their relative motion; calculation of the trajectory of the center of the journal; the calculation of the temperature of the lubricating film.

Mathematical models used in the calculation must reflect the nature of the live load, lubricant properties, geometry and elastic properties of a construction. The choice of models is built on the working conditions of tribounits in general and the properties of the lubricant. This will allow on the early stages of the design of tribounits to evaluate their bearing capacity, thermal stress and longevity.

References

1. Elrod, H. (1981). A Cavitation Algorithm. *Journal of lubrication Technology*, Vol.103, No.3, (July 1981), pp. 354-359, ISSN 0201-8160
2. Prokopiev, V., Rozhdestvensky, Y. et al. (2010). The Dynamics and Lubrication of Tribounits of Piston and Rotary Machines: Ponograph the Part 1, *South Ural State University*, ISBN 978-5-696-04036-3, Chelyabinsk
3. Prokopiev, V. & Karavayev, V. (2003). The Thermohydrodynamic Lubrication Problem of Heavy-loaded Journal Bearings by Non-Newtonian Fluids, *Herald of the SUSU. A series of "Engineering"*, Vol.3, No 1(17), pp. 55-66
4. Whilkinson, U. (1964). Non-Newtonian fluids, *Moscow: Mir*
5. Gecim, B. (1990). Non-Newtonian Effect of Multigrade Oils on Journal Bearing Perfomance, *Tribology Transaction*, Vol. 3, No 3, pp. 384-394.
6. Mukhortov, I., Zadorozhnaya, E., Levanov, I. et al. (2010). Improved Model of the Rheological Properties of the Boundary layer of lubricant, *Friction and lubrication of machines and mechanisms*, No 5, pp. 8-19
7. Oh, K. & Genka, P. (1985). The Elastohydrodynamic Solution of Journal Bearings Under Dynamic Loading, *Journal of Tribology*, No 3, pp. 70-76.

8. Bonneau, D. (1995). EHD Analysis, Including Structural Inertia Effect and Mass-Conserving Cavitation Model, *Journal of Tribology*, Vol. 117, (July 1995), pp. 540-547

9. Zakharov, S. (1996). Calculation of unsteady-loaded bearings, taking into account the deviation of the shaft and the regime of mixed lubrication, *Friction and Wear*, Vol.17, No 4, pp. 425-434, ISSN 0202-4977

10. Zakharov, S. (1996). Tribological Evaluation Criteria of Efficiency of Sliding Bearings of Crankshafts of Internal Combustion Engines, *Friction and Wear*, Vol.17, No 5, pp. 606 – 615, ISSN 0202-4977

4

The Bearing Friction of Compound Planetary Gears in the Early Stage Design for Cost Saving and Efficiency

1. Introduction

The efficiency of planetary gearboxes mainly depends on the tooth- and bearing friction losses. This work shows the new mathematical model and the results of the calculations to compare the tooth and the bearing friction losses in order to determine the efficiency of different types of planetary gears and evaluate the influence of the construction on the bearing friction losses and through it on the efficiency of planetary gears. In order to economy of energy transportation it is very important to find the best gearbox construction for a given application and to reach the highest efficiency.

In transmission system of gas turbine powered ships, power stations, wind turbines or other large machines in industry heavy-duty gearboxes are used with high gear ratio, efficiency of which is one of the most important issues. During the design of such equipment the main goal is to find the best constructions fitting to the requirements of the given application and to reduce the friction losses generated in the gearboxes. These heavy-duty tooth gearboxes are often planetary gears being able to meet the following requirements declared against the drive systems:

- High specific load carrying capacity
- High gear ratio
- Small size
- Small mass/power ratio in some application
- High efficiency.

There are some types of planetary gears which ensure high gear ratio, while their power flow is unbeneficial, because a large part of the rolling power (the idle power) circulate inside the planetary gearbox decreasing the efficiency. In the simple planetary gears there is no idle power circulation. Therefore heavy-duty planetary drives are set together of simple planetary gears in order to transmit megawatts or even more power, while they must be compact and efficient.

2. Planetary gearbox types

The two- and three-stage planetary gears consisting of simple planetary gears are able to meet the requirements mentioned above [Fig. 1(a)–1(d).].

Varying the inner gear ratio (the ratio of tooth number of the ring gear and of the sun gear) of each simple planetary gear stage KB the performance of the whole combined planetary gear can be changed and tailored to the requirements.

There are special types of combined planetary gears containing simple KB units (differential planetary gears), which can divide the applied power between the planetary stages thereby increasing the specific load carrying capacity and efficiency of the whole planetary drives [Fig. 1(b)-1.(d)]. Proper connections between the elements of the stages in these differential planetary gears do not result idle power circulation.

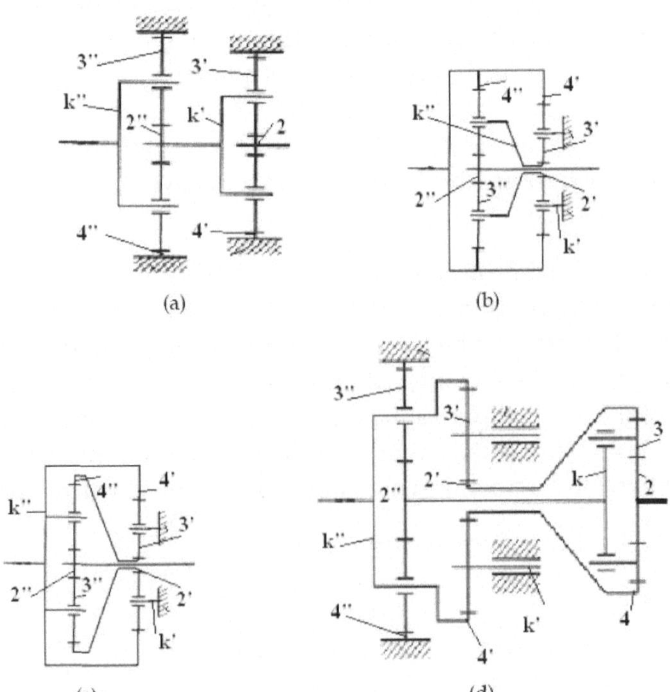

Fig. 1. (a) Gearbox KB+KB; (b). Planetary gear PKG; (c). Planetary gear PV; (d). Planetary gear GPV

The efficiency of planetary gears depends on the various sources of friction losses developed in the gearboxes. The main source of energy loss is the tooth friction of meshing gears, which mainly depends on the arrangements of the gears and the power flow inside the planetary gear drives. The tooth friction loss is influenced by the applied load, the entraining speed and the geometry of gears, the roughness of mating tooth surfaces and the viscosity of lubricant. Designers of planetary gear drives can modify the geometry of tooth profile in order to lower the tooth friction loss and to reach a higher efficiency [Csobán, 2009].

3. Friction loss model of roller bearings

It is important to find the parameters (such as inner gear ratio, optimal power flow) of a compound planetary gear drive which result its highest performance for a given application. The power flow and the power distribution between the stages of a compound gearbox is also a function of the power losses generated mainly by the friction of mashing teeth and the bearing friction.

This is why it is beneficial, when, during the design of a planetary gear beside the tooth friction loss also the friction of rolling bearings is taken into consideration even in the early stage of design. In this work a new method is suggested for calculate the rolling bearing friction losses without knowing the exact sizes and types of the bearings.

In this model first the torque and applied loads (loading forces and, if possible, bending moments) originated from the tooth forces between the mating teeth have to be determined. Thereafter the average diameter of the bearing d_m can be calculated as a function of the applied load and the prescribed bearing lifetime. Knowing the average diameter d_m, the friction loss of bearings can be counted using the methods suggested by the bearing manufacturers based on the Palmgren model [SKF 1989].

For determining the functions between the bearing average diameter and between the basic dynamic, static load, inner and outer diameter [Fig. 2-6.], the data were collected from SKF catalog [SKF 2005].

The functions loss between the bearing parameters (inner diameter d_b, dynamic basic load C) and the average diameters d_m being necessary for calculation of the friction moment and the load can be searched in the following form:

$$Y = \tilde{c} \cdot \tilde{d}_m^{\tilde{d}} \tag{1}$$

The equations of the diagrams [Fig. 2-6.] give the values of c and d for the inner diameter of the bearings d_b and for the basic dynamic loads C of the bearings.

Knowing the torque M_{24} and the strength of the materials of the shafts (t_m, s_m) the mean diameter of the bearing for central gears (sun gear, ring gear) necessary to carry the load can be calculated using the following formula:

$$d_{m_{2;4}}(d) = \sqrt[\tilde{d}]{\frac{\sqrt[3]{\frac{16 \cdot M_{2;4}}{\tau_m \cdot \pi}}}{\tilde{c}}} \tag{2}$$

Calculating the tangential components of the tooth forces the applied radial loads of the planet gear shafts F_r can be determined (which are the resultant forces of the two tangential components F_{t2} and F_{t4}). The shafts of the planet gears are sheared and bended by the heavy radial forces, this is why, in this analysis, at the calculation of shaft diameter, once the shear stresses, then the bending stresses are considered. Calculating the maximal bending moment $M_h max$ of the planet gear shafts, and the allowable equivalent stress s_m of planet gear pins, the bearing inner diameter d_b necessary to carry the applied load of the planet gear shaft and the average bearing diameter d_{m3} can be calculated:

$$d_{m_3}(d) = \sqrt[\tilde{d}]{\frac{\sqrt[3]{\frac{32 \cdot M_{h_{max}}}{\sigma_m \cdot \pi}}}{\tilde{c}}} \tag{3}$$

The functions between the bearing geometry and load carrying capacity for deep groove ball bearings [Fig. 2(a)-2(d)]. The points are the average data of the bearings taken from SKF Catalog [SKF 2005] and the continuous lines are the developed functions between the parameters.

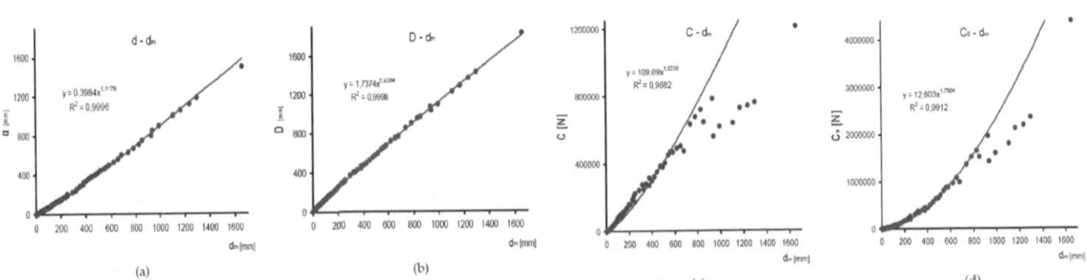

Fig. 2. (a) The average inner diameter of the deep groove ball bearing as a function of its average diameter. (b). The average outer diameter of deep groove ball bearing as a function of the average diameter. (c). The average basic dynamic load of deep groove ball bearing as a function of the average diameter. (d). The average static load of deep groove ball bearing as a function of the average diameter

The functions between the bearing geometry and load carrying capacity for cylindrical roller bearings [Fig. 3(a)-3(d)].

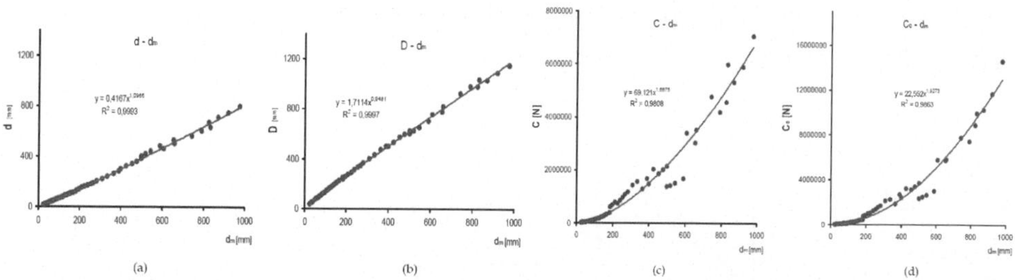

Fig. 3. (a) The average inner diameter of the cylindrical roller bearing as a function of its average diameter. (b). The average outer diameter of cylindrical roller bearing as a function of the average diameter. (c). The average basic dynamic load of the cylindrical roller bearing as a function of its average diameter. (d). The average static load of different types of cylindrical roller bearing as a function of the average diameter

The functions between the bearing geometry and load carrying capacity for full complement cylindrical roller bearings [Fig. 4(a)-4(d)].

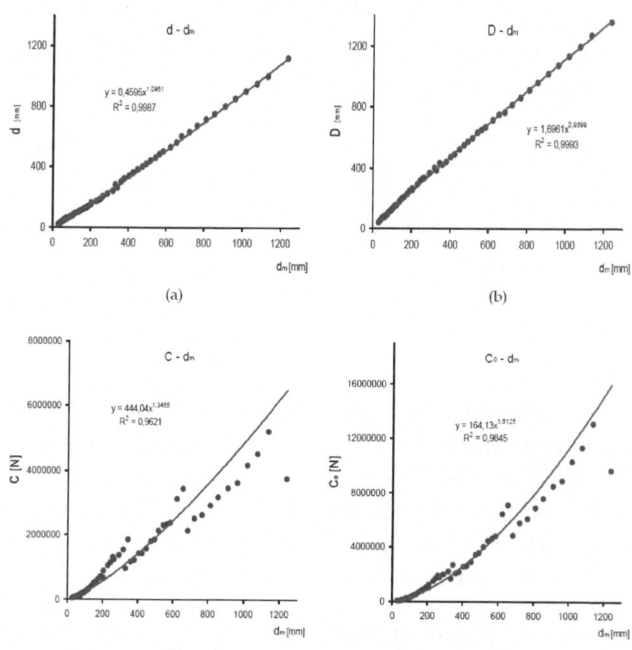

Fig. 4. (a) The average inner diameter of the full complement cylindrical roller bearing as a function of its average diameter. (b). The average outer diameter of the full complement cylindrical roller bearing as a function of the average diameter. (c). The average basic dynamic load of the full complement cylindrical roller bearing as a function of its average diameter. (d). The average static load of different types of full complement cylindrical roller bearing as a function of the average diameter

The functions between the bearing geometry and load carrying capacity for spherical roller bearings [Fig. 5(a)-5(d)].

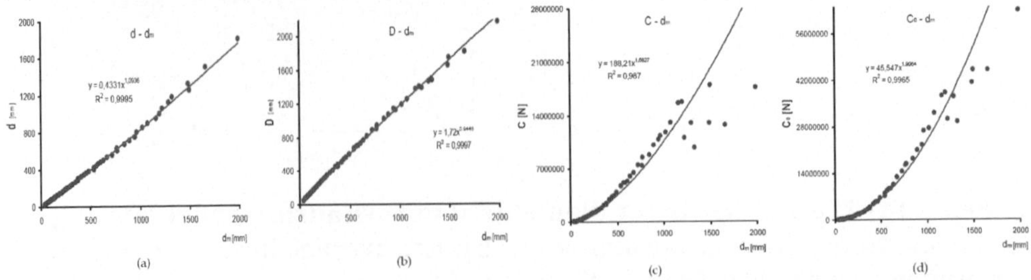

Fig. 5. (a) The average inner diameter of the spherical roller bearing as a function of its average diameter. (b). The average outer diameter of spherical roller bearing as a function of the average diameter. (c). The average basic dynamic load of spherical roller bearing as a function of its average diameter. (d). The average static load of spherical roller bearing as a function of its average diameter

The functions between the bearing geometry and load carrying capacity for of CARB toroidal roller bearings [Fig. 6(a)-6(d)].

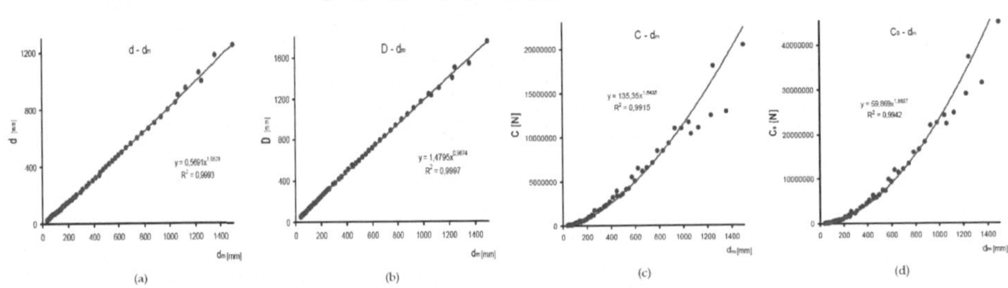

Fig. 6. (a) The average inner diameter of CARB toroidal roller bearing as a function of its average diameter. (b). The average outer diameter of CARB toroidal roller bearing as a function of the average diameter. (c). The average basic dynamic load of CARB toroidal roller bearing as a function of its average diameter. (d). The static load of CARB toroidal roller bearing as a function of the average diameter

The V shear load of the planet gear shaft is equal with the applied load F_r divided by the number of sheared areas A of the shaft. Knowing the V shear load and the allowable equivalent stress t_m of planet gear pins, the bearing inner diameter d_b necessary to carry the applied load of the planet gear shaft and the average bearing diameter d_{m3} can be calculated:

$$d_{m_3}(d) = \sqrt[\tilde{d}]{\frac{\sqrt{\dfrac{16 \cdot V}{3 \cdot \tau_m \cdot \pi}}}{\tilde{c}}} \qquad (4)$$

The average diameters of bearings necessary to reach the prescribed lifetime L_{1h} was determined using the SKF modified lifetime equation [SKF 2005] (C is the basic dynamic load, F_r is the radial bearing load and a_1 is the bearing life correction factor) as follows:

$$d_m(L_h) = \sqrt[a]{\frac{\sqrt[p]{\frac{L_h \cdot 60 \cdot n}{10^6 \cdot a}} \cdot F_r}{\tilde{c}}} \tag{5}$$

From the two calculated average diameters of bearings ($d_m(d)$ and $d_m(L_h)$) the larger ones have to be chosen. This biggest average diameter can be called resultant average (ball or roller) bearing diameter ($d_{m\,re}$

$$d_{m_{res}} = \left[d_m(d) + \frac{1}{2} \cdot \left(\left| \left(d_m(L_h) - d_m(d) \right) \right| + \left(d_m(L_h) - d_m(d) \right) \right) \right] \tag{6}$$

3.1 Calculating the friction losses and efficiency of roller bearings

The sun gears and the ring gears are well balanced by radial components of tooth forces; the friction losses of their bearings are not depending on the applied load. The energy losses of these bearings are determined by the entraining speed of the bearings, the viscosity of lubricant and the bearing sizes.

The calculation of the component of friction torque M_0 being independent of the bearing load can be performed using the following equations [SKF 1989].

When

$$\begin{gathered} \nu \cdot n \geq 2000 \\ M_0 = 10^{-7} \cdot f_0 \cdot (\nu \cdot n)^{2/3} \cdot d_m^3 \end{gathered} \tag{7}$$

And when

$$\begin{gathered} \nu \cdot n < 2000 \\ M_0 = 160 \cdot 10^{-7} \cdot f_0 \cdot d_m^3 \end{gathered} \tag{8}$$

At bearings of planet gears the component of friction torques M_1 depending on the bearing loads was calculated using the following simple equation [SKF 1989]:

$$M_1 = f_1 \cdot P_1^a \cdot d_m^b \tag{9}$$

Using the average bearing diameters the friction torques of the bearings can be determined:

$$M_v\left(d_{m_{res}}\right) = M_0\left(d_{m_{res}}\right) + M_1\left(d_{m_{res}}\right) + \ldots \tag{10}$$

Knowing the friction torques of the sun gear its bearing efficiency can be calculated using the following equation:

$$\eta 2_{Bearing} = \frac{\left[M_2 - M_v(d_{m_{res}})\right] \cdot \omega_2}{M_2 \cdot \omega_2} = 1 - \frac{M_v(d_{m_{res}})}{M_2} \tag{11}$$

The bearing efficiency of planet gears can be determined with the following equation:

$$\eta 3_{Bearing} = \frac{\left[M_3 - M_v(d_{m_{res}})\right] \cdot \omega_{3g}}{M_3 \cdot \omega_{3g}} = 1 - \frac{M_v(d_{m_{res}})}{M_3} \tag{12}$$

The power loss generated only by the bearings in the gearbox can be calculated as (the rolling efficiency of a simple stage and the gearbox efficiency is a function of only the bearing efficiencies):

$$\begin{aligned} &\eta_g = \eta 2_{Bearing} \cdot \eta 3_{Bearing} \rightarrow \eta Gearbox_{Bearing} \\ &\upsilon_{Bearing} = P_{in} \cdot \left(1 - \eta Gearbox_{Bearing}\right) \end{aligned} \tag{13}$$

The power loss generated by the tooth friction can be calculated with the following equations (the rolling efficiency of a simple stage and the gearbox efficiency is a function of only the tooth efficiencies):

$$\begin{aligned} &\eta_g = \eta_{z_{23}} \cdot \eta_{z_{34}} \rightarrow \eta Gearbox_{Tooth} \\ &\upsilon_{Tooth} = P_{in} \cdot \left(1 - \eta Gearbox_{Tooth}\right) \end{aligned} \tag{14}$$

The rolling efficiency of a simple planetary gear stage can be calculated as:

$$\eta_g = \eta_{z_{23}} \cdot \eta_{z_{34}} \cdot \eta 2_{Bearing} \cdot \eta 3_{Bearing} \rightarrow \eta Gearbox \tag{15}$$

The total power loss generated in the planetary gear drive as a function of the gearbox efficiency:

$$\Sigma\upsilon = P_{in} \cdot \left(1 - \eta_{Gearbox}\right) \tag{16}$$

The power loss ratios show the dominant power loss component. The tooth power loss ratio is the tooth power loss component divided by the total power loss:

$$\frac{\upsilon_{Tooth}}{\Sigma\upsilon} \tag{17}$$

The bearing loss ratio is the power loss generated by the bearing friction divided by the total power loss:

$$\frac{\upsilon_{Bearing}}{\Sigma\upsilon} \tag{18}$$

The bearing selecting and efficiency calculation algorithm can be seen in figure 10.

4. Comparing the properties of planetary gears

The performance of a planetary gear drive depends on its kinematics, its inner gear ratios and the connections between the planetary stages. Only detailed calculations can reveal the behavior of planetary gears and show their best solutions for given applications. To calculate the gear ratios and the gearbox efficiencies of various planetary gears (Fig. 1(a).- 1(d).) the following equations were developed:

The gear ratio of planetary gear KB+KB (Fig. 1(a).) (sun gears drive and carriers are driven):

$$i_{KB+KB} = (1 - i_{b''}) \cdot (1 - i_{b'}) \tag{19}$$

The efficiency of planetary gear KB+KB:

$$\eta_{KB+KB} = \frac{(1 - i_{b''} \cdot \eta_{g''}) \cdot (1 - i_{b'} \cdot \eta_{g'})}{(1 - i_{b''}) \cdot (1 - i_{b'})} \tag{20}$$

The gear ratio of planetary gear PKG (Fig. 1(b).):

$$i_{PKG} = (i_{b''} + i_{b'} - i_{b''} \cdot i_{b'}) \tag{21}$$

The efficiency of planetary gear PKG:

$$\eta_{PKG} = \frac{i_{b''} \cdot \eta_{g''} + i_{b'} \cdot \eta_{g'} - i_{b''} \cdot i_{b'} \cdot \eta_{g''} \cdot \eta_{g'}}{i_{b''} + i_{b'} - i_{b''} \cdot i_{b'}} \tag{22}$$

Power distribution between the stages (the power of the driven element of the first stage $P_{4''}$ divided by the output power P_{out}):

$$\frac{P_{4''}}{P_{out}} = \frac{1}{\left[1 + \dfrac{i_{b'} \cdot \eta_{g'}}{i_{b''} \cdot \eta_{g''}} - i_{b'} \cdot \eta_{g'}\right]} \tag{23}$$

The gear ratio of planetary gear PV (Fig. 1(c).):

$$i_{PV} = 1 + i_{b''} \cdot i_{b'} - i_{b''} \tag{24}$$

The efficiency of planetary gear PV:

$$\eta_{PV} = \frac{1 - i_{b''} \cdot \eta_{g''} + i_{b''} \cdot i_{b'} \cdot \eta_{g''} \cdot \eta_{g'}}{1 - i_{b''} + i_{b''} \cdot i_{b'}} \tag{25}$$

Power distribution between the stages (the power of the driven element of the first stage $P_{k''}$ divided by the output power P_{out}):

$$\frac{P_{k''}}{P_{out}} = \frac{1}{\left[1 + \dfrac{i_{b''} \cdot i_{b'} \cdot \eta_{g''} \cdot \eta_{g'}}{(1 - i_{b''} \cdot \eta_{g''})}\right]} \tag{26}$$

Table 1. Parameters for the bearing selection

Bearing Types/function	$c^{\sim}$	$d^{\sim}$
Deep groove ball bearing		
$d - d_m$	0,3984	1,1179
$D - d_m$	1,7374	0,9364
$C - d_m$	109,09	1,3236
$C_0 - d_m$	12,603	1,7564
Cylindrical roller bearings	$c^{\sim}$	$d^{\sim}$
$d - d_m$	0,4167	1,0966
$D - d_m$	1,7114	0,9481
$C - d_m$	69,121	1,6675
$C_0 - d_m$	22,552	1,9273
Full complement cylindrical roller bearings	$c^{\sim}$	$d^{\sim}$
$d - d_m$	0,4595	1,0951
$D - d_m$	1,6961	0,9399
$C - d_m$	444,04	1,3465
$C_0 - d_m$	164,13	1,6125
Spherical roller bearings	$c^{\sim}$	$d^{\sim}$
$d - d_m$	0,4331	1,0936
$D - d_m$	1,72	0,9448
$C - d_m$	188,21	1,5827
$C_0 - d_m$	45,547	1,9064
CARB® toroidal roller bearings	$c^{\sim}$	$d^{\sim}$
$d - d_m$	0,5691	1,0529
$D - d_m$	1,4795	0,9674
$C - d_m$	135,35	1,6435
$C_0 - d_m$	59,869	1,8637

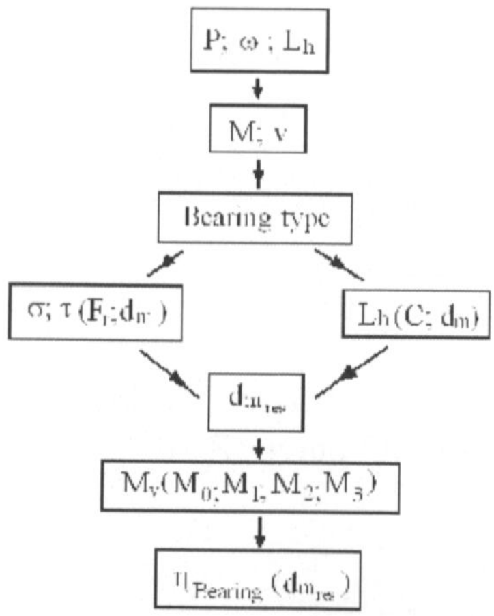

Fig. 7. The bearing selecting and efficiency calculation algorithm

The gear ratio of planetary gear GPV (Fig. 1(d).):

$$i_{GPV} = 1 - i_{b''} - i_b + i_{b''} \cdot i_b + i_b \cdot i_{b'} \quad (27)$$

The efficiency of planetary gear GPV:

$$\eta_{GPV} = \frac{\left[(1 - i_b \cdot \eta_g) \cdot (1 - i_{b''} \cdot \eta_{g''}) + i_b \cdot i_{b'} \cdot \eta_g \cdot \eta_{g'}\right]}{1 - i_b - i_{b''} + i_b \cdot i_{b''} + i_b \cdot i_{b'}} \quad (28)$$

Power distribution between the stages (the power of the driver element of the first stage $P_{2''}$ divided by the power of the driver element of the second stage $P_{2'}$):

$$\frac{P_{2''}}{P_{2'}} = \frac{\left(\frac{1}{i_b \cdot \eta_g} - 1\right) \cdot (1 - i_{b''})}{i_{b'}} \quad (29)$$

5. Results of calculations

Calculations were to compare the tooth and the bearing friction losses in order to determine the efficiency of different types of planetary gears and evaluate the influence of the construction on the bearing friction losses and the efficiency of planetary gears. Comparing the calculated power losses caused by only the friction of tooth wheels or only by the bearing friction with the total power losses of the gearboxes, it is obvious that the bearing friction loss is a significant part of the whole friction losses. Behavior of various types of two- and three-stage and differential planetary gears were investigated and compared using the derived equations, following a row of systematical procedures. If the input power, the input speed and lubricant viscosity are known, the calculation can be performed. The first step is to choose various inner gear ratios for every stage and to combine them creating as many planetary gear ratios as possible. Using the equations presented above (1-29) the efficiency and the bearing power loss of every gear can be calculated. Some results are presented in diagrams (Fig. 8-17). Comparing the calculated values of efficiency and power loss ratios the optimal gearbox construction can be selected. The beneficial inner gear ratio of each stage and the power ratios were determined for all the four types of planetary gears. When the optimal inner gear ratios are known, the tooth profile ensuring the lowest tooth friction can be calculated for every planetary gear stage by varying the addendum modification of tooth wheels [Csobán 2009]. The calculations were performed for all planetary gears presented above for transmitting a power of 2000 kW at a driving speed of 1500 rpm. In the calculations the parameters of Table 2 and 3 were used.

Table 2. Other important parameters for the analyses

σ_F [MPa]	η_M [mPas]	R_{a23} [mm]	R_{a34} [mm]	P_{in} [kW]	n_{in} [1/min]	β [°]	x_2 [-]	N [-]	b/d_w [-]
500	63	0,63	1,25	2000	1500	0	0	3	0,8

Table 3. Parameters for calculate the bearing friction losses

a	b	f_0	f_1	$c(d_b)$	$d(d_b)$	$c(L_{1h})$	$d(L_{1h})$
1	1	7,5	0,00055	0,4595	1,0951	444,04	1,3465

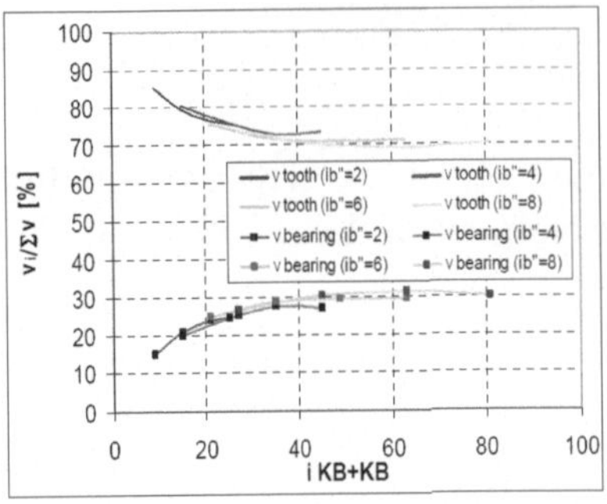

Fig. 8. The power loss ratio of planetary gear KB+KB as a function of gear ratio. Prescribed gearbox lifetime=5000[h]

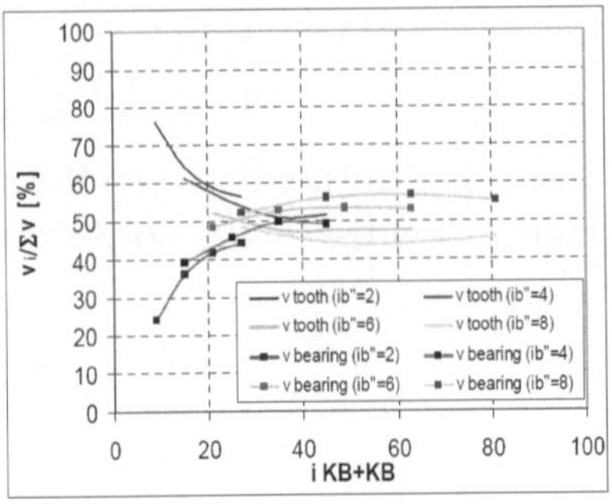

Fig. 9. The power loss ratio of planetary gear KB+KB as a function of gear ratio. Prescribed gearbox lifetime=50000[h]

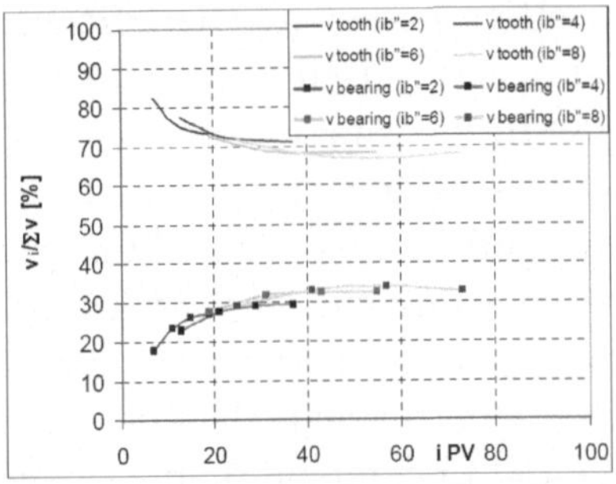

Fig. 10. The power loss ratio of planetary gear PV as a function of gear ratio. Prescribed gearbox lifetime=5000[h]

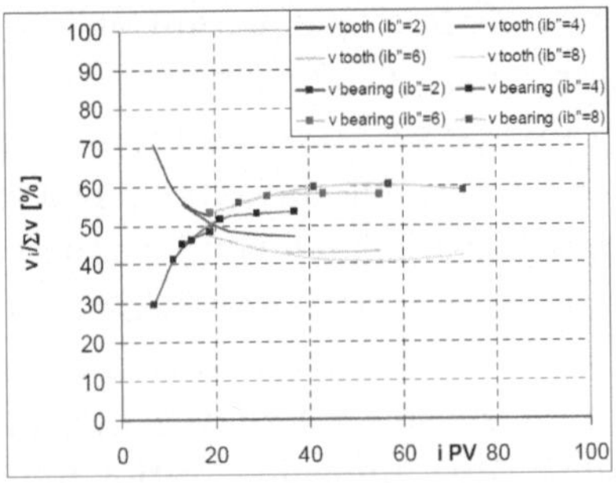

Fig. 11. The power loss ratio of planetary gear PV as a function of gear ratio. Prescribed gearbox lifetime=50000[h]

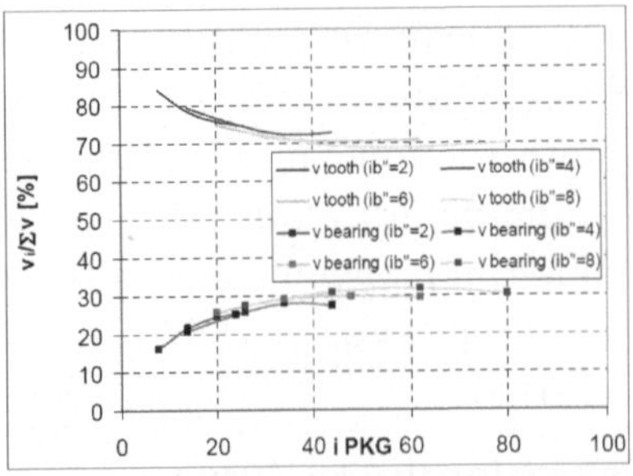

Fig. 12. The power loss ratio of planetary gear PKG as a function of gear ratio. Prescribed gearbox lifetime=5000[h]

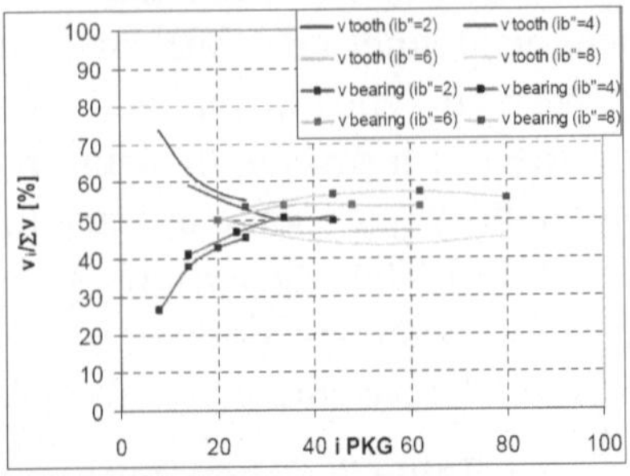

Fig. 13. The power loss ratio of planetary gear PKG as a function of gear ratio. Prescribed gearbox lifetime=50000[h]

The power loss ratios of the three-stage GPV planetary gearbox were investigated at the same gear ratio range as the two-stage differential gears have (Figure 14-15).

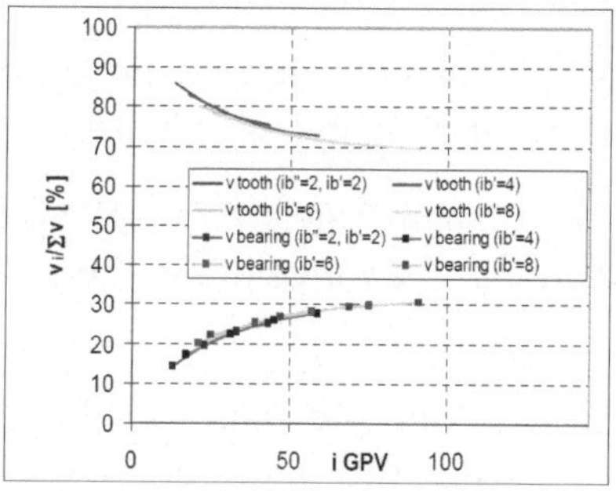

Fig. 14. The power loss ratio of planetary gear GPV as a function of gear ratio. Prescribed gearbox lifetime=5000[h], ib"=2

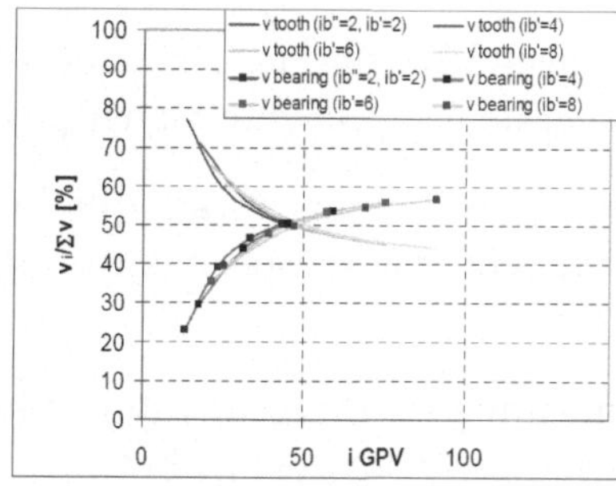

Fig. 15. The power loss ratio of planetary gear GPV as a function of gear ratio. Prescribed gearbox lifetime=50000[h], ib"=2

The GPV gearbox can operate with higher gear ratios than the two-stage gears. The gear ratio range was changed with increasing the inner gear ratio of the first stage while the inner gear ratios of the second and third stage were changed and combined (Figure 16-17).

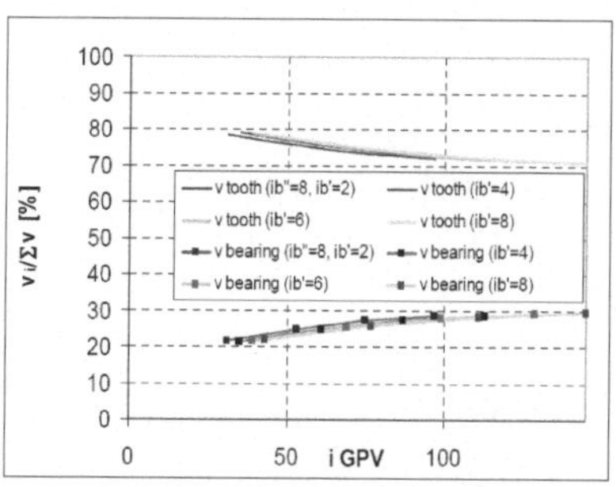

Fig. 16. The power loss ratio of planetary gear GPV as a function of gear ratio. Prescribed gearbox lifetime=5000[h], ib"=8

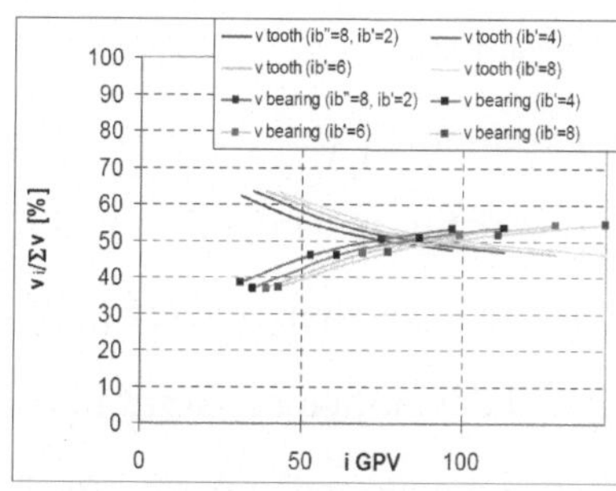

Fig. 17. The power loss ratio of planetary gear GPV as a function of gear ratio. Prescribed gearbox lifetime=50000[h], ib"=8

6. Conclusions

Comparing the results of the analysis the following can be stated:

- It is obvious that the bearing friction loss is a significant part of the friction losses.
- Higher gear ratios can be realized with the planetary gear PKG than with planetary gear PV.

- ❖ It can be stated that thanks to relatively low predicted lifetime, smaller bearings have to be build in the gearbox. Having smaller bearings, the tooth power loss ratio will be higher (at the same types of bearings).
- ❖ If longer bearing life is needed larger bearings have to build in the gearbox. Larger bearings lead to higher power losses (at the same types of bearings).
- ❖ Varying the inner gear ratios of the investigated planetary gear drives the values of the power loss rates change significantly only in the range of the lower gear ratios.
- ❖ Depending on the gear ratios and prescribed lifetime the values of the tooth power loss ratio change between 80% to 40% while the values of the bearing power loss ratio change between 20% to 60%.

Using the bearing power loss model presented above all types of bearings can be considered for a given planetary gearbox optimization and application and all the important parameters like efficiency, size and even cost can be compared easily.

7. Nomenclature

2 sun gear,

3 planet gear,

4 ring gear,

a, b exponents [-],

a_1 factor for bearing life correction (a_1=0,21...1) [-],

C is the basic dynamic load [N],

$c(d_b;L_{1h})$ developed constant for bearing calculation

$d(d_b;L_{1h})$ developed constant for bearing calculation, c˜ ; d˜ constant and exponent (table 1.),

d_m average diameter of bearing [mm],

$d_{m\ res}$ resultant average bearing diameter [mm],

f_0 coefficient (which is a function of bearing type and size) [-], f_1 coefficient (which is a function of bearing type and load) [-], F_r is the radial bearing load [N],

h_g is the rolling efficiency of a simple planetary gear stage *KB*,

h_M viscosity at operating temperature [Pas],

I is the gear ratio,

i_b is the ratio of the number of teeth of sun gear and ring gear at the third stage,

$i_{b'}$ is the ratio of the number of teeth of sun gear and ring gear at the second stage,

$i_{b''}$ is the ratio of the number of teeth of sun gear and ring gear at the first stage,

k planetary carrier,

L_{1h} prescribed lifetime [h],

M_0 load independent friction torque [Nmm], M_1 load dependent friction torque [Nmm], $M_{2;4}$ sun or ring gear torque [Nm],

M_3 planet gear torque [Nm], n bearing velocity [rpm], nin driving speed [rpm],

n kinematical viscosity at operating temperature [mm2/s],

P_1 load of the bearing [N],

P_{in} driving power [W],

R_a average surface roughness (CLA),

s_F bending strength of teeth [MPa],

s_m, t_m allowable equivalent and shear stress components [MPa],

Sv total power loss [W],

v entraining speed [m/s],

V shear load of planet gear pin [N], $v_{Bearing}$ bearing friction loss component [W], v_{tooth} tooth power loss component [W],

w_{3g} angle velocity of planet gear [rad/s], Y calculated parameter (d_f, C).

References

1. Bartz, W.J.: Getriebeschmierung. Expert Verlag. Ehningen bei Böblingen, 1989
2. Csobán Attila., Kozma Mihály: Comparing the performance of heavy-duty planetary gears. Proceedings of fifth conference on mechanical engineering Gépészet 2006 ISBN 9635934653
3. Csobán Attila, Kozma Mihály: Influence of the Power Flow and the Inner Gear Ratios on the Efficiency of Heavy-Duty Differential Planetary Gears, 16th International Colloquium Tribology, Technische Akademie Esslingen, 2008
4. Csobán Attila, Kozma Mihály: A model for calculating the Oil Churning, the Bearing and the Tooth Friction Generated in Planetary Gears, World Tribology Congress 2009, Kyoto, Japan, September 6 – 11, 2009

5. Duda, M.: Der geometrische Verlustbeiwert und die Verlustunsymmetrie bei geradverzahnten Stirnradgetrieben. Forschung im Ingenieurwesen 37 (1971) H. 1, VDI-Verlag
6. Erney György: Fogaskerekek, Műszaki Könyvkiadó, Budapest, 1983
7. Klein, H.: Bolygókerék hajtóművek. Műszaki Könyvkiadó, Budapest, 1968.
8. Kozma, M.: Effect of lubricants on the performance of gears. Proceedings of Interfaces'05. 15-17 September, 2005. Sopron 134-141
9. Müller, H.W.: Die Umlaufgetriebe. Springer-Verlag, Berlin. 1971
10. Niemann G.- Winter, H.: Maschinenelemente. Band II. Springer-Verlag, Berlin, 1989 Shell Co: The Lubrication of Industrial Gears. John Wright & Sons Ltd. London 1964. SKF Főkatalógus, Reg. 47.5000 1989-12, Hungary
11. SKF General catalogue, 6000 EN, November 2005

5

Three-Dimensional Stress-Strain State of a Pipe with Corrosion Damage Under Complex Loading

1. Introduction

In studies of the stress-stain state of models of pipeline sections without corrosion defects of a pipe, in the two-dimensional statement (cross section), pipes are usually modeled by a ring, whereas in the three-dimensional statement – by a thick-wall cylindrical shell [Ponomarev et al., 1958; Seleznev et al. 2005]. Usually, internal pressure or temperature is considered as a load applied to the pipe. The solution of the problems stated in this manner yields not bad results when a relatively not complicated procedure of calculation, both analytical and numerical, is adopted.

The presence of corrosion damage at the inner surface of the pipe (Figures 1, 2), being a particular three-dimensional concentrator of stresses, requires a special approach to defening the stress-strain state. In addition, account should be taken of a simultaneous compound action of such loading factors as internal pressure and friction of the mineral oil flow over the inner surface of the pipe, as well as of soil.

The analysis of the known references to articles shows that the problem of investigating the spatial stress-strain states of the pipe with regard to its corrosion damage with the account of various types of loading has not been stated up to now. In essence, the problems of determining individual stress-strain states under the action of internal pressure ($\sigma_{ij}^{(p)}$, $\varepsilon_{ij}^{(p)}$) or temperature ($\sigma_{ij}^{(T)}$, $\varepsilon_{ij}^{(T)}$) [Ainbinder et al., 1982; Borodavkin et al., 1984; Grachev et al., ij 1982; Dertsakyan et al., 1977; Mirkin et al., 1991; O'Grady et al., 1992] are under consideration. The problem of determining stress-strain state caused by wall friction due to viscous fluid motion ($\sigma_{ij}^{(\tau)}$, $\varepsilon_{ij}^{(\tau)}$), as well as the most general problems of determining

$\sigma_{ij}^{(p+\tau)}$, $\varepsilon_{ij}^{(p+\tau)}$; $\sigma_{ij}^{(p+T)}$, $\varepsilon_{ij}^{(p+T)}$; $\sigma_{ij}^{(p+\tau+T)}$ $\varepsilon_{ij}^{(p+\tau+T)}$ has not been stated. In addition, the problems of stress-strain state determination are usually being solved for shell models of a pipe. Although, for example, in [Seleznev et al. 2005] $\sigma_{ij}^{(p)}$ is described for the three- dimensional model of the section of the pipe with corrosion damage.

Therefore the statement and solutions of the problem of determining three-dimensional stress-strain state of the models of pipes with corrosion defects under the action of internal pressure, friction caused by oil flow and temperature discussed in the present chapter are important for pipeline systems and such related disciplines as solid mechanics, fluid mechanics and tribology.

2. Statement of the problem

The present Chapter deals with some of the results of investigation of the three-dimensional stress-strain state of the model of a pipe with corrosion damage (Figure 2).

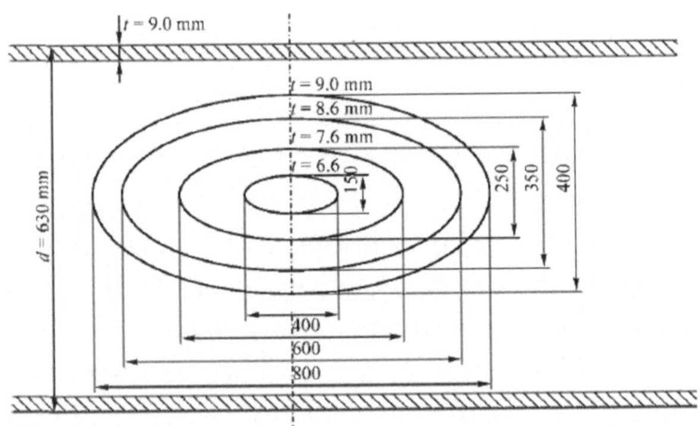

Fig. 1. Simplified scheme of elliptical corrosion damage at the inner surface of the pipe displaying reduced wall thickness inside the damage

In calculations, the following basic loads applied to the pipe were taken into consideration:

- internal pressure

$$\sigma_r\big|_{r=r_1} = p, \tag{1}$$

where r_1 is the inner radius of the pipe;

- mineral oil friction over the inner surface of the pipe, thus exciting wall tangential stresses

$$\tau_{rz}\big|_{r=r_1} = \tau_0, \tag{2}$$

t_0 - tangential forces modeling the viscous fluid friction force over the inner surface of the pipe;

❖ change of the thermodynamic state (temperature) of the pipe

$$\left|T_{r_1} - T_{r_2}\right| = \Delta T, \tag{3}$$

where r_2 is the outer radius of the pipe.

It should be emphasized that in the presence of corrosion damage

$$r_1 = r_1(\varphi, z), \tag{4}$$

where j and *z* are the components of the cylindrical coordinate system (*r*, j, *z*).

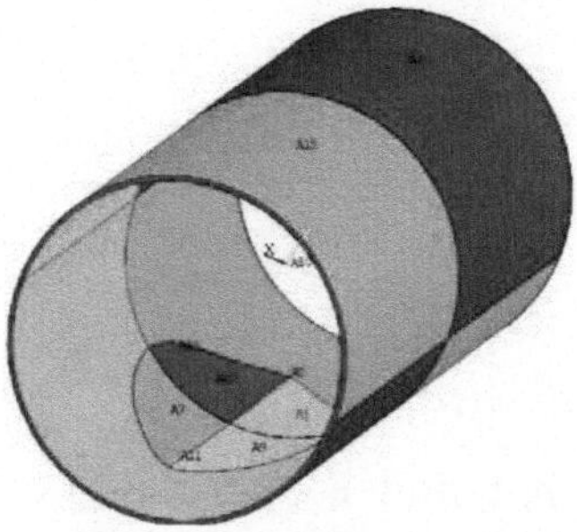

Fig. 2. Solid model of the neighborhood of the pipe with corrosion damage

The distinctive feature of computer finite-element modeling of pipes of trunk pipelines with corrosion damage lies in the opportunity to assign different boundary conditions for the outer surface of the pipe. So, for loads (1)–(3), consideration is made of different restrictions on displacements of the outer surface of the pipe:

a. no pipeline fixing

$$\sigma_r\big|_{r=r_2} = 0, \tag{5}$$

b. no displacements of the outer surface of the pipe in the *x* and *y* directions and in the *z* direction on the right end

$$u_x\big|_{r=r_2} = u_y\big|_{r=r_2} = 0,\ u_z\big|_{z=L} = 0, \tag{6}$$

c. no displacements of the outer surface of the pipe in all directions

$$u_x\big|_{r=r_2} = u_y\big|_{r=r_2} = u_z\big|_{r=r_2} = 0, \tag{7}$$

d. contact between the pipe and soil where no displacements of the soil outer surface take place

$$\begin{aligned} &\sigma_r^{(1)}\Big|_{r=r_2} = -\sigma_r^{(2)}\Big|_{r=r_2}, \\ &\sigma_\tau^{(1)}\Big|_{r=r_2} = -\sigma_\tau^{(2)}\Big|_{r=r_2} = f\sigma_n^{(1)}\Big|_{r=r_2}, \\ &u_x\big|_{r=r_3} = u_y\big|_{r=r_3} = 0 \end{aligned} \tag{8}$$

where the superscript 1 means the pipe, whereas 2 – soil, s_t is the tangential component of the stress vector, f is the friction coefficient, and r_3 is the soil outer radius.

With the use of the pipeline fixing (type a), tests of the pipe dug out of soil (in the air) are modeled. The stress-strain state of a pipe lying in hard soil without friction in the axial direction is modeled by means of pipe fixing (type b) while that of a pipe lying in hard soil and rigidly connected with it – by means of pipe fixing (type c). Subject to boundary conditions (8) (type d), a pipe lying in soil having particular mechanical characteristics is modeled.

Thus, the problem has been stated to make a comparative analysis of the stress-strain states of the pipe with corrosion damage for different combinations of boundary conditions (1)– (3), (6)–(8):

$$\begin{gathered}\sigma_{ij}^{(p)},\varepsilon_{ij}^{(p)};\sigma_{ij}^{(\tau)},\varepsilon_{ij}^{(\tau)};\sigma_{ij}^{(T)},\varepsilon_{ij}^{(T)};\\ \sigma_{ij}^{(p+\tau)},\varepsilon_{ij}^{(p+\tau)};\sigma_{ij}^{(p+T)},\varepsilon_{ij}^{(p+T)};\sigma_{ij}^{(p+\tau+T)},\varepsilon_{ij}^{(p+\tau+T)}.\end{gathered} \tag{9}$$

where the superscripts p, t, and T correspond to the stress states caused by internal pressure, friction force over the inner surface of the pipe, and temperature.

In the case of the elastic relationship between stresses and strains, the stress states in (9) are connected by the following relations

$$\begin{aligned}\sigma_{ij}^{(p+\tau)}&=\sigma_{ij}^{(p)}+\sigma_{ij}^{(\tau)},\\ \sigma_{ij}^{(p+T)}&=\sigma_{ij}^{(p)}+\sigma_{ij}^{(T)},\\ \sigma_{ij}^{(p+\tau+T)}&=\sigma_{ij}^{(p)}+\sigma_{ij}^{(\tau)}+\sigma_{ij}^{(T)}.\end{aligned} \tag{10}$$

Further, some of the solutions to more than 70 problems of studying the stress-strain state of the pipe cross section in the damage area (dot-and-dash line in Figure 1) [Kostyuchenko et al., 2007a; 2007b; Sherbakov et al., 2007b; 2008a; 2008b; Sherbakov, 2007b; Sosnovskiy et al., 2008] are analyzed. These two-dimensional problems mainly describe the stress-strain states of straight pipes with different-profile damage along the axis. Also, with the use of the finite-element method implemented in the software ANSYS, the essentially three- dimensional stress-strain state of the pipe in the three-dimensional damage area (Figure 1) was investigated.

3. Wall friction in the turbulent mineral oil flow in the pipe with corrosion damage

Within the framework of the present work, hydrodynamic calculation was made of the motion characteristics of a viscous, incompressible, steady, isothermal fluid in a cylindrical channel that models a pipe and in a cylindrical channel with geometric characteristics with regard to the peculiarities of a pipe with corrosion damage (see, Sect. 2). Calculations were performed for the initial incoming flow velocities u_0: 1 m/sec and 10 m/sec.

The kinematic viscosity of fluid was taken equal to v_K = 1.4 10-4 m2/sec, the viscous fluid density – 865 kg/m3. The calculated Reynolds numbers will be, respectively,

$$\mathrm{Re}_{1\,m/sec} = \frac{\upsilon_0 D}{\nu_K} = \frac{1m/sec*0.612m}{1.4*10^{-4}m^2/sec} = 4371.43, \quad (11)$$

$$\mathrm{Re}_{10\,m/sec} = \frac{\upsilon_0 D}{\nu_K} = \frac{10m/sec*0.612m}{1.4*10^{-4}m^2/sec} = 43714.3. \quad (12)$$

The critical Reynolds number (a transition from a laminar to a turbulent flow) for a viscous fluid moving in a round pipe is Re_{cr} » 2300. Thus, the turbulent flow motion should be considered in our problem. The software Fluent calculations used the turbulence k – e model for modeling turbulent flow viscosity [Launder et al., 1972; Rodi, 1976].

As boundary conditions the following parameters were used: at the incoming flow surface the initial turbulence level equal to 7% was assigned; at the pipe walls the fixing conditions and the logarithmic velocity profile were predetermined; in the pipe the fluid pressure equal to 4 MPa was set.

Calculations of the steady regime of the fluid flow (quasi-parabolic turbulent velocity profile of the incoming flow) and of the unsteady regime (rectangular velocity profile of the incoming flow) were made.

In the problems with a rectangular velocity profile of the incoming flow

$$\upsilon_x\big|_{x=0} = \upsilon_{r1}, \quad (13)$$

The unsteady regime of the fluid flow was considered.

In the problems with a quasi-parabolic turbulent velocity profile, at the entrance surface of the pipe the empirically found profile of the initial velocity was assigned, which is determined by the formula:

- for the two-dimensional case

$$\upsilon_x\big|_{x=0} = \upsilon_{\max}\left(1 - \frac{|r - 2r_0|}{2r_0}\right)^{\frac{1}{7}}, \upsilon_{\max} = 1.1428\,\upsilon_0, 0 \le r \le 2r_1, \quad (14)$$

- for the three-dimensional case

$$\upsilon_x\big|_{x=0} = \upsilon_{\max}\left(1 - \frac{r}{r_0}\right)^{\frac{1}{7}}, \upsilon_{\max} = 1.2244\,\upsilon_0, r = \sqrt{y^2 + z^2}, 0 \le r \le r_1. \quad (15)$$

The calculation results have shown that the motion becomes steady (as the flow moves in the pipe, the quasi-parabolic turbulent profile of the longitudinal velocity V_x develops) at some distance from the entrance (left) surface of the pipe (Figure

3). So, from Figure 4 it is seen that for the quasi-parabolic velocity profile of the incoming flow the zone of the steady motion begins earlier than for the rectangular profile.

Further, we will consider the results obtained for the velocity profiles of the incoming flow calculated in accordance to (14) and (15).

Consider the flow turbulence intensity being the ratio of the root-mean-square fluctuation velocity *u*¢ to the average flow velocity u_{avg} (Figure 5).

$$I = \frac{u'}{u_{avg}}, \tag{16}$$

At the surface of the incoming flow, the turbulence intensity is calculated by the formula

$$I = 0.16\left(\mathrm{Re}_{D_H}\right)^{-\frac{1}{8}}, \quad \mathrm{Re}_{D_H} = \frac{\upsilon_0 D_H}{\nu}, \tag{17}$$

where D_H is the hydraulic diameter (for the round cross section: $D_H = 2r_1 = 0.612$ m), u_0 is the incoming flow velocity, and ν is the kinematic viscosity of oil (ν = 1.4×10–4 m2/sec).

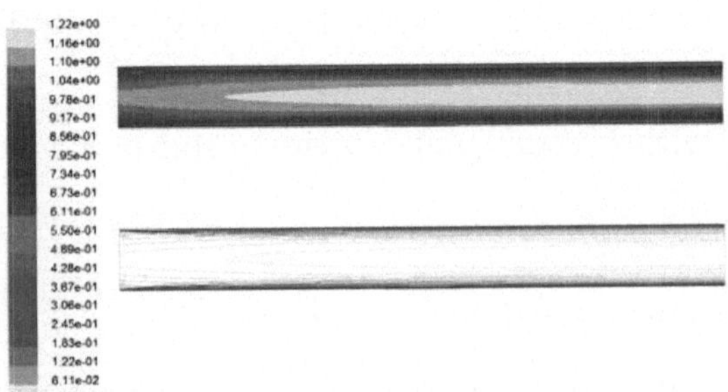

Fig. 3. Longitudinal velocity V_x (two-dimensional flow) for the quasi-parabolic turbulent velocity profile of the incoming flow at u_0 = 1 m/sec

Fig. 4. Profiles of the longitudinal velocity V_x, over the pipe cross sections (three-dimensional flow) for the quasi-parabolic turbulent velocity profile of the incoming flow at u_0 = 1 m/sec

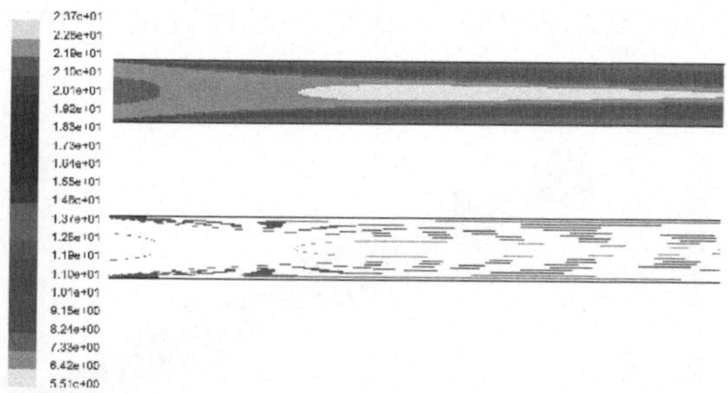

Fig. 5. Turbulence intensity (two-dimensional pipe flow, quasi-parabolic turbulent velocity profile, u0 = 1 m/sec)

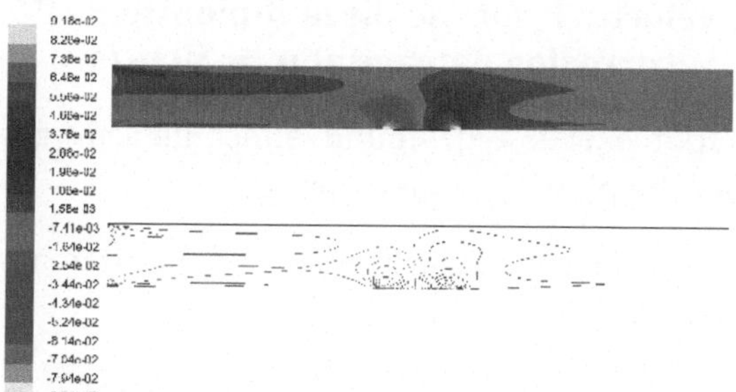

Fig. 6. Transverse velocity V_y for the two-dimensional flow in the pipe with corrosion damage at u0 = 10 m/sec

The zone of the unsteady turbulent motion is characterized by the higher turbulence intensity (vortex formation) in comparison with the remaining region of the pipe (Figure 5). The highest intensity is observed in the steady motion zone, which is especially noticeable in the calculations with the initial velocity of 1 m/sec in the pipe wall region, whereas the lowest one – at the flow symmetry axis.

At high initial flow velocity values the vortex formation rate is higher.

It should be emphasized that at a higher value of the initial flow velocity, the instability region is longer: at u_0 = 1 m/sec its length is about 2 m, while at u_0 = 10 m /sec its length is about 5 m.

The behavior of the motion (steady or unsteady) exerts an influence on the value of wall stresses. In the unsteady motion zone, they are essentially higher as against the appropriate stresses in the identical steady motion zone.

These figures illustrate that at that place of the pipe, where the fluid motion becomes steady, the value of tangential stress at u_0 = 1 m/sec is approximately equal to 8 Pa, whereas at u_0 = 10 m/sec it is about 240 Pa.

The results as presented above are peculiar for a pipe with corrosion damage and without it. At the same time, the presence of corrosion damage affects the kinematics of the moving flow in calculations with both the rectangular profile of the initial flow velocity and the quasi-parabolic turbulent one. In this domain of geometry, there appear transverse displacements that form a recirculation zone (Figure 7).

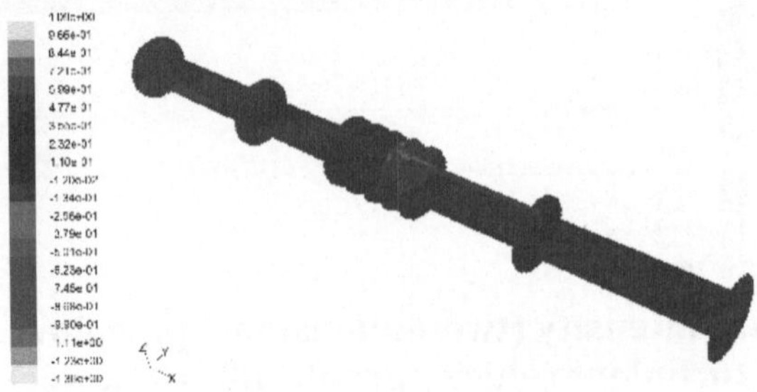

Fig. 7. Transverse velocity V_z for the three-dimensional flow in the pipe with corrosion damage at u_0 = 10 m/sec

The corrosion spot exerts a profound effect on changes in wall tangential stresses in the area of the pipe corrosion damage.

Figures 8 and 9 demonstrate that in the corrosion damage area, the values of wall tangential stresses undergo jumping.

For the laminar fluid motion, the value of tangential stresses at the pipe wall is calculated by the following formula [Sedov, 2004]:

$$\tau_0 = \mu\left(\frac{\partial v_x}{\partial y} + \frac{\partial v_y}{\partial x}\right) = \mu\frac{dv}{dr} = \frac{4\mu v_0}{r_0}, \qquad (18)$$

where m = u×r = 1.4×10–4×865 = 0.1211 kg/(m*sec) is the molecular viscosity, r_0 = 0.306 m is the pipe radius.

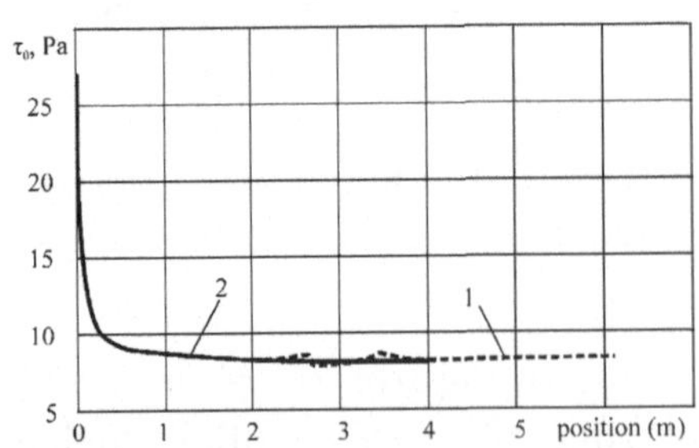

Fig. 8. Wall tangential stresses at the pipe wall: stresses at $y = f(x)$, stresses at $y = 2r_1$ for the two-dimensional flow in the pipe with corrosion damage at u_0 = 1 m/sec

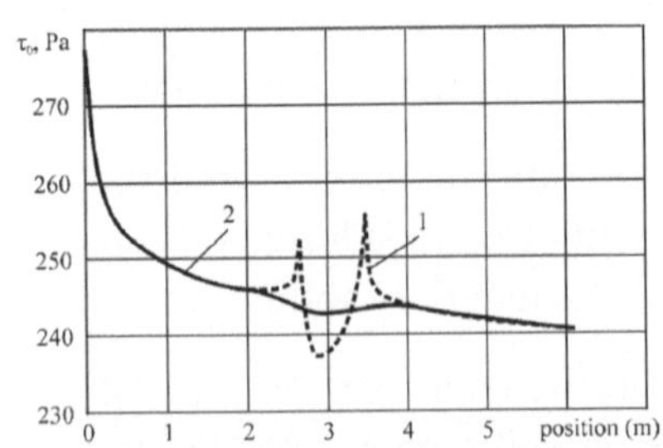

Fig. 9. Wall tangential stresses at the pipe wall: stresses at $y = f(x)$, stresses at $y = 2r_1$ for the two-dimensional flow in the pipe with corrosion damage at u_0 = 10 m/sec

Then t_0 for the velocities u_0 = 10 m/sec and u_0= 1 m/sec will be

$$\tau_0^{10} = \frac{4\cdot 0.1211\cdot 10}{0.306} = 15.83\ \text{Pa},\ \tau_0^{1} = \frac{4\cdot 0.1211\cdot 1}{0.306} = 1.58\ \text{Pa}. \qquad (19)$$

The expression for the tangential stresses with regard to the turbulence is of the form [Sedov, 2004]:

$$\tau_{xy} = \tau_0 + \tau'_{xy} = \mu\left(\frac{\partial \bar{\upsilon}_x}{\partial y} + \frac{\partial \bar{\upsilon}_y}{\partial x}\right) - \rho\overline{\upsilon'_x \upsilon'_y} = (\mu + \mu_t)\left(\frac{\partial \bar{\upsilon}_x}{\partial y} + \frac{\partial \bar{\upsilon}_y}{\partial x}\right). \tag{20}$$

The last formula and the analysis of the calculations enable evaluating the turbulence influence on the value of tangential stresses at the pipe wall. As indicated above, at different profiles and initial velocity values the tangential stresses were obtained: at $u_0 = 1$ m/sec: $t_{xy} = t_w$ » 8 Pa, at $u_0 = 10$ m/sec: $t_{xy} = t_w$ » 240 Pa. The value of the turbulent stress (Reynolds stress):

at $u_0 = 1$ m/sec :

$$\tau'_{xy} = -\rho\overline{\upsilon'_x \upsilon'_y} = \mu_t\left(\frac{\partial \bar{\upsilon}_x}{\partial y} + \frac{\partial \bar{\upsilon}_y}{\partial x}\right) = \tau_{xy} - \tau_0 = 8 - 1.58 = 6.42\,\text{Pa}, \tag{21}$$

at $u_0 = 10$ m/sec :

$$\tau'_{xy} = -\rho\overline{\upsilon'_x \upsilon'_y} = \mu_t\left(\frac{\partial \bar{\upsilon}_x}{\partial y} + \frac{\partial \bar{\upsilon}_y}{\partial x}\right) = \tau_{xy} - \tau_0 = \tag{22}$$

$$= 240 - 15.83 = 224.17\,\text{Pa},$$

The results obtained are evident of the fact that the turbulence much contributes to the formation of wall tangential stresses. At the higher turbulence intensity (it is especially high in the pipe wall region), Reynolds stresses increase, too. I.e., the turbulence stresses are:

$$\text{at } \upsilon_0 = 1\ \text{m/sec}: \frac{\tau_{xy} - \tau_0}{\tau_{xy}} 100\% = \frac{8 - 1.58}{8} 100\% = 80.25\%; \tag{23}$$

$$\text{at } \upsilon_0 = 10\ \text{m/sec}: \frac{\tau_{xy} - \tau_0}{\tau_{xy}} 100\% = \frac{240 - 15.83}{240} 100\% = 93.4\%. \tag{24}$$

The analysis as made above shows that the calculation of the motion of a viscous fluid in the pipe as laminar can result in a highly distorted distribution pattern of the tangential stresses at the inner surface of the pipe. It can be concluded that the analysis of viscous fluid friction, when the flow interacts with the pipe wall, must be performed on the basis of the calculation of flow motion as essentially turbulent one.

4. Analytical solutions for the stress-strain state of the pipeline model under the action of internal pressure and temperature difference

In the simplified analytical statement, the problem of calculating the stress-strain state of a long cylindrical pipe reduces to the problem of the strain of a thin ring loaded with a pressure p_1 uniformly distributed over its inner wall and also with a pressure p_2 uniformly distributed over the outer surface of the ring (Figure 10).

Operating conditions of the ring do not vary depending on whether it is considered either as isolated or as a part of the long cylinder.

Work [Ponomarev et al., 1958] and many other publications contain the classical solution to this problem based on solving the following differential equation for radial displacements:

$$\frac{d^2u_r}{dr^2}+\frac{1}{r}\frac{du_r}{dr}-\frac{1}{r^2}u_r=0. \tag{25}$$

The general solution of this equation is of the form:

$$u_r=C_1 r+C_2\frac{1}{r}. \tag{26}$$

With the use of the relationship between stresses and strains, and also of Hook's law, it is possible to determine integration constants $C_1$and C_2 under the boundary conditions of the form:

$$\begin{aligned}\sigma_r\big|_{r=r_1}&=-p_1,\\ \sigma_r\big|_{r=r_2}&=-p_2.\end{aligned} \tag{27}$$

where p_1 is the internal pressure; p_2 is the external pressure.

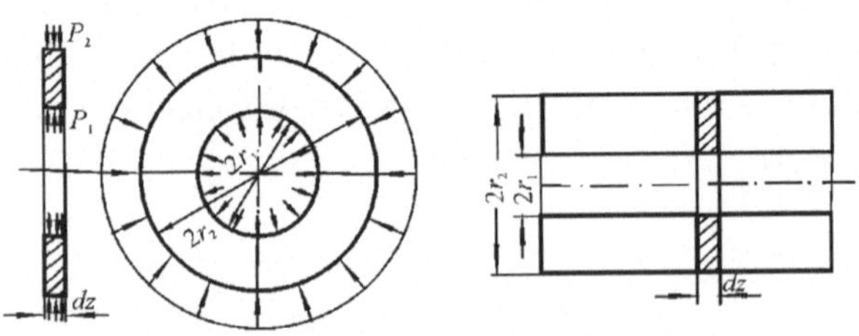

Fig. 10. Loading diagram of the circular cavity of the pipe

In such a case, the general formulas for stresses at any pipe point have the following form:

$$\begin{aligned}\sigma_r&=\frac{p_1 r_1^2-p_2 r_2^2}{r_2^2-r_1^2}-\frac{(p_1-p_2)\,r_1^2r_2^2}{r_2^2-r_1^2}\frac{1}{r^2},\\ \sigma_\varphi&=\frac{p_1 r_1^2-p_2 r_2^2}{r_2^2-r_1^2}+\frac{(p_1-p_2)\,r_1^2r_2^2}{r_2^2-r_1^2}\frac{1}{r^2}.\end{aligned} \tag{28}$$

Assuming that the cylinder is loaded only with the internal pressure ($p_1 = p$, $p_2 = 0$), the following expressions are obtained for the stresses based on the internal pressure:

$$\frac{\sigma_r^{(p)}}{p}=\frac{1}{k_{r2}^2}\left(\frac{k_{r12}^2}{1-k_{r12}^2}-1\right),\quad \frac{\sigma_\varphi^{(p)}}{p}=\frac{1}{k_{r2}^2}\left(\frac{k_{r12}^2}{1-k_{r12}^2}+1\right), \tag{29}$$

where $kr2 = r/r2$, $kr12 = r1/r2$

To analyze the rigid fixing of the outer surface of the pipeline, as one of the equations of the boundary conditions we choose expression (26) for displacements, the value of which tends to zero at the outer surface of the model. As the secondary boundary condition we use an expression for stresses at the inner surface of the cylinder from (27):

$$\sigma_r\Big|_{r=r_1}=-p_1,\ u_r\Big|_{r=r_2}=0. \tag{30}$$

Then, the expressions for the stresses will assume the form:

$$\begin{aligned}\frac{\sigma_r^{(p)}}{p}&=-\frac{k_{r12}^2}{k_{r2}^2}\frac{k_{r2}^2(1+\nu_1)-(\nu_1-1)}{k_{r12}^2(1+\nu_1)-(\nu_1-1)},\\ \frac{\sigma_\varphi^{(p)}}{p}&=-\frac{k_{r12}^2}{k_{r2}^2}\frac{k_{r2}^2(1+\nu_1)+(\nu_1-1)}{k_{r12}^2(1+\nu_1)-(\nu_1-1)}.\end{aligned} \tag{31}$$

Consider a long thick-wall pipe, whose wall temperature t varies across the wall, but is constant along the pipe, i. e., $t = t(r)$ [Ponomarev et al., 1958].

If the heat flux is steady and if the temperature of the outer surface of the pipe is equal to zero and that of the inner surface is designated as T, then from the theory of heat transfer it follows that the dependence of the temperature t on the radius r is given by the formula

$$t=\frac{T}{\ln k_{r12}}\ln k_{r2}, \tag{32}$$

Any other boundary conditions can be obtained by making uniform heating or cooling, which does not cause any stresses. Thus, the quantity T in essence represents the temperature difference DT of the inner and outer surfaces of the pipe.

As the temperature is constant along the pipe, it can be considered that cross sections at a sufficient distance from the pipe ends remain plane, and the strain ε_z is a constant quantity. The temperature influence can be taken into account if the strains due to stresses are added with the uniform temperature expansion De = aDT where a is the linear expansion coefficient of material.

The stress-strain state in the presence of the temperature difference between the pipe walls can be determined by solving the differential equation [Ponomarev et al., 1958]:

$$\frac{d^2u}{dr^2}+\frac{1}{r}\frac{du}{dr}-\frac{u}{r^2}=\frac{1+\nu_1}{1-\nu_1}\alpha\frac{dt}{dr}. \tag{33}$$

Subject to the boundary conditions

$$\sigma_r\Big|_{r=r_1}=0,\ \sigma_r\Big|_{r=r_2}=0. \tag{34}$$

Having solved boundary-value problem (33), (34), the expressions for stresses are of the form:

$$\sigma_r^{(T)} = \frac{E_1\alpha\Delta T}{2(1-\nu)}\frac{1}{\ln k_{r12}}\left[-\ln k_{r2} - \frac{k_{r12}^2}{1-k_{r12}^2}\left(1-\frac{1}{k_{r2}^2}\right)\ln k_{r12}\right],$$

$$\sigma_\varphi^{(T)} = \frac{E_1\alpha\Delta T}{2(1-\nu)}\frac{1}{\ln k_{r12}}\left[-1-\ln k_{r2} - \frac{k_{r12}^2}{1-k_{r12}^2}\left(1+\frac{1}{k_{r2}^2}\right)\ln k_{r12}\right], \quad (35)$$

$$\sigma_z^{(T)} = \frac{E_1\alpha\Delta T}{2(1-\nu)}\frac{1}{\ln k_{r12}}\left[-1-2\ln k_{r2} - \frac{2k_{r12}^2}{1-k_{r12}^2}\ln k_{r12}\right],$$

Figures 11–14 show the distribution of dimensionless stresses (29), (31), (35) along *r* and their sums

$$\sigma_i^{(p+T)} = \sigma_i^{(p)} + \sigma_i^{(T)}, i = r, \varphi, z, \quad (36)$$

for $k_{r12} = 0.8, v = 0.3, E_1 aT / p = 10$ (for example, at $E_1 = 2×1011$ Pa, a = 10-5°C-1, $DT = 20$ °C). These figures well illustrate the essential influence of the temperature and the procedure of fixing the pipe on its stress-strain state.

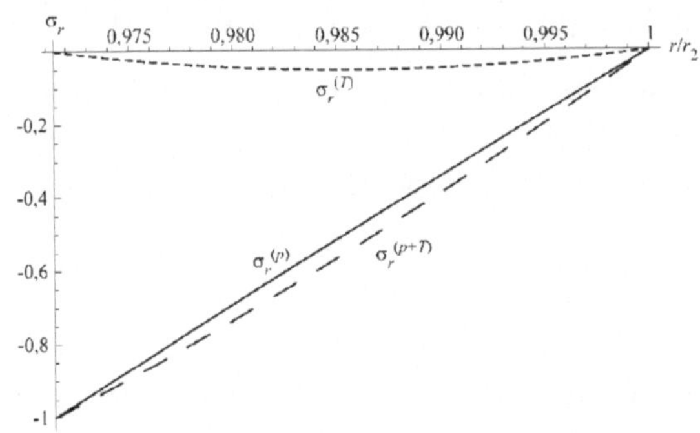

Fig. 11. Radial stresses for problems (25), (27) and (33), (34) at *r*1 £ *r* £ *r*2

Compare the distribution of the stresses calculated analytically with the use of (31) for a non-damaged pipe with the finite-element calculation results by plotting the graphs of the pipe thickness stress distribution (Figures 1.15–1.16). To make calculation, take the following initial data: inner and outer radii r_1= 0.306 m and r_2= 0.315 m, p_1= 4M Pa, p_2= 0, E = 2×1011 Pa, ν = 0.3.

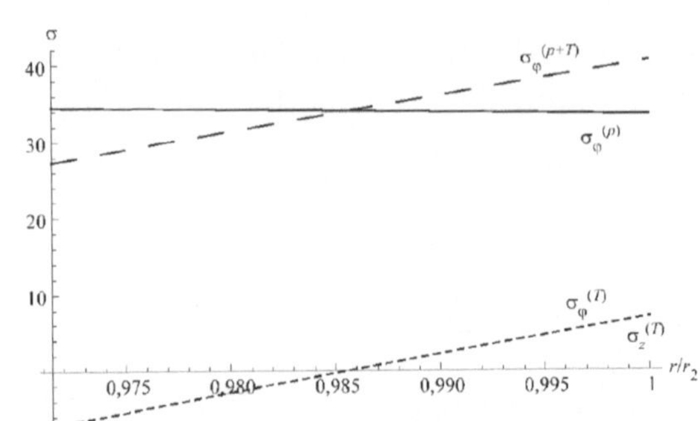

Fig. 12. Circumferential stresses for problems (25), (27) and (33), (34) at *r*1 £ *r* £ *r*2

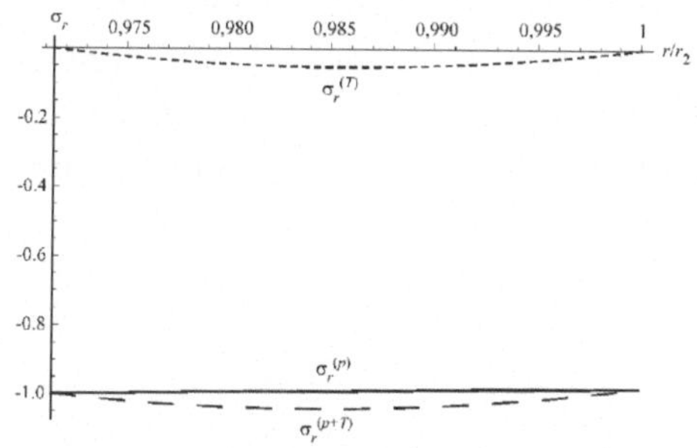

Fig. 13. Longitudinal stresses for problems (25), (30) and (33), (34) at *r*1 £ *r* £ *r*2

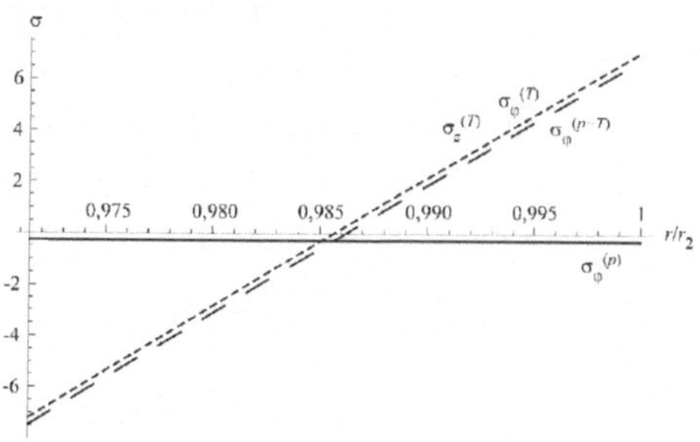

Fig. 14. Circumferential stresses for problems (25), (30) and (33), (34) at *r*1 £ *r* £ *r*2

As seen from Figures 15–16, the s_r and s_j distributions obtained from the analytical calculation practically fully coincide with those obtained from the finite-element calculation, which points to a very small error of the latter.

5. Stress-strain state of the three-dimenisonal model of a pipe with corrosion damage under complex loading

Consider the problem of determining the stress-strain state of a two-dimenaional model of a pipe in the area of three-dimensional elliptical damage.

In calculations we used a model of a pipe with the following geometric characteristics (Figure 2): inner (without damage) and outer radii r_1 = 0.306 m and r_2= 0.315 m, respectively, the length of the calculated pipe section L=3 m, sizes of elliptical corrosion damage length ´ width ´ depth – 0.8 m ´ 0.4 m ´ 0.0034 m.

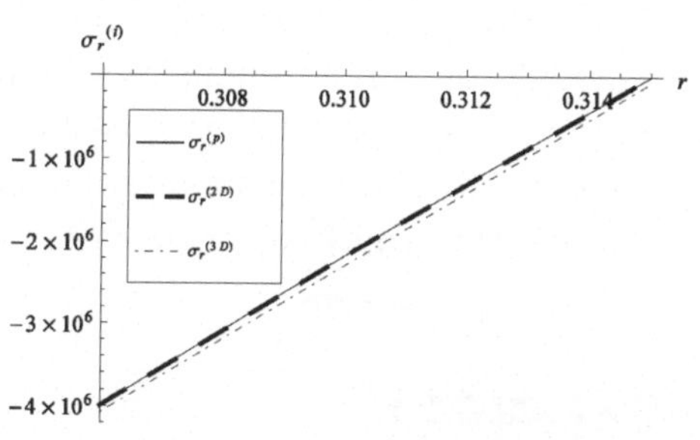

Fig. 15. Radial stress distribution for the analytical calculation (s (p)), for the two-dimensional computer model (s (2*D*)), for the three-dimensional computer model (

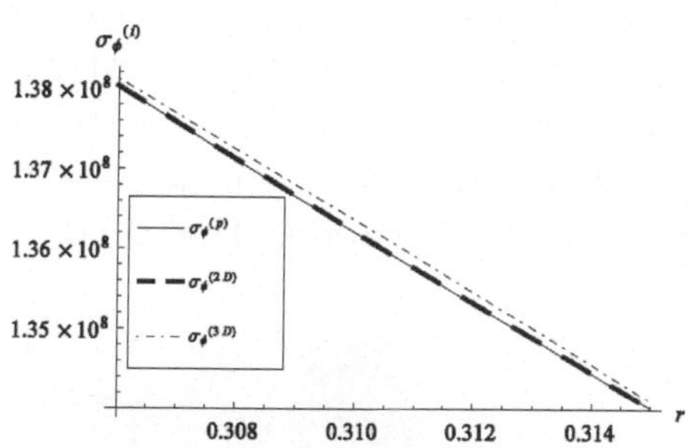

Fig. 16. Circumferential stress distribution for the analytical calculation (s (*p*)), for the two-dimensional computer model (s (2*D*)), for the three-dimensional computer model (s (3*D*))

The pipe mateial had the following characteristics: elasticity modulus E_1 = 2×1011 Pa, Poisson's coefficient v_1 = 0.3, temperature expansion coefficient a = 10-5 °C-1, thermal conductivity k = 43 W/(m°C), and the soil parameters were: E_2 = 1.5×109 Pa, Poisson's coefficient v_2 = 0.5. The coefficient of friction between the pipe and soil was m = 0.5.

The internal pressure in the pipe (1) is:

$$\sigma_r\big|_{r=r_1} = p = 4 \text{ MPa.} \tag{37}$$

The temperature diffference between the pipe walls is (3)

$$\left|T_{r_1} - T_{r_2}\right| = \Delta T = 20^oC. \tag{38}$$

The value of internal tangential stresses (wall friction) (2) is determined from the hydrodynamic calculation of the turbulent motion of a viscous fluid in the pipe.

Calculations in the absence of fixing of the outer surface of the pipe and in the presence of the friction force over the inner surface (2) were made for 1/2 of the main model (Figure 2), since in this case (in the presence of friction) the calculation model has only one symmetry plane. In the absence of outer surface fixing, calculations were made for 1/4 of the model of the pipeline section since the boundary conditions of form (2) are also absent and, hence, the model has two symmetry planes.

The investigation of the stress state of the pipe in soil is peformed for 1/4 of the main model of the pipe placed inside a hollow elastic cylinder modeling soil (Figure 17).

In calculations without temperature load, a finite-element grid is composed of 20-node elements SOLID95 (Figure 17) meant for three-dimensional solid calculations. In the presence of temperature difference, a grid is composed of a layer of 10-node finite elements SOLID98 intended for three-dimensional solid and temperature calculations. The size of a finite element (fin length) a_{FE} =10-2 m.

Fig. 17. General view and the finite-element partition of ¼ of the pipe model in soil

Thus, the pipe wall is composed of one layer of elements since its thickness is less than centermeter. During a compartively small computer time such partition allows obtaining the results that are in good agreement with the analytical ones (see, below).

Calculations for boundary conditions (8) with a description of the contact between the pipe and soil use elements CONTA175 and TARGE170.

As seen from Figure 17, the finite elements are mainly shaped as a prism, the base of which is an equivalateral triangle. The value of the tangential stresses t *rz r* =*r* applied to each node of the inner surface will then be calculated as follows:

$$\tau_{rz}^{(node)}\Big|_{r=r_1} = \tau_0 S, \qquad (39)$$

where *S* is the area of the romb with the side a_{FE} and with the acute angle b_{FE} = p/3. Thus, the value of the tangential stress applied at one node will be

$$\tau_{rz}^{(node)}\Big|_{r=r_1} = \tau_0 a_{FE}^2 \sin\beta_{FE} = 260\cdot 10^{-4}\sqrt{3}\,/\,2 = 2.25\cdot 10^{-2}\text{Pa}. \qquad (40)$$

The analysis of the calculation results will be mainly made for the normal (principal) stresses s_x, s_y, s_z in the Cartesian system of coordinates. It should be noted that for axis- symmetrical models, among which is a pipe, the cylindrical system of coordinates is natural, in which the normal stresses in the radial s_r, circumferential s_t, and axial s_z directions are principal. Since the software ANSYS does not envisage stresses in the polar system of coordinates, the analysis of the stress state will be made on the basis of s_x, s_y, s_z in those domains where they coincide with s_r, s_t, s_z corresponding to the last principal stresses s_1, s_2, s_3 and also to the tangential stresses s_{yz}.

Make a comparative analysis of the results of numerical calculation for boundary conditions (1), (6) and (1), (7) with those of analytical calculation as described in Sect. 1.4. Consider pipe stresses in the circumferential s_t and radial s_r directions.

Figures 18 and Figure 19 show that in the case of fixing $u_x\big|_{r=r_2} = u_y\big|_{r=r_2} = 0$ corrosion damage exerts an essential influence on the s_t distribution over the inner surface of the pipe. At the damage edge, the absolute value of circumferential s_t is, on average, by 15% higher than the one at the inner surface of the pipe with damage and, on average, by 30 % higher than the one inside damage. In the case of fixing $u_x\big|_{r=r_2} = u_y\big|_{r=r_2} = u_z\big|_{r=r_2} = 0$ distributions are localized just in the damage area. The additional key condition *uz r* =*r* = 0 (coupling along the *z*-axis) is expressed in increasing $|s_t|$ at the inner surface without damage in the calculation for (1), (7) approximately by 60% in comparison with the calculation for (1), (6). However in the calculation for (1), (7), the $|s_t|$ differences between the damage edge, the inner surface without damage, and the inner surface with damage are, on average, only 6 and 3% , respectively. Maximum and minimum values of s_t in the calculation for (1), (6) are: $\sigma_t^{\min} = -1.27\cdot 10^6$ Pa and $\sigma_t^{\max} = -7.96\cdot 10^5$ Pa; in the calculation for (1), (7) are: $\sigma_t^{\min} = -1.72\cdot 10^6$ Pa and $\sigma_t^{\max} = -1.61\cdot 10^6$ Pa. The analysis of the stress distribution reveals a good coincidence of the results of the analytical and finite-element calculations for s_t. At r_1 £ y £ r_2, x=z=0 in the vicinity of the pipe without damage, the error is at $r = r_1$

$$e = \frac{1.093 - 1.082}{1.093} \cdot 100\% = 1.03\%, \tag{41}$$

$$e = \frac{1.175 - 1.165}{1.175} \cdot 100\% = 0.94\%. \tag{42}$$

Thus, at the upper inner surface of the pipe the damage influence on the st variation is inconsiderable. A comparatively small error as obtained above is attributed to the fact that the three-dimensional calculation subject to (1), (6) was made at the same key conditions as the analytical calculation of the two-dimensional model. At the same time, owing to the additonal condition *uz r =r* = 0 the difference between the results of the analytical calculation and the calculation for (1), (7) is much greater – about 45 %.

Fig. 18. Distribution of the stress s,(s,) at $\sigma_r\big|_{r=r_1} = p\,,\, u_x\big|_{r=r_2} = u_y\big|_{r=r_2} = 0$

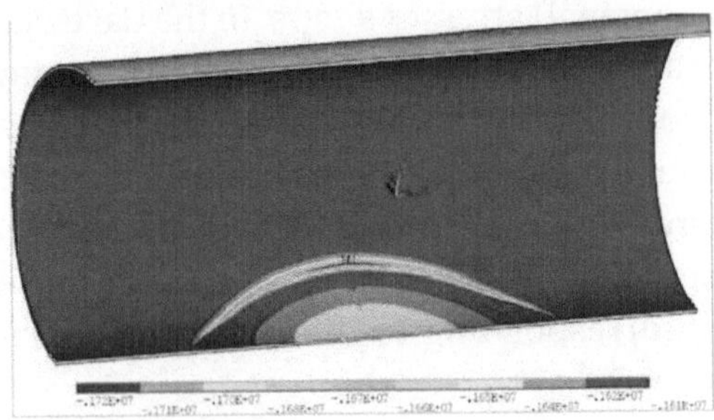

Fig. 19. Distribution of the stress s, (s,) at $\sigma_r\big|_{r=r_1} = p\,,\, u_x\big|_{r=r_2} = u_y\big|_{r=r_2} = u_z\big|_{r=r_2} = 0$

A more detailed analysis of the stress-strain state can be made for distributions along the below paths.

For 1/2 of the pipe model:

Path 1. Along the straight line r_1 £ y £ r_2 at x=z=0: from $P_{11}(0, r_1, 0)$ to $P_{12}(0, r_2, 0)$.

Path 2. Corrosion damage center (– r_1 – h £ y £ – r_2 at x=z=0): from $P_{21}(0, -r_1 - h, 0)$ to $P_{22}(0, -r_2, 0)$.

Path 3. Cavity boundary over the cross section z=0:

from $P_{31}(0.186, -0.243, 0)$ to $P_{32}(0.192, -0.25, 0)$.

Path 4. Cavity boundary over the cross section x=0:

from $P_{41}(0, -r_1, d/2)$ to $P_{42}(0, -r_2, d/2)$.

Path 5. Along the straight line of the upper inner surface of the pipe – $0.8L/2$ £ z £ $0.8L/2$ at $x = 0, y = r_1$: from $P_{51}(0, r_1, - 0.8L/2)$ to $P_{52}(0, r_1, 0.8L/2)$.

Path 6. Along the curve of the lower inner surface of the pipe – $0.8L/2$ £ z £ $0.8L/2$ at x=0, $y = \begin{cases} -r_1 = f(z), 0 \le |z| \le d/2 \\ -r_1, d/2 \le |z| \le 0.8L/2 \end{cases}$ through the points:

$P_{64}(0, - r_1, - 0.8L/2)$, $P_{63}(0, - r_1, - d/2)$, $P_{62}(0, - r_1, - 0.0025, -0.2)$, P_{61}**(0, –** r_1, – h, 0), $P_{62}(0, - r_1, - 0.0025, 0.2)$, $P_{63}(0, - r_1, d/2)$, $P_{64}(0, - r_1, 0.8L/2)$.

For 1/4 of the pipe model, paths 1–4 are the same as those for 1/2, whereas paths 5 and 6 are of the form:

Path 5. Along the strainght upper inner surface of the pipe 0 £ z £ $0.8L/2$ at x=0, y=r_1: from

$P_{51}(0, r_1, 0)$ to $P_{52}(0, r_1, 0.8L/2)$.

Path 6. Along the curve of the lower inner surface of the pipe 0 £ z £ $0.8L/2$ at x=0, $y = \begin{cases} -r_1 = f(z), 0 \le z \le d/2 \\ -r_1, d/2 \le z \le 0.8L/2 \end{cases}$ through the points:

$P_{61}(0, - r_1, - h, 0)$, $P_{62}(0, - r_1, - 0.0025, 0.2)$, $P_{63}(0, - r_1, d/2)$, $P_{64}(0, - r_1, 0.8L/2)$.

In the above descriptions of the paths, d=0.8 m is the length of corrosion damage along the z axis of the pipe. The function $f(z)$ describes the inhomogeneity of the geometry of the inner surface of the pipe with corrosion damage.

The analysis of the distributions shows that $|s_t|$ increases up to 10% from the inner to the outer surface along paths 1, 2, 4 and decreases up to 2% along path 3. Thus, it is seen that at the corrosion damage edge over the cross section (path 3), the $|s_t|$ distribution has a specific pattern. It should also be mentioned that if in the calculation for (1), (6), $|s_t|$ inside the damage is approximately by 20% less than the one at the inner surface without damage, then in the calculation for (1), (7) this stress is approximately by 2% higher.

Figure 20 shows the sr distribution that is very similar to those in the calculations for (1), (6) and for (1), (7). I.e., the procedure of fixing the outer surface of the pipe practically does not influencesthe s_r distribution. At the corrorion damage edge of the inner surface of the pipe, the s_r distribution undergoes small variation (up to 1%). Maximum and minimum values of s_r in the calculation for (1), (6) are: $\sigma_r^{\min} = -4.02 \cdot 10^6$ Pa and $\sigma_r^{\max} = -3.91 \cdot 10^6$ Pa; in the calculation for (1), (7): $\sigma_r^{\min} = -4.02 \cdot 10^6$ Pa and $\sigma_r^{\max} = -3.92 \cdot 10^6$ Pa.

The numerical analysis of the results reveals a good agreement between the results of analytical and finite-element calculations for s_r ((1), (6)). For r_1 £ y £ r_2, x=z=0 in the region of the pipe without damage at $r = r_1$*e is* >>1%, whereas at $r = r_2$*e is* »1% for (1), (6).

Make a comparative analysis of the results of these numerical calculations for (1), and (1), (8) with those of the analytical calculation described in Sect. 1.4 for the boundary conditions of the form $\sigma_r|_{r=r_1} = p$, $\sigma_r|_{r=r_2} = 0$. Consider pipe stresses in the circumfrenetial s_t and radial s_r directions under the action of internal pressure (1) for fixing absent at the outer surface and at the contact between the the pipe and soil (1), (8).

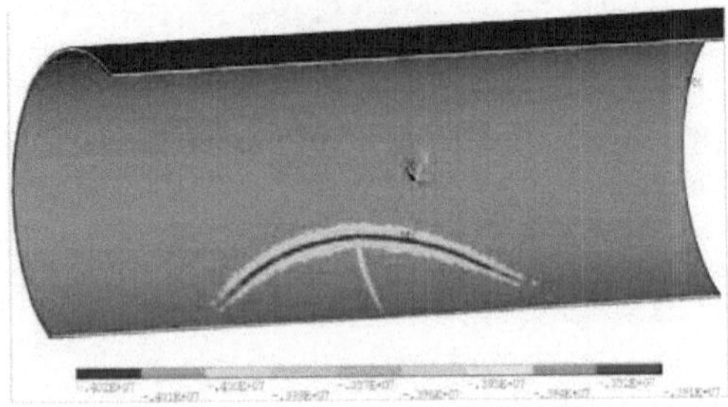

Fig. 20. Distribution of the stress s_3 (s_r) at s_r $r=r=p$, ux $r=r$ = $uy = 0$

From Figures 21 and 22 it is seen that in the case of pipe fixing $u_x|_{r=r_2} = u_y|_{r=r_2} = 0$ the corrosion damage exerts an essential influence on the s_t distribution over the inner surface of the pipe. The minimum of the tensile stress s_t is at the damage edge over the cross section, whereas the maximum – inside the damage. The s_t value at the damage edge is, on average, by 30% less than the one at the inner surface of the pipe without damage and by 60% less than the one inside the damage. The stress s_t is approximately by 50% less at the surface without damage as against the one inside the damage. At the contact between the pipe and soil, the s_t disturbances are localized just in the damage area. In the calculation for (1), (8), the s_t differences between the damage edge, the inner surface without damage, and the damage interior are, on average, 60 and 70%, respectively. The stress s_t is approximately by 30% less at the surface without damage as against the one inside the damage. In this calculation there appear essential end disturbances of s_t. Such a disturbance is the drawback of the calculation involvingh the modeling of the contact between the pipe and soil. Additional investigations are needed to eliminate this disturbance. On the whole, s_t at the inner surface of the pipe in the calculation for (1) is, on average, by 70% larger than the one in the calculation for (1), (8). Maximum and minimum values of s_r in the calculation for (1) are: $s^{min} = 8.39 \times 10^7$ Pa $s^{max} = 6.65$ and $\times 10^8$ Pa; in the calculation for (1), (8): $s^{min} = 7.66 \times 10^6$

The numerical analysis of the results shows not bad coincidence of the results of the analytical and finite-element calculations for s_t, (1). At r_1 £ y £ r_2, $x = z = 0$ in the region of the pipe without damage the error at $r = r_1$ is approximately equal to

$$e = \frac{1.38 - 1.45}{1.38} \cdot 100\% = -6.71\%, \qquad (43)$$

at $r = r_2$

$$e = \frac{1.34 - 1.305}{1.34} \cdot 100\% = 2.61\%. \qquad (44)$$

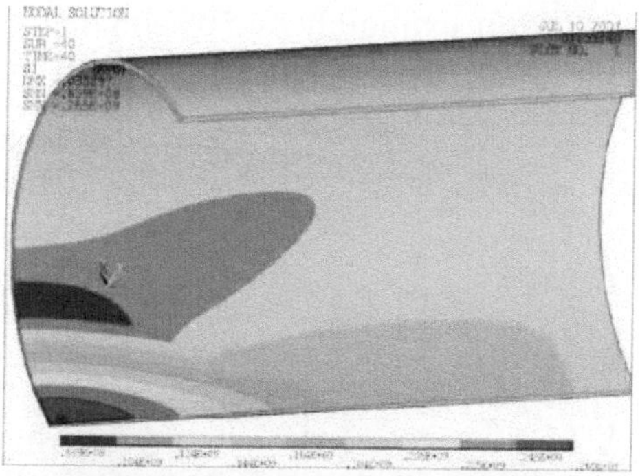

Fig. 21. Distribution of the stress s$_1$ (s$_t$) at s$_r$ $r=r=p$

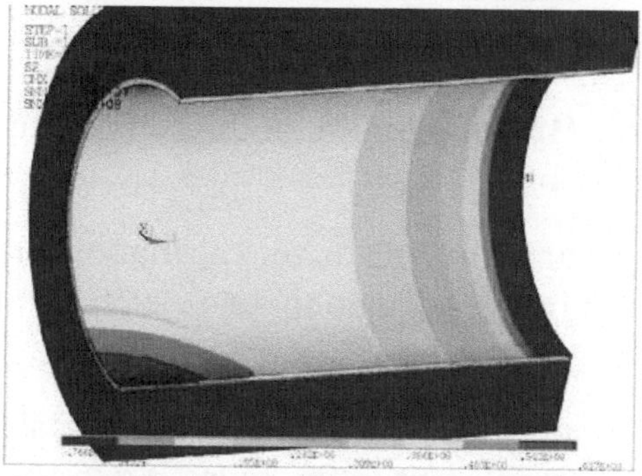

Fig. 22. Distribution of the stress s$_2$ (s$_t$) at $\sigma_r\big|_{r=r_1}=p$, $\sigma_r^{(1)}\big|_{r=r_2}=-\sigma_r^{(2)}\big|_{r=r_2}$

$$\sigma_\tau^{(1)}\Big|_{r=r_2}=-\sigma_\tau^{(2)}\Big|_{r=r_2}=f\,\sigma_n^{(1)}\Big|_{r=r_2},\ u_x\big|_{r=r_3}=u_y\big|_{r=r_3}=0$$

Thus, at the upper inner surface of the pipe, the damage influence on the st variation is inconsiderable. A comparatively small error obtained says about the fit of the key condition $\sigma_r\big|_{r=r_1}=p$ in the three-dimensional calculation with the key condition for the two- dimensional model $\sigma_r\big|_{r=r_1}=p$, $\sigma_r\big|_{r=r_2}=0$ in the analytical calculation. For (1), (8), because of the presence of elastic soil the difference between the results of the analytical and finite- element calculations and the calculation for (1), (7) is much larger – about 70 %.

The analysis shows that from the inner to the outer surface along paths 1, 2, 4, the stress s$_t$ decreases approximately by 7, 36 and 43%, respectively, and increases approximately by 120% along path 3. Thus, it is seen that at the corrosion damage edge over cross section (path 3) the s$_t$ distribution has an essentially peculiar pattern. The s$_t$ variations in the calculation for (1), (8) along paths 1, 2, 3 are identical to those in the calculation for (1) and are approximately 3, 1.5 and 15 %, respectively. However unlike the calculation for (1), in the calculation for (1), (8) s$_t$ increases a little (up to 1%) along path 4.

The stress s$_r$ distributions shown in Figures 23 and 24 illustrate a qualitative agreement of the results of the analytical and finite-element calculations for (1). In the calculation for (1) $|s_r|$ is approximately by 70% higher at the damage edge than the one at the inner surface without damage.

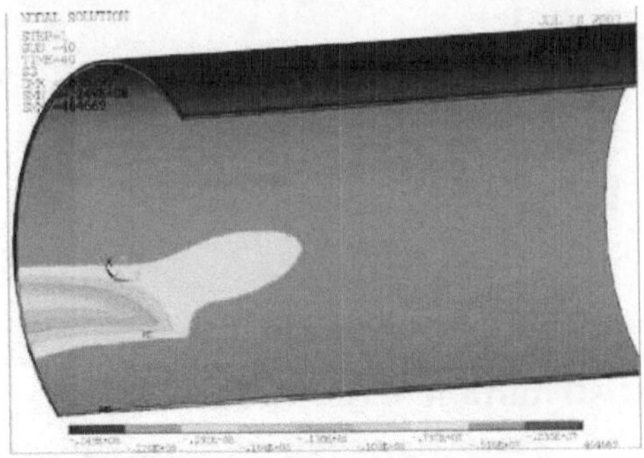

Fig. 23. Distribution of the stress s_3 (s_r) at s_r $r = r = p$

In the calculation for (1), (8), because of the soil pressure, $|s_r|$ practically does not vary in the damage vicinity.

Maximum and minimum values of s_r in the calculation for (1) are: s^{min}=-2.49×10^7 Pa and s^{max}=4.64×10^5 Pa; in the calculation for (1), (8): s^{min}=-1.62×10^7 Pa and s^{max} =1.09×10^6

Figures 1.18– 1.28 plot the distributions of the principal stresses corresponding to the sresses st, sr, sz for different fixing types. From the comparison of theses distributions it is seen that four forms of boundary conditions form two qualitatively different types of the stress st distributions. So, in the case of rigid fixing of the outer surface of the pipe (at $u_x|_{r=r_2} = u_y|_{r=r_2} = 0$ or $u_x|_{r=r_2} = u_y|_{r=r_2} = u_z|_{r=r_2} = 0$) $\sigma_t<0$. In the case, fixing is absent and contact is present, $s_t>0$. At the contact interaction between the pipe and soil, the level due to the pressure soil in s_t is approximately three times less than in the absence of fixing. The

Fig. 24. Distribution of the stress σ3 (σr) at $\sigma_r|_{r=r_1} = p$, $\sigma_r^{(1)}|_{r=r_2} = -\sigma_r^{(2)}|_{r=r_2}$,

Fig. 25. Distribution of the stress σz $\sigma_r|_{r=r_1} = p$, $u_x|_{r=r_2} = u_y|_{r=r_2} = 0$

Fig. 26. Distribution of the stress $\sigma_r\big|_{r=r_1} = p$, $u_x\big|_{r=r_2} = u_y\big|_{r=r_2} = u_z\big|_{r=r_2} = 0$

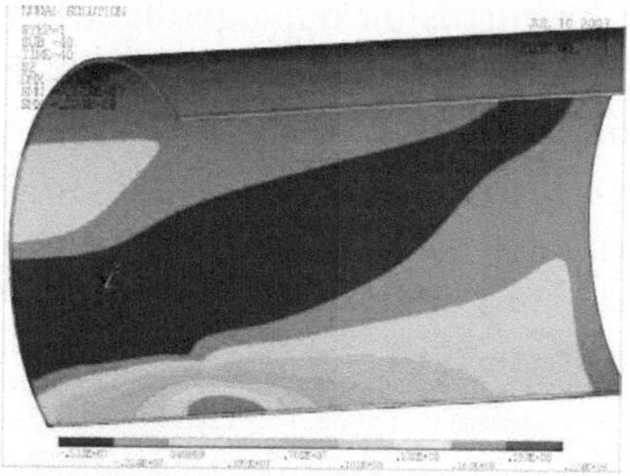

Fig. 27. Distirbution of the stress s_2 (s_z) at s_r $r = r = p$

st<0 distributions over the inner surface of the pipe are qualitatively and quantitatively indentical in all calculations. The sz distributions are essensially different for the considered calculations. In the calculations for $u_x\big|_{r=r_2} = u_y\big|_{r=r_2} = 0$ and in the absence of fixing, there exist regions of both tensile and compressive stresses sz. In the calculation for $u_x\big|_{r=r_2} = u_y\big|_{r=r_2} = u_z\big|_{r=r_2} = 0$, the peculiarities of the sz<0 distributions manefest themselves just in the damage region (fixing influence in all directions). At the contact interaction between the pipe and soil, the sz>0 distribution in the damage region is similar to the distribution for $u_x\big|_{r=r_2} = u_y\big|_{r=r_2} = 0$. The bulk analysis of the stress distributions has shown that the results of calculation of the contact interaction of the pipe and soil are intermediate between the calculation results for the extreme cases of fixing. So, the sr<0 distribution has a similar pattern in all calculations. By the st distribution, the case of the contact between the pipe and soil is close to that of absent fixing since in these calculations the boundary conditions allow the pipe to be expanded in the radial direction. By the sz distributions, the case of the contact between the pipe and soil is close for $u_x\big|_{r=r_2} = u_y\big|_{r=r_2} = 0$, since in these calculations for the outer surface of the pipe, displacements along the *z* axis of the pipe are possible and at the same time displacements in the radial direction are limited.

Fig. 28. Distribution of the stress s_1 (s_z) at $\sigma_r\big|_{r=r_1} = p$, $\sigma_r^{(1)}\big|_{r=r_2} = -\sigma_r^{(2)}\big|_{r=r_2}$,

$$\sigma_\tau^{(1)}\big|_{r=r_2} = -\sigma_\tau^{(2)}\big|_{r=r_2} = f\sigma_n^{(1)}\big|_{r=r_2},\ u_x\big|_{r=r_3} = u_y\big|_{r=r_3} = 0$$

The corrosion damage disturbance of the strain state of the pipe as a whole corresponds to the disturbance of the stress state (Figures 29–34). The exception is only e_t (Figures 29, 30) that is tensile at the entire inner surface of the pipe, except for the damage edge where it becomes essentially compressive. This effect in principle corresponds to the effect of developing compressive strains inside the damage in a total compressive strain field. This effect was reaveled during full-scale pressure tests of pipes.

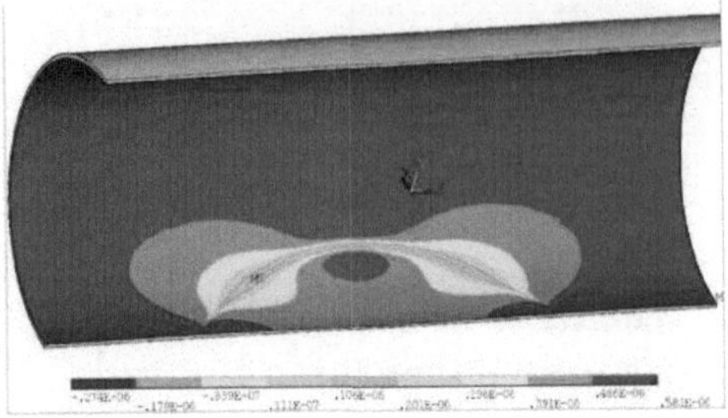

Fig. 29. Strains e_t at

$$\sigma_r\big|_{r=r_1} = p\,,\ u_x\big|_{r=r_2} = u_y\big|_{r=r_2} = 0$$

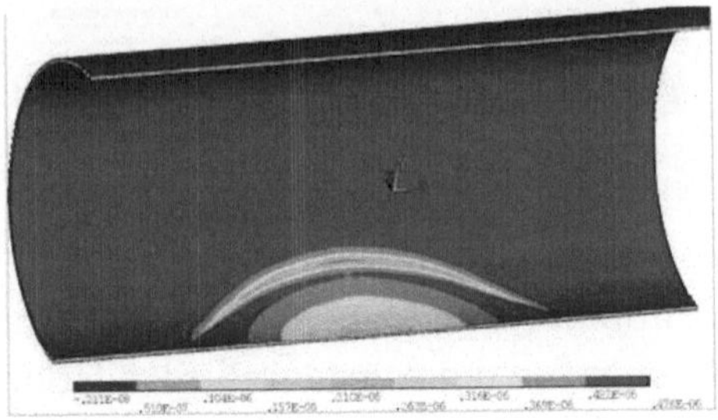

Fig. 30. Strains e_t at

$$\sigma_r\big|_{r=r_1} = p\,,\ u_x\big|_{r=r_2} = u_y\big|_{r=r_2} = u_z\big|_{r=r_2} = 0$$

Fig. 31. Strains e_r at

$$\sigma_r\big|_{r=r_1} = p\,,\ u_x\big|_{r=r_2} = u_y\big|_{r=r_2} = 0$$

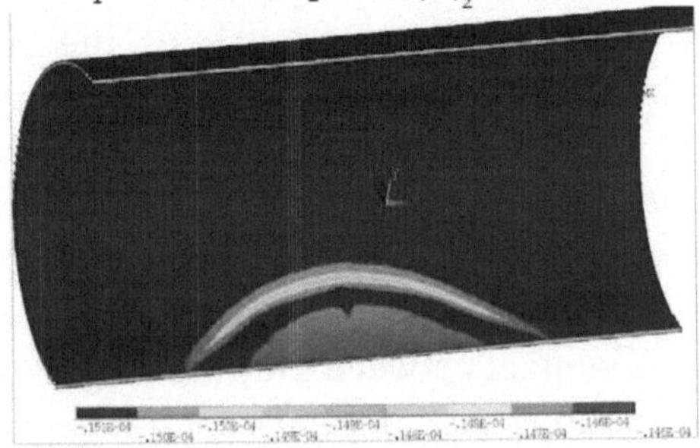

Fig. 32. Strains e_r at

$$\sigma_r\big|_{r=r_1} = p\,,\ u_x\big|_{r=r_2} = u_y\big|_{r=r_2} = u_z\big|_{r=r_2} = 0$$

Fig. 33. Strains e_z at

$$\sigma_r\big|_{r=r_1} = p\,,\ u_x\big|_{r=r_2} = u_y\big|_{r=r_2} = 0$$

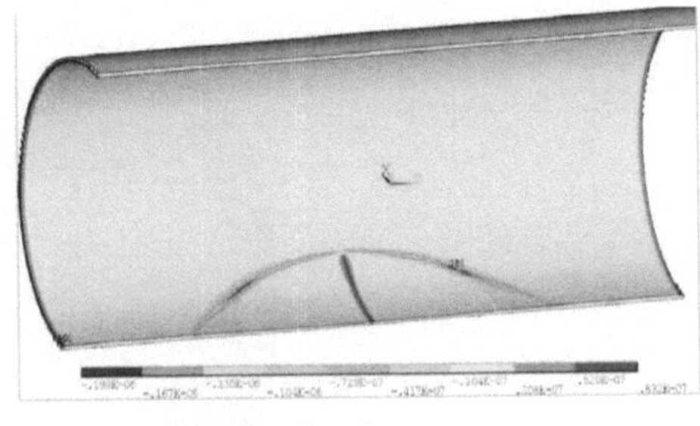

Fig. 34. Strains e_z at s_r

$$\sigma_r\big|_{r=r_1} = p\,,\ u_x\big|_{r=r_2} = u_y\big|_{r=r_2} = u_z\big|_{r=r_2} = 0$$

6. Influnce of different loading types on the stress-strain state of three- dimensional pipe models

Figures 35, 36 present the distributions of the principal stresses corresponding to the stresses s_t for different loading types in the absence of fixing of the outer surface of the pipe. From the comparison of these distributions it is seen that three loading types form three characteristic distribution types of the stresses $\sigma_{ij}^{(p)}$, $\sigma_{ij}^{(T)}$, $\sigma_{ij}^{(p+T)}$ such that according to (10) $\sigma_{ij}^{(p+T)} = \sigma_{ij}^{(p)} + \sigma_{ij}^{(T)}$.

Fig. 35. Distribution of the stress s_1 (s_t) in the absence of the outer surface fixing for

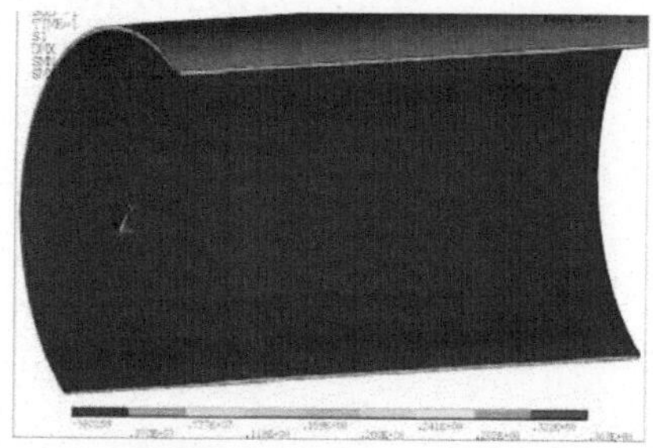

Fig. 36. Distribution of the stress s_1 (s_t) in the absence of the outer surface fixing for $\left|T_{r_1} - T_{r_2}\right| = \Delta T$

A comparative analysis of the stress distributions along the assigned paths shows that at the corrosion damage center (path 2) there is an almost two-fold increase of the stresses (s_t), as compared to the surface of the pipe without damage (path 1). The disturbing effect of corrosion damage (path 6) on the stress state is clearly seen.

Figures 37–39 plot the distributions of the principal stresses corresponding to the stresses st for different loading types when displacements are absent along the x and y axes of the outer surface of the pipe $u_x|_{r=r_2} = u_y|_{r=r_2} = 0$ and along the z axis at the right end $u_z|_{z=L} = 0$ when friction is present at the inner surface $\tau_{rz}|_{r=r_1} \neq 0$. From the comparison of these figures it is possible to single out several characteristic distribution types of the stresses $\sigma_{ij}^{(p)}$, $\sigma_{ij}^{(\tau)}$, $\sigma_{ij}^{(T)}$, $\sigma_{ij}^{(p+\tau)}$, $\sigma_{ij}^{(p+T)}$, $\sigma_{ij}^{(p+\tau+T)}$ related by (10).

Figures 1.37–1.38 illustrate a noticeable influence of the viscous fluid (oil) pipe wall friction (s $^{(t)}$) on the s $^{(p+t)}$ formation. From Figure 39 it is seen that temeprature stresses are dominant, exceeding by no less than 2-3 times the stresses developed by the action of s_r $r = r = p$ =4 MPa, t_{rz} $r = r$ = t_0 =260 Pa. In view of the fact that the temperature difference 1 1 Tr - Tr = DT =20°C exerts a dramatic influence on the formation of the stress state of the pipe, the distributions of $\sigma_{ij}^{(p+T)}$ and $\sigma_{ij}^{(p+\tau+T)}$ are qualitatively similar to the $\sigma_{ij}^{(p+\tau+T)}$ distribution, slightly differing in numerical values.

Fig. 37. Distribition pf the stress s_1 ($s^{(p)}$) at *ux*

Fig. 38. Distribution of the stress $\sigma_1\,(\sigma_{ij}^{(p+\tau)})$ at $u_x\big|_{r=r_2}=u_y\big|_{r=r_2}=0$, $u_z\big|_{z=L}=0$ for $\sigma_r\big|_{r=r_1}=p$, $\tau_{rz}\big|_{r=r_1}=\tau_0$

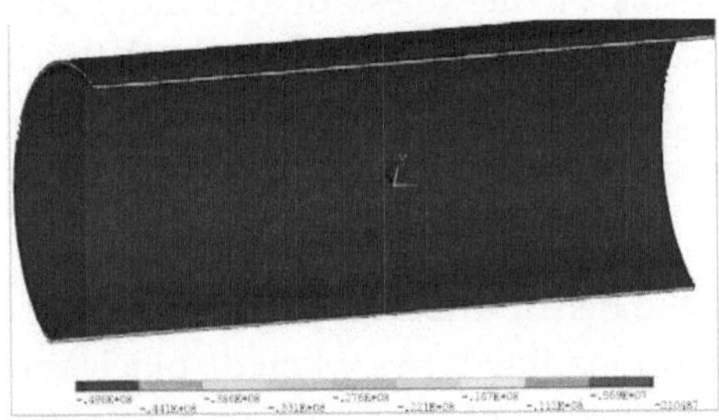

Fig. 39. Distribution of the stress $\sigma_1\,(\sigma_{ij}^{(p+\tau+T)})$ at $u_x\big|_{r=r_2}=u_y\big|_{r=r_2}=0$, $u_z\big|_{z=L}=0$ for $\sigma_r\big|_{r=r_1}=p$, $\tau_{rz}\big|_{r=r_1}=\tau_0$, $\left|T_{r_1}-T_{r_2}\right|=\Delta T$

A comparative analysis of the stress distributions shows that at the corrosion damage center the stresses grow (almost two-fold increase for s_t) in comparison with the surface of the pipe without damage.

7. Conclusion

Within the framework of the investigations made, the method for evaluation of the influence of the process of friction of moving oil on the damage of the inner surface of the pipe has been developed. The method involves analytical and numerical calculations of the motion of the two-and three-dimensional flow of viscous fluid (oil) in the pipe within laminar and turbulent regimes, with different average flow velocities at some internal pipe pressure, in the presence or the absence of corrosion damage at the inner surface of the pipe.

The method allows defining a broad spectrum of flow motion characteristics, including: velocity, energy and turbulence intensity, a value of tangential stresses (friction force) caused by the flow motion at the inner surface of the pipe.

The method for evaluation of the stress-strain state of two-and three-dimensional pipe models as acted upon by internal pressure, uniformly distributed tangential stresses over the inner surface of the pipe (pipe flow friction forces), and temperature with regard to corrosion-erosion damages of the inner surface of the pipe has been developed, too. For finite-element pipe models with boundary conditions of type (1)–(7) mainly the circumferential stresses, being the largest, were considered.

The meth of allows defining the variation in the values of the tensor components of stresses and strains in the pipe with corrosion damage for assigned pipe fixing under individual loading (temperature, pressure, fluid flow friction over the inner surface of the pipe) and their different combinations.

References

1. Ainbinder A.B., Kamershtein A.G. Strength and stability calculation of trunk pipelines. M: Nedra, 1982. – 344 p.
2. Borodavkin P.P., Sinyukov A.M. Strength of trunk pipelines. M: Nedra, 1984. – 286 p.
3. Grachev V.V., Guseinzade M.A., Yakovlev E.I. et al. Complex pipeline systems. M: Nedra, 1982. – 410 p.
4. Handbook on the designing of trunk pipelines / Ed, by A.K. Dertsakyan. L: Nedra, 1977. – 519 p.
5. Kostyuchenko A.A. Influence of friction due to the oil flow on the pipe loading / A.A. Kostyuchenko, S.S. Sherbakov, N.A. Zalessky, P.A. Ivankin, L.A. Sosnovskiy // Reliability and safety of the trunk pipeline transportation: Proc. VI International Scientific-Technical Conference, Novopolotsk, 11–14 December 2007 / PSU; eds: V.K. Lipsky et al. – Novopolotsk, 2007 a. – P. 76-78.
6. Kostyuchenko A.A. Wall friction in the turbulent oil flow motion in the pipe with corrosion defect / A.A. Kostyuchenko, S.S. Sherbakov, N.A. Zalessky, P.S. Ivankin, L.A. Sosnovskiy // Reliability and safety of the trunk pipeline transportation: Proc. VI International Scientific-Technical Conference, Novopolotsk, 11–14 December 2007 / PSU; eds: V.K. Lipsky et al. –Novopolotsk, 2007 b. – P. 78-80.
7. Launder B.E., Spalding D.B. Mathematical Models of Turbulence. London: Academic Press, 1972.
8. Mirkin A.Z., Usinysh V.V. Pipeline systems: Handbook Edition. M: Khimiya, 1991. – 286 p.

9. O'Grady T.J., Hisey D.T., Kiefner J. F. Pressure calculation for corroded pipe developed // Oil & Gas J. 1992. Vol. 42. – P. 64-68.
10. Ponomarev S.D. Strength calculations in engineering industry / S.D. Ponomarev, V.D. Biderman, K.K. Likharev, V.M. Makushin, N.N. Malinin, V.I. Fedosiev. M: State Scientific-Technical Publishing House of Engineering Literature, 1958. Vol. 2. – 974 p.
11. Rodi W. A new algebraic relation for calculating the Reynolds stresses //ZAMM 56. 1976.
12. Sedov L.I. Continuum mechanics: in 2 volumes. 6th edition, Saint-Petersburg: Lan', 2004. 2nd vol.
13. Seleznev V.E., Aleshin V.V., Pryalov S.N. Fundamentals of numerical modeling of trunk pipelines / Ed. by V.E. Seleznev. – M: KomKniga, 2005. – 496 p.
14. Sherbakov S.S. Influence of fixing of a pipe with a corrosion defect on its stress-strain state / S.S. Sherbakov, N.A. Zalessky, P.A. Ivankin, V.V. Vorobiev // Reliability and safety of the trunk pipeline transportation: Proc. VI International Scientific-Technical Conference, Novopolotsk, 11–14 December 2007 / PSU; eds: V.K. Lipsky et al. – Novopolotsk, 2007 a. – P. 52-55.
15. Sherbakov S.S. Modeling of the three-dimensional stress-strain state of a pipe with a corrosion defect under complex loading / S.S. Sherbakov, N.A. Zalessky, P.S. Ivankin, L.A. Sosnovskiy// Reliability and safety of the trunk pipeline transportation: Proc. VI International Scientific-Technical Conference, Novopolotsk, 11–14 December 2007 / PSU; eds: V.K. Lipsky et al. – Novopolotsk, 2007 b. – P. 55-58.
16. Sherbakov S.S. Modeling of the stress-strain state of a pipe with a corrosion defect under complex loading / S.S. Shcherbakov, N.A. Zalessky, P.S. Ivankin // X Belarusian Mathematical Conference: Abstract of the paper submitted to the International Scientific Conference, Minsk, 3–7 Novermber 2008 – Part 4. – Minsk: Press of the Institute of Mathematics of NAS of Belarus, 2008. – P. 53- 54.
17. Sherbakov S.S. Influence of wall friction in the turbulent oil flow motion in the pipe with a corrosion defect on the stress-strain state of the pipe / S.S. Sherbakov // Strength and reliability of trunk pipelines (Abstracts of the papers submitted to the International Scientific-Technical Conference "MT-2008", Kiev, 5–7 June 2008). – Kiev: IPS NAS Ukraine, 2008. – P.120-121.
18. Sosnovskiy L.A. Modeling of the stress-strain state of pipes of trunk pipelines with corrosion defects with regard to pressure, temperature, and interaction between the oil flow and the inner surface / L.A. Sosnovskiy, S.S. Sherbakov // Strength and safety of trunk pipelines (Abstracts of the papers submitted to the InternationalScientific-Technical Conference "MT-2008", Kiev, 5–7 June 2008). – Kiev: IPS NAS Ukraine 2008. – Pp. 107-108.

Part 2

Lubrication Tests and Biodegradable Lubricants

6

Experimental Evaluation on Lubricity of RBD Palm Olein Using Fourball Tribotester

1. Introduction

Tribology is defined as "the science and technology of surface interacting in motion". Thus it is important for us to understand the surface interaction when they are loaded together as to understand the tribology process occurring in the system. The physical, chemical and mechanical properties not only cause the effects to the surface material in tribology behavior but also the near surface material. Apart from that, on the surface of the bulk material, lies a layer formed as a result from the manufacturing process. This deformed layer is covered by a compound layer resulting of chemical reaction of metal with the environmental substance such as air. In addition, the machining process such as cutting lubricants to be trapped may also cause the deformed regions of the surface. The regions on the surface material can critically affect both friction and wear of metals. In addition, the forces which arise from the contact of solid bodies in relative motion also affect both friction and wear. Thus, it is important for us to understand the mechanics contact of solid bodies in order to evaluate the friction and wear on solid bodies. Solid bodies are subjected to an increasing load deform elastically until the stress reaches a limit or maximum yield stress then deform plastically (Gohar and Rahnejat, 2008).

Friction is known as resistance to motion. Friction can be categorized into five types; which are dry friction, fluid friction, lubricated friction, skin friction and internal friction. The friction forces are divided into two types; static friction force which is required to initiate sliding, and kinetic friction force which is required to maintain sliding. Coefficient of friction is known as the constant of proportionality in which the typical two materials may be similar or dissimilar, sliding against each other under a given set of surfaces and environmental conditions (Arnell and Davies, 1991).

The first laboratory test device for determining lubricant quality was known as fourball tribotester is proposed by Boerlage in the year of 1993 (Ivan, 1980). The concept of friction for this machine is three stationary balls pressed against a rotating ball. The quality and the characteristics of the lubricant were established by the size of the wear scar or the seizure load and the value of friction obtained. The main elements of fourball machine are vertical driving shaft which hold the moving ball at the lower end with conical devices. Besides that, three stationary balls which are fixed by a conical ring and lock nut are pressed by the moving ball. The stationary ball holder is mounted on an axial bearing so that it can rotate and displace in the vertical direction freely. In addition, a lever device is used to apply load on stationary balls. The friction occurring on the fixed stationary balls by the rotating ball is transmitted by means of a lever to the measuring device. The wear is viewed based on the size of the wear scar on the stationary balls. 12.7mm diameter of balls is usually used. These are specially processed to ensure high dimensional accuracy as well as uniform hardness and surface quality. The tested lubricant was immersed into the stationary balls cup hold with desire volume. Apart from that, the speed for rotating ball depends on the type of machine and the experiment conditions. There are several standards and specifications for fourball machine: such as Socialist Republic of Romania State Standard 8618-70; FTM no. 791 a/6503; ASTM D2596-67 and DIN 51350 (Ivan, 1980).

Boundary lubrication is defined as a condition of lubrication in which the friction and wear between two surfaces in relative motion are determined by the properties of lubricant. Lubrication is critical for minimizing the wear in mechanical systems that operate for extended time period. Developing lubricants that can be used in engineering systems without replenishment is very important for increasing the functional lifetime of mechanical components. The additives usually to be added in to the base oil to improve its performance. Joseph Perez stated that the number of additives and the amount present depends on the application (Joseph and Waleska, 2005). They are selected to enhance the base oil performance so that they will meet the system requirement.

The increasing and wide usage of petro and synthetic based oil overwhelm the lubricant industry because the major damage to the environment and the rise of concern about health and environmental damage caused by the mineral oil based lubricant; have created a growing worldwide trend of promoting vegetable oil as based oil in industries. Biodegradable oils are becoming an important alternative to conventional lubricants as a result of awareness of ecological pollution and their detrimental effects on our lives. The use of vegetable oils in industrial sector is not a new idea. They had been used in the construction of monuments in Ancient Egypt (Nosonovsky, 2000). Vegetable oil with high stearic acid content is considered to be potential candidates as the substitute for conventional mineral oil based lubricants because they are biodegradable and non toxic. Besides that, they have better intrinsic boundary lubricant properties because of the presence of long chain fatty acids in their composition (Carcel and Palomares, 2004). Other advantages include very low volatility due to the high molecular weight of triglyceride molecule and excellent temperature viscosity properties. Their polar ester groups are able to adhere to metal surface and therefore possess good lubricating ability. In addition, vegetable oils have high solution power for polar contaminants and additive molecules (Sevim et al, 2006). Vegetable oils show good lubricating abilities as they give rise to low coefficient of friction. However, many researchers report that although the co-efficiency of friction is low with vegetable oil as boundary lubricant, the wear rate is high. This behavior is possible due to the chemical attack on the surface by the fatty acid present in vegetable oil. The metallic soap film is rubbed away during sliding and producing the non-reactive detergents increase in wear (Bowden and Tabor, 2001).

In western country, the common vegetable oils that have been widely used in the tribology test are sunflower oil, rapeseed oil and corn oil. For this research, the authors used RBD palm olein as test oil and evaluated its friction and wear performance using fourball tribotester. Nowadays, palm oil has been widely tested for engineering applications. The potential of palm oil as fuels for diesel engines (Kinoshita et al, 2003; Bari et al, 2002), hydraulic fluid (Wan Nik et al, 2002), and lubricants (Syahrullail et al., 2011) has been confirmed in previous studies. In addition, Malaysia is one of the world's largest palm oil producers.

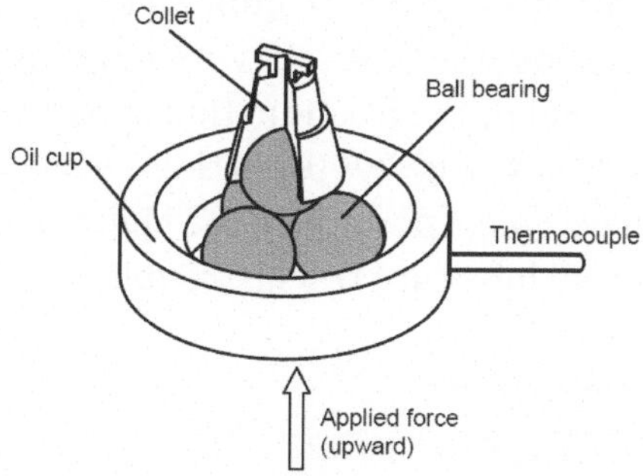

Fig. 1. A schematic sketch of the fourball tribotester

Throughout all the previous studies, the characteristics of RBD palm olein were investigated using fourball tribotester. The objective of this experiment is to study the lubricity characteristics of vegetable oils compared to the petroleum based oil. RBD palm olein and additive free paraffinic mineral oil were used as lubricants in this experiment. RBD palm olein is a refined, bleached and deodorized palm olein product and it exists in liquid state at room temperature. Fourball tester was used in this experiment to evaluate the lubricity of those lubricants. The lubricity performance of RBD palm olein and non-aditive paraffinic mineral oil were compared mutually. The experiments were carried out at the temperature of 75°C for one hour duration. Besides that, the load applied on the fourball tester was 40 kg (392.4N). Apart from that, the speed of spindle was set to 1200 rpm. At the end of the experiments, the evaluations of lubricants focused on the friction and wear of each lubricant. From the experiments, the authors confirmed that RBD palm olein showed satisfactory lubrication performance as compared to additive free paraffinic mineral oil, especially in terms of friction reduction.

2. Experimental procedures

2.1 Experimental apparatus

The fourball wear tester machine was first described by Boerlage to have acquired the status of an established institution in the fundamental investigation of lubricants characteristics (Boerlage, 1933). In this research, the fourball wear tester was used. This instrument uses four balls, three at the bottom and one on top. The bottom three balls are held firmly in a ball pot containing the lubricant under test and pressed against the top ball. The top ball is made to rotate at the desired speed while the bottom three balls are pressed against it. The important components are ballpot (oil cup) assembly, collet, locknut adaptor and standard steel balls. The components surface needs to be clean with acetone before the tests. The amount of lubricant test is 10 ml.

2.2 Test lubricants

The tested lubricants for this experiment were RBD palm olein and additive free paraffinic mineral oil (written as paraffinic mineral oil). The RBD is an abbreviation for refined, bleached and deodorized. As shown in Figure 2, RBD palm olein is the liquid fraction that is obtained by the fractionation of palm oil after crystallization at a controlled temperature (Pantzaris, 2000). In these experiments, a standard grade of RBD palm olein, which was incorporated in the Malaysian Standard MS 816:1991, was used. The amount for all lubricant tests is 10 ml.

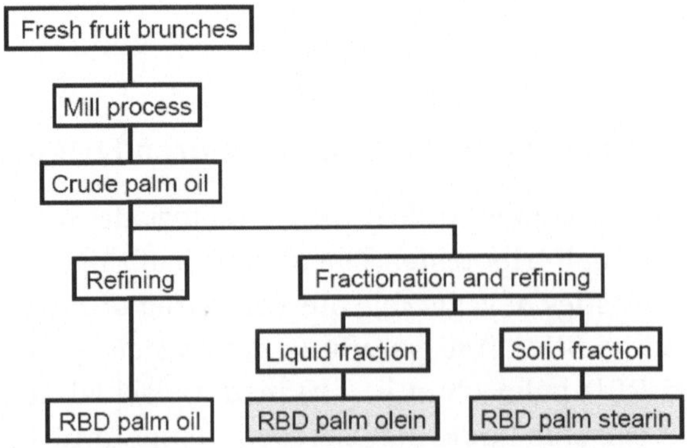

Fig. 2. Refining method of RBD palm olein

2.3 Experimental procedures

The wear tests were carried out under the ASTM method D-4172 method B with the applied load of 392.4 N (40 kg) at a spindle speed of 1200 revolution per minute (rpm). The experiment was carried out for duration of one hour and conducted under the temperature of 75 degree Celsius. The three bottoms stationary balls in the wear test were evaluated the average diameter of the circular scar formed. Besides that, the lubricating ability of the RBD palm olein was also being evaluated based on the friction torque produced compared with the additive free paraffinic mineral oil. All parts in fourball (upper ball, lower balls and oil cup) were cleaned thoroughly using acetone and wiped using a fresh lint free industrial wipe. There should not be any trace of solvent remain when the test oil was introduced and the parts were assembled. The steel balls were placed into the ballpot assembly and to be tightened using torque wrench. This purpose was to prevent the bottoms steel balls from moving during the experiment. The top spinning ball was locked inside the collector and tightened into the spindle. 10 ml of test lubricant (RBD palm olein or paraffinic mineral oil) was to be poured into the ballpot assembly. Apart from that, researcher should note or observe that this oil level filled all the voids in the test cup assembly. The ballpot assembly components were installed onto the non-friction disc in the four-ball machine and avoided shock loading by slowly applying the test load up to 392.4 N. After that, the lubricant used was heated up to 75 degree Celsius. When the set temperature was reached, researcher started the drive motor which had been set to drive the top ball at 1200 rpm. For the duration of one hour, the heater was turn off and the oil cup assembly was removed from the machine. Then, the test oil in the oil cup was drained off and wear scar area was wiped using tissue. The wear scars on the bottom balls were put on a special base of a microscope that has been designed for the purpose. All tests were repeat several times.

3. Results and discussions

3.1 Density and viscosity

Density of fluids is defined as the unit of mass per volume. A laboratory experiment had been carried out to measure the density of RBD palm olein and paraffinic mineral oil. The result was shown in Table 1. Dynamic viscosity is a measure of the resistance of a fluid which is formed by either shear stress or tensile stress of the fluids. It is also known as the internal friction of the fluids. A viscometer was used to measure the viscosity for both lubricants. Viscometer rotor was immerged into the lubricants to evaluate it fluidity by turning the rotor for 99 seconds. The viscosity of the RBD palm olein and paraffinic mineral oil was shown in the Figure 3. The viscosity of both lubricants dropped as the temperature of the lubricants increase. The lesser the viscosity of the fluids, the easier the particles will move in the fluids.

Table 1. Properties of RBD palm olein and paraffinic mineral oil

Test oil	RBD palm olein	Paraffinic mineral oil
Density at 25ºC (kg/m3)	915	848
Flash point (ºC)	315-330	140-180
Pour point (ºC)	18-24	-20

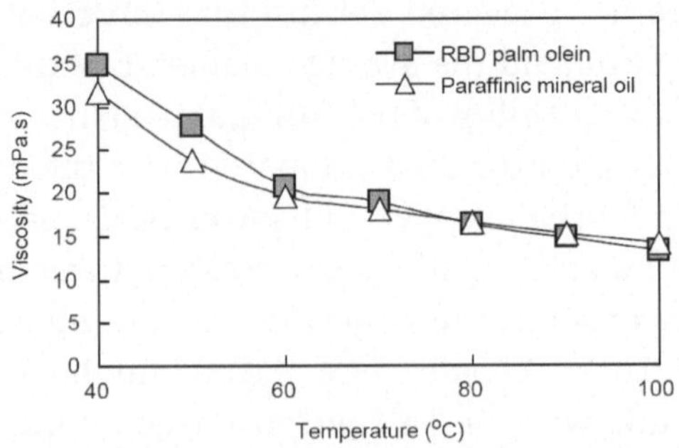

Fig. 3. Viscosity curves of RBD palm olein and paraffinic mineral oil

3.2 Friction

$$F = \mu \cdot N$$

The friction coefficient (μ) between two solid surfaces is defined as the ratio of the tangential force (*F*) which required sliding, and is divided by the normal force between the surfaces (*N*) (Jamal, 2008). Coefficient of friction for RBD palm olein and paraffinic mineral oil had been obtained using the relevant software. Figure 4 shows the value of coefficient of friction at steady state for both lubricants in the fourball tribotester. The coefficient of friction for RBD palm olein is lower than paraffinic mineral oil. As shown in Figure 5 the steady state friction torque for RBD palm olein is lower than paraffin mineral oil, thus the steady state coefficient of friction also shows the same trend of result. From the experiment, the value of coefficient of friction for RBD palm olein is 0.065 while the value of coefficient of friction for paraffinic mineral oil is 0.075.

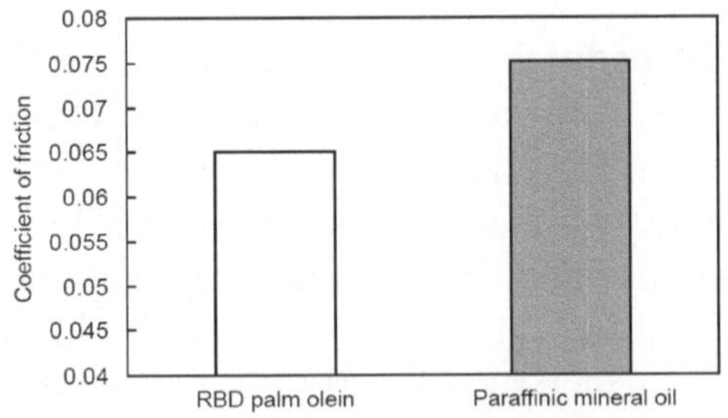

Fig. 4. Coefficient of friction for RBD palm olein and paraffinic mineral oil at steady state condition

Few series of wear tests had been conducted using fourball tribotester. Figure 4 illustrates the friction torque obtained for RBD palm olein and paraffinic mineral oil using fourball tribotester along the period of experiments. The trend of graph for both lubricants was similar to each other. The friction torque for both lubricants was increased along the period of experiments. In Figure 5 the friction torque of RBD palm olein is lower than paraffinic mineral oil. The value of friction torque at steady state for RBD palm olein and paraffinic mineral oil is 0.12 Nm and 0.14 Nm respectively. Based on the previous study, the long chain of fatty acids present in the palm oil has the potential to reduce the friction constraint (Abdulquadir and Adeyemi, 2008).

3.3 Wear

The wear scar on the surface of balls bearing was obtained and measured using the CCD microscope and its specific software. The measured wear scar diameter on the balls bearing

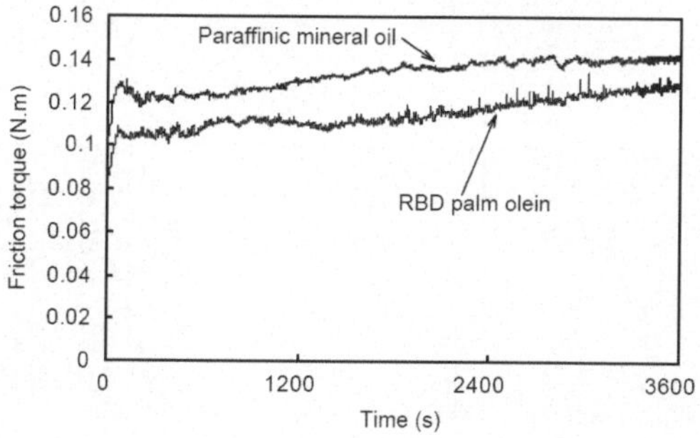

Fig. 5. Friction torque curves for RBD palm olein and paraffinic mineral oil

was recalculated to obtain the mean or average wear scar diameter for each lubricant test. Figure 6 illustrates the average wear scar diameter of fourball tribotester for RBD palm olein and paraffinic mineral oil. The average wear scar diameter measured for RBD palm olein is larger than paraffinic mineral oil in this experiment. RBD palm olein shows 0.828 mm and paraffinic mineral oil shows 0.764 mm in wear scar diameter. In addition, this result is totally opposite with the result of friction. The wear increases as the friction decreases as shown in Figure 4 and Figure 5. This due to the increased shear strength of the adsorbed oil on the surface of the balls and affected chemical attack on the surface by the fatty acid present in vegetable oil (Bowden and Tabor, 2001).

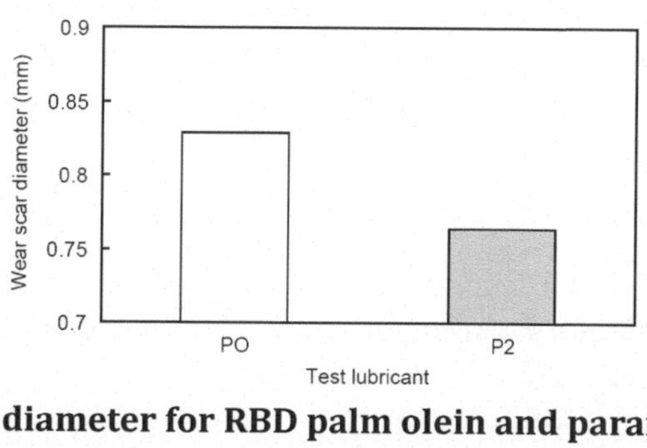

Fig. 6. Wear scar diameter for RBD palm olein and paraffinic mineral oil

Wear scar track that was lubricated with RBD palm olein and paraffinic mineral oil had been viewed and captured using the microscope. The enlargement of wear scar track for both tested oils is shown in Figure 6. The wear scar track on the ball bearing lubricated with RBD palm olein shows smoother surface than wear scar track lubricated with paraffin mineral oil on the surface of ball bearing. The wear scar worn on the ball bearing lubricated with paraffin mineral oil has more ploughed traces or grooves as the result of material transfer. The narrower and deeper of groove on the wear traces would be the sources of roughening the surface of ball bearing after the experiments (Meng and Jian, 2008).

Fig. 7. Observation of the wear scar condition for RBD palm olein and paraffinic mineral oil

4. Conclusion

The lubricating ability of RBD palm olein had been evaluated using the fourball tribotester. All the results were compared mutually with the additive free paraffinic mineral oil. For the reduction in friction, RBD palm olein shows better result compared to the additive free paraffinic mineral oil. RBD palm olein shows lower coefficient of friction and friction torque compared to the paraffinic mineral oil. This behavior is related to the long chain fatty acid in the RBD palm olein. However in wear, due to the increasing shear strength of the RBD palm olein on the surface of the balls, it shows larger wear scar diameter compared to the paraffin mineral oil. Besides that, from the observation of scar view using CCD microscope, the scar surface of balls lubricated with RBD palm olein looks smoother than paraffinic mineral oil.

References

1. Abdulquadir, B.A. and Adeyemi, M.B., 2008, "Evaluations of Vegetable Oil-Based as Lubricants for Metal-Forming Processes," Industrial Lubricant and Tribology, Vol. 60, pp.242-248.
2. Arnell, R.D., Davies, P.B. and Halling, J., 1991, "Tribology-Principles and Design Applications", Macmillan Education Ltd, First edition.
3. Bari, S., Lim, T.H. and Yu, C.W., 2002, "Effect of Preheating of Crude Palm Oil (CPO) on Injection System, Performance and Emission of a Diesel Engine", Renewable Energy, Vol. 27, pp.339-351.

4. Boerlage, G.D., 1933, "Four-ball Testing Apparatus for Extreme-pressure Lubricants," Engineering, Vol. 136, pp.46-47.
5. Bowden, F.P. and Tabor, D., 2001, "The Nature of Metallic Wear. The Friction and Lubrication of Solids," Oxford Classic Texts. New York: Oxford University Press; pp.285-98.
6. Carcel, A.D. and Palomares, D., 2004, "Evaluation of Vegetable Oils as Pre-Lube Oils for Stamping", Materials and Design, Vol. 26, pp.587-593.
7. Gohar, R and Rahnejat, H., 2008, "Fundamentals of Tribology", Imperial College Press. Ivan Iliuc, 1980, Tribology of Thin Layers", Elsevier Scientific Publishing Company.
8. Jamal Takadoum, 2008, "Materials and Surface Engineering in Tribology," John Wiley & Sons, Inc.
9. Joseph, M.P. and Waleska, C., 2005, "The Effect of Chemical Structure of Basefluids on Antiwear Effectiveness of Additives", Tribology International, Vol. 38, pp.321-326.
10. Kinoshita, E., Hamasaki, K. and Jaqin, C., 2003, "Diesel Combustion of Palm Oil Methyl Ester", SAE, 2003, Paper No. 2003-01-1929.
11. Meng Hua and Jian Li, 2008, "Friction and Wear Behavior of SUS 304 Austenitic Stainless Steel against AL2O3 Ceramic Ball under Relative High Load," Wear, Vol. 265, pp.799-810.
12. Nosonovsky, M., 2000, "Oil as a Lubricant in the Ancient Middle East", Tribology Online, Vol. 2-2, pp.44-49.
13. Syahrullail, S., Zubil, B.M., Azwadi, C.S.N. and Ridzuan, M.J.M., 2011. Experimental evaluation of palm oil as lubricant in cold forward extrusion process. International journal of mechanical Sciences, 53, 549-555.
14. Pantzaris, T.P., 2000, "Pocketbook of Palm Oil Uses," Malaysian Palm Oil Board.
15. Sevim, Z.E, Brajendra, K.S. and Joseph, M.P., 2006, "Oxidation and Low Temperature Stability of Vegetable Oil-Based Lubricants", Industrial Crops and Products, Vol. 24, pp.292-299.
16. Wan Nik, W.B., Ani, F.N. and Masjuki, H., 2002, "Thermal Performances of Bio-fluid as Energy Transport Media", The 6th Asia Pacific International Symposium on Combustion and Energy Utilization, Kuala Lumpur, Malaysia, pp.558-563.

7

Biodegradable Lubricants and Their Production Via Chemical Catalysis

1. Introduction

The primary purpose of this chapter is to describe the differences among biolubricants and petroleum-based lubricants, especially their production and physical and chemical properties. Established production methodology will be described, especially those using chemical catalysis that have been developed at the laboratories of the Petrobras Research Center (CENPES), in Rio de Janeiro, Brazil.

Today there is growing concern about the future availability of petroleum-based products. In addition, millions of tons of lubricants are dumped into the environment through leakage, exhaust gas and careless disposal. Some of these wastes are resistant to biodegradation and are threats to the environment. Thus, there are two major issues confronting the lubricant industries today: the search for raw materials that are renewable and products that are biodegradable.

The oleochemistry represents a significant challenge to biolubricants production by petroleum companies. All the required technologies from seed crushing to oil refining, fractionation and chemical transformation are in place. The main research emphasis has been placed on ways to produce biolubricants with suitable viscosity and liquid-state temperature range. In addition, these lubricants must not corrode the machinery they lubricate and they must be stable under the conditions of their use. These requirements eliminate many simple fatty acid esters. Saturated esters with long enough chains to not be too volatile or lacking in viscosity are solids in the temperature range required by many lubricant applications. Double bonds will lower their melting point but introduce instability to oxygen attack, especially for the typical polyunsaturated fatty acid found in most vegetable oils. Branching will reduce the melting point but such fatty acids are relatively rare in nature. The solution to these problems that will be described on this chapter emphasizes the use of Brazilian raw materials. The well-developed Brazilian program of biodiesel production from soybean and castor oils has led to the choice of ricinoleate esters as potential biolubricant ingredients.

2. Castor and its derivatives

Castor oil is produced in the seed of the castor oil plant, *Ricinus communis*, and has been used for medicinal purposes for many years. During the 20th century, a number of industrial uses were developed including its use as a lubricant (Azevedo & Lima, 2001).

Castor oil was introduced into Brazil by the Portuguese for use as in illumination and as a carriage shaft lubricant. The climate of Brazil is suitable for growing castor plants and it can be found today among the wild flora in many parts of Brazil as well as a drought resistant cultivated plant.

From its seeds industrialization is obtained, as main product, the oil (47%) and, as by- product, the castor waste that may be used as a fertilizer.

Castor oil posses unusual and has greater density, viscosity, ethanol solubility and lubricity compared with other vegetable oil. This oil also has a wide chemical versatility inside the industry, due to be used as raw material to the synthesis of a large amount of products.

Furthermore, we can obtain biodiesel from castor oil, which replaces the petroleum-derived diesel as fuel. Besides, this oil posses the unusually fatty acid, ricinoleic acid, which makes about 90% of its composition. Ricinoleic acid is similar to the common fatty acid, oleic acid, except it has a hydroxyl group on the 12th carbon of its 18 carbon chain. Like oleic acid, ricinoleic acid has a *cis* double bond between the 9th and 10th carbon, as can be seen in figure 1.

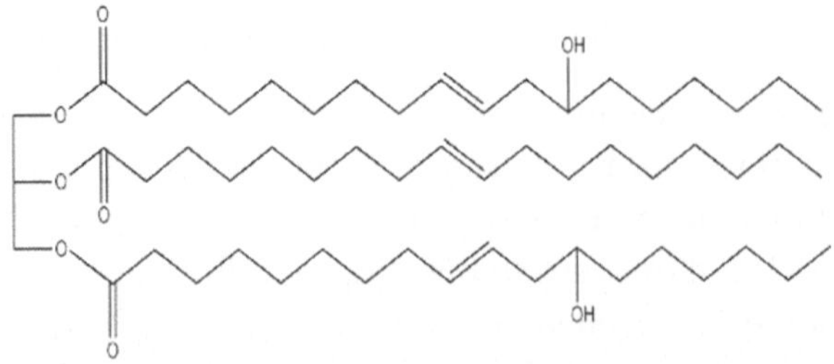

Fig. 1. Castor oil molecular structure (*Ricinus Communis*)

Table 1 presents the main physical-chemical characteristics of this oil.

The hydroxyl group of castor oil increases its polarity and makes it a better solvent for lubricant additives than other vegetable or mineral oils. Besides, castor oil presents high viscosity and low pour point, but its viscosity index is lower than the others vegetable oils, which means that its viscosity changes more with temperature than the other oils.

Table 1. Typical castor oil physical-chemical characteristics

Property	Value
Iodine Index	84-88
Viscosity at 100°C	20.00 cSt
VI (Viscosity Index)	90
Melting Point	-23°C
Ricinoleic Acid Content	90%
Linoleic Acid Content	4.2%
Oxidative Stability by RPVOT	25 Min.
RPVOT: Rotary pressure vessel oxidation test.	

Castor oil has been used on the manufacturing of more than 800 products, ranging from bullet-proof glasses, contact lenses, lipsticks, metal soaps, special engine and high rotation reactors lubricants, high resistance plastics, polyurethanes, etc. Its odd properties give lubricity to the mineral diesel, like sulfur, becoming a special oil in the current world market.

The major castor seeds and oil producing nations in order of their production are India, China and Brazil. Germany and Thailand are the greatest castor beans importers (94%), but the United States consumes the most castor oil.

The state of Bahia produces 85% of Brazil's production of castor oil, being together with the state of Minas Gerais, the states where are located the main oil extraction companies. Brazil produces about 160,000 metric tons of beans per year. As the internal consumption of castor oil is small (10,000-15,000 metric tons per year), there is an excess of about 45,000-50,000 metric tons per year for export.

3. Base oils

The term "base oils" refers to the various oils used in the world's technological applications. This chapter will focus on lubricant oils. The base oils are the larger proportion constituents at the lubricants formulations and most of them are derived from petroleum. They can be classified as mineral or synthetic oils, depending on their production history (Lastres, 2003). The first known lubricants used by humans were animal and vegetable based oils. In the 19th Century, the natural triglycerides were replaced by petroleum based oils, called mineral oils. In some lubricants applications, certain performance standards are required that cannot be met by conventional mineral oils. Alternate processes have been devised for their production usually to achieve greater durability or lower environmental impact. Vegetable oils are less expensive than minerals and are produced from renewable resources.

Mineral oils are produced through the petroleum distillation and refining. They are classified in paraffinics, naphthenics and aromatics, depending on the hydrocarbon type predominant in its composition. They possess 20 to 50 carbon atoms, on average, per molecule, and these can be paraffinic chains (linear or branching alkanes), naphthenic chains (cicloalkanes with side chains) or aromatic chains (alkyl benzenes), as illustrated on the figure 2.

The paraffinic base oils owe high pour point and viscosity index. To produce them, the dewax step is very important and the product, even dewaxed, still needs to be additivated with a pour point depressor to avoid the wax crystals growth at low temperatures and to reduce the product flow temperature.

The naphthenic base oils possess higher levels of carbons in cycle chains (naphthenics) than the paraffinics. The cut of a naphthenic petroleum has low linear wax levels and does not need to be dewaxed. Its pour point can achieve -51°C (base oil NH-10). On the other hand, they have low VI values (becoming very hard their usage on the engine oil formulations). They are more used on the formulations of cutfluids, shock absorbers oils and as isolation fluid to electrical transformers. The aromatic oils are used as extensor oils at the rubber industry.

Oil type	Carbon chains type
Paraffinics	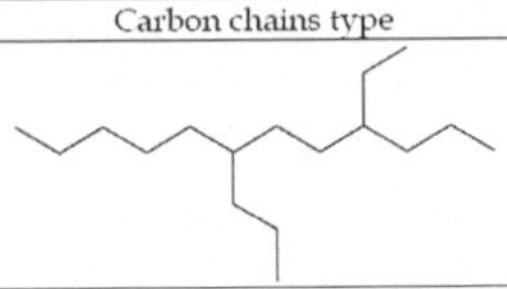
Naphthenics	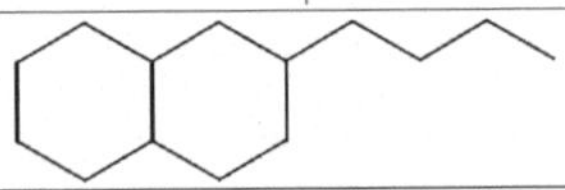
Aromatics	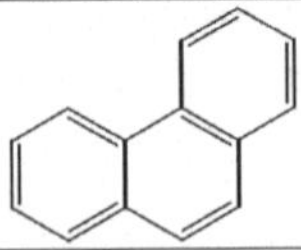

Fig. 2. Structure of the mineral oils composition

Table 2. Base oils API classification

CATEGORY	SATURATES (1)		SULFUR, %P (2)	VISCOSITY INDEX (3)
GROUP I	< 90	and / or	> 0.03	80 - 120
GROUP II	≥ 90	and	≤ 0.03	80 - 120
GROUP III	≥ 90	and	≤ 0.03	> 120
GROUP IV	POLYALPHAOLEFINS (PAO)			
GROUP V	OTHER BASE OILS NOT INCLUDED ON THE GROUPS I, II, III and IV			
(1) ASTM D 2007 (2) ASTM D 2622 or ASTM D 4294 or ASTM D 4927 or ASTM D 3120 (3) ASTM D 2270				

Mineral base oils can also be classified by the production process. The most common is the solvent extraction, or conventional process, where compounds like aromatics and compounds that contain heteroatoms, as nitrogen and sulfur, are removed, increasing the VI and improving the products stability. This process also includes dewax steps, in order to reach the specified pour point, and hydrotreatment, to improve the products specifications. The non conventional process includes more severe steps of hydrocracking, where the molecules are cracked and saturated, with very stable and high VI final products.

On the other hand, synthetic base oils are produced through chemical reactions. Approximately 80% of the synthetic lubricant world market is composed by: polyalphaolefins (45%), organic esters (25%) and polyglycols (10%) (Murphy et al., 2002). The most used synthetic base oils are the polyalphaolefins, and the synthetic oils have as an advantage, in general, higher thermal and oxidative stability, better low temperature properties and lower volatility when compared to mineral oils. However, these base oils are more expensive than mineral oils.

Applications that require high level of biodegradability need to use vegetable based synthetic base oils.

Regarding the automotive oils, the American Petroleum Institute, API, classifies the base oils in five categories as illustrated on the table 2.

The lubricant's performance is evaluated by their friction reduction, oxidation resistance, deposits formation minimization, corrosion and wear avoiding abilities. The main problem with lubricants is related to the oil degradation and its contamination by the engine combustion by-products (automotives). Thus, the main causes of engine bad working, regarding the lubricant quality, are due to deposit formation, viscosity increase, high consumption, corrosion and wear.

Deposit formation occurs when the detergent/dispersant power of the lubricant is not enough to keep the contaminants in suspension. The oil thickness results from the lubricant oxidation and the insolubles material accumulation. The viscosity increases due to the oxygenated compounds polymerization and to the insoluble products in suspension, derived from the irregular fuel burning. The sulfur level in the diesel mays cause corrosion and wear on the cylinders and rings, because of the sulfur acids or organic acids attack on the iron surfaces. To avoid this attack, lubricants with a good alkaline reserve must be used. To minimize such problems, lubricants are obtained from the mixture between base oils and additives. These additives have antioxidant, antiwear, detergent and dispersant, and others functions. Therefore, to design a lubricant to play all these roles is a hard task which involves a careful evaluation of the base oils and additives properties.

4. Biolubricants

The world final lubricants market is about 38,000,000 tons/year (Whitby, 2005). The US market is about 9.5 millions tones, from which 32% are discarded on the environment (Lal & Carrick, 1993). On the other hand, the European biodegradable lubricants market is 172,000 tons/year, concentrated on Germany and Scandinavian countries (Whitby, 2006).

From the 1.3 million tons German lubricants market, 53% are collected as used oil, which is equivalent to 100% of all oil collection of the several applications. These used oils are recycled or used as thermal energy source. The remainder is lost to the environment as leakages, total loss applications or specific systems. Only 5% of all lubricants from the German market are biodegradable (Wagner et al., 2001). To increase this market, one must increase the acceptance and the trust on the biodegradable lubricants and decrease its price. Nearly 13% (Europe) and 32% (USA) of all commercialized lubricants return to the environment with properties and appearance modified (Bartz, 1998). These are used on total loss wear contacts, approximately 40,000 tons/year in Germany, and on circulation systems, which are not collected neither disposed. Besides, one must take account of the lubricants from leakages and the remainder amounts in filters and recipients. Thus, the German environment is exposed to nearly 150,000 tons/year, based on the 13% previously cited. A calculation based on the current lubricants consumption in Germany and on the discard rates for the different lubricants results in nearly 250,000 tons/year. Including the not defined amount (leakages, etc.), the lubricants discarded amount on the German environment may reach 300,000 tons/year. Taking account of the lubricants market share represented by Germany, as well as the fact that in many places around the world the collect and recycling rates of used lubricants are lower than in Europe, the total amount of lubricants returning to the environment is about 12 million tons/year.

Only 10-50% of the lubricants used on the world market are recycled (Kolwzan & Gryglewicz, 2003). The remainder, which represents millions of tons, is disposed irreversibly on the environment through leakages, oil-water emulsions, components exhaust gases, etc. Some of them are carcinogenic and resistant to biodegradation, representing a serious menace to the environment. One of the solutions to modify this situation is replacing mineral oils with biodegradable synthetic lubricants.

In the last decades, there has been an increased worldwide concern about the environmental impact from the petroleum derivatives usage. Although only approximately 1% of all consumed petroleum be used on the lubricants formulations, the most part of these products are disposed in the environment without any treatment and this concern has driven the biodegradable lubricants development.

The pollution potential of the mineral oil is extremely high. For example, 1 liter of mineral oil contaminates 1 million liters of water for the human consumption (Ravasio et al., 2002). Regarding the 2 strokes engines (currently, the main use of biolubricants), the lubrication mechanism results in the release of unburned oil, together with exhaust gases, promoting the possibility of environmental pollution. Furthermore, when using these engines in rivers, lakes or oceans, the unburned oil, released in the water, can become a possible pollution source. Tractors, agricultural machines, chain-saws, and other forest equipments, may pollute forests and rivers, as well due to the unburned released oil.

Measures to reduce the environmental impact of lubricants, that means to eliminate or decrease the problems caused by lubricant contact, are driven by the following forces: environmental facts, public awareness, government rules, market globalization and economic incentives.

A biolubricant is a biodegradable lubricant. A substance is called biodegradable when it presents the proved capacity of being decomposed within 1 year, through natural biological processes in carbonaceous land, water and carbon dioxide (Whitby, 2005).

In general, biodegradability means a lubricant trend to be metabolized by microorganisms within 1 year. When it is complete, it means that the lubricant has essentially been back to Nature, but when it partially decomposes, one or more lubricant compounds are not biodegradable.

Some of the readily biodegradable lubricants are based on pure unmodified vegetable oils (Wagner et al., 2001), that present a biodegradability of about 99% (CEC L-33-A-93) (Birova et al., 2002). In Europe there is a predominance of sunflower and rapeseed oils, which are esters of glycerin and long chain fatty acids (triglycerides). The fatty acids are specific for each plant, being variable. The fatty acids found in natural vegetable oils differ in chain length and in their double carbon bond number. Moreover, function groups may be present. Natural triglycerides are highly biodegradable and efficient as lubricants. However, their thermal, oxidative and hydrolytic stabilities are limited. Thus, pure vegetable oils are used only on applications with low thermal requirements, as unmolding and chain-saws.

The reasons for the thermal and oxidative instabilities of the vegetable oils are the double bonds in the fatty acid molecule and the group β-CH in the alcohol counterpart (figure 3). Double bonds are especially reactive and react immediately with the air oxygen, while the hydrogen β atom is easily eliminated from the molecule structure. This results in the ester breakage in olefins and acids. A further weak point of the esters is its trend to undergo hydrolysis in the presence of water. Chemical modifications may improve the thermal, oxidative and hydrolytic stabilities of the vegetable oils. The most important modifications occur on the carboxyl groups of the fatty acids, approximately 90%, while oleochemical reactions on the fatty acid chain are approximately 10%.

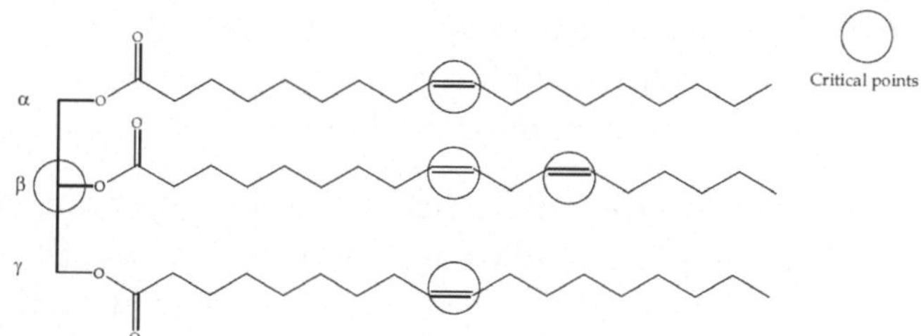

Fig. 3. Typical structure of a vegetable oil and its instability critical points

Esters, similar substances to triglycerides in terms of chemical structure, are excellent replacements for mineral oils, which possess only 20% of biodegradability (CEC-L-33-A-93). The organic esters are a growing interest in the base lubricants industry and its advantages compared to mineral base oil are (Lal & Carrick, 1993):

- Low toxicity;
- Higher biodegradability;
- Obtained from renewable sources;
- High flash point;
- Low volatility;
- High additives solvency power;
- High added value;
- Good lubricity (due to molecule polarity);
- High viscosity index (VI), due to the double bonds and the molecule linearity; However, the main disadvantages of these compounds are:
- Oxidative instability;
- Hydrolitic instability;
- Low temperature properties.

These disadvantages can be minimized by additives, but the biodegradability, the toxicity and the price can be endangered. Thus, the chemical synthesis of these compounds seems to be the best choice to overcome these disadvantages.

The additives used traditionally are antioxidant, antiwear, anticorrosion, etc. These agents have low biodegradability. However, the additives industry has increased efforts on the development of biodegradable additives.

5. Esters

Organic esters have a wide diversity of applications in the lubricant industry because of the growing awareness of health and environment beneficial aspects, besides the benefits from better products performance: chain-saw, drilling fluids, food industry equipments, hydraulic fluids, boat engines, 2 stroke engines, tractors, agriculture equipments, cut fluids, cooling fluids, etc (Erhan & Asadauskas, 2000).

Esters have been used as lubricants since the beginning of the 19th Century, in the form of natural esters in pig fat and whale oil (Whitby, 1998). During World War II, a large number of synthetic fluids were developed such as alcohol and long chain acids esters, that presented excellent low temperature properties.

Nowadays, the esters represent only 0.8% of the world lubricants market. However, while the global consumption of lubricants has been stagnant, the consumption of synthetic oils has grown approximately 10% per year. This growing esters consumption is due to performance reasons and also to changes on the environmental laws of several European Community countries, mainly Germany.

Esters have a low environmental impact and its metabolization consists of the following steps: ester hydrolysis, beta-oxidation of long chain hydrocarbons and oxygenases attack to aromatic nucleus. The main characteristics that reduce the microbial metabolization or degradability are:

- Branching position and degree (that reduce the beta-oxidation);
- Molecule saturation degree;
- Ester molecular weight increase.

The strongest effect of the ester group on the lubricant physical properties is a decrease in its volatility and increase in its flash point. This is due to the strong dipole moment (London forces) that keeps the ester molecules together. The ester group affects other properties, too such as: thermal and hydrolytic stabilities, solvency, lubricity and biodegradability. Besides, esters, mainly from polyalcohols, as trimethylolpropane (TMP), produce a unimolecular layer on the metal surface, protecting it against wear. This layer is produced by the oxygen atoms which are presents in the ester molecules.

The ester's most important physical-chemistry properties are viscosity, viscosity index (VI), pour point, lubricity, thermal and hydrolytic stabilities and solvency.

The main esters used as biolubricants are: diesters, phthalates, trimethilates, C_{36} dimerates and polyolesters. The polyolesters are formed from polyols with one quaternary carbon atom (neopentylalcohols), as trimethylolpropane, neopentylglycol and pentaerythritol. This class of compounds is very stable due to the absence of a secondary hydrogen on the β position and to the presence of a central quaternary carbon atom (Wagner et al., 2001). The main applications to the esters are: engine oil, 2 stroke engine oils, compressor oils, cooling fluids, aviation fluids and hydraulic fluids.

5.1 Synthesis of biolubricant esters

According to (Solomons, 1983), the carboxylic acids react with alcohols to produce esters, through a condensation reaction called esterification (figure 4). This reaction is catalyzed by acids and the equilibrium is achieved in a few hours, when an alcohol and an acid are heated under reflux with a small amount of sulfuric acid or hydrochloric acid. Since the equilibrium constant controls the amount of produced ester, an excess of the carboxylic acid or of the alcohol increases the yield of the ester. The compound choice to use in excess will depend on its availability and cost. The yield of a esterification reaction may be increased also through the removal of one of the products, the water, as it is formed.

The typical mechanism of esterification reactions is the nucleophilic substitution in acyl-carbon, as illustrated on figure 5.

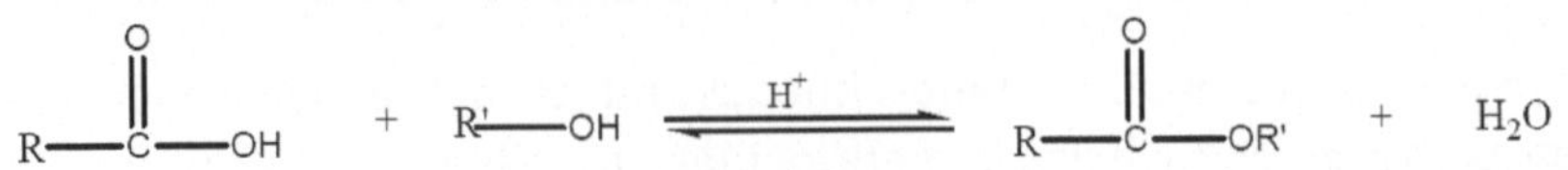

Fig. 4. Esterification reaction scheme between a carboxylic acid and an alcohol

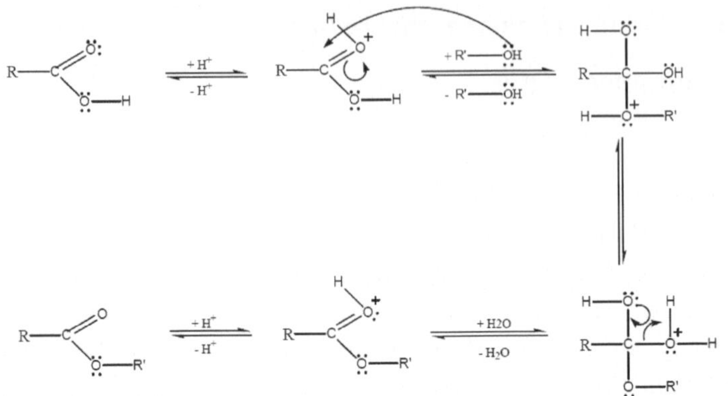

Fig. 5. Esterification reaction mechanism

When one follows the reaction clockwise, this is the direction of a carboxylic acid esterification, catalyzed by acid. If, however, one follows the counterclockwise, this is the mechanism of an ester hydrolysis, catalyzed by acid. The final result will depend on the choice conditions to the reaction. If the goal is to ersterify an acid, one uses an alcohol excess and if it is possible, one promotes the water removal as it is formed. However, if the goal is the hydrolysis, one uses a large water excess.

The steric hindrance strongly affects the reaction rates of the ester hydrolysis catalyzed by acids. The presence of large groups near to the reaction center in the alcohol component or in the acid component retards the reaction.

Esters can be synthesized through transesterification reactions (figure 6). In this process, the equilibrium is shifted towards the products, allowing the alcohol, with the lower boiling point, to be distilled from the reactant mixture. The transesterification mechanism is similar to the one of a catalyzed by acid esterification (or to the one of a catalyzed by acid ester hydrolysis).

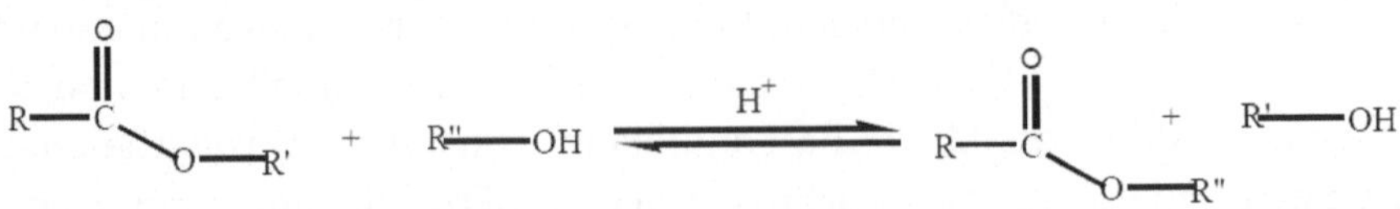

Fig. 6. Transesterification reaction between an ester and an alcohol

The methylricinoleate, from a transesterification reaction of the castor oil with methanol, is the main constituent of castor biodiesel. The transesterification of this compound with superior alcohols (TMP, Pentaerythritol or Neo-pentylglycol) (figure 7) allows the production of poliolesters, important synthetic base oils precursors.

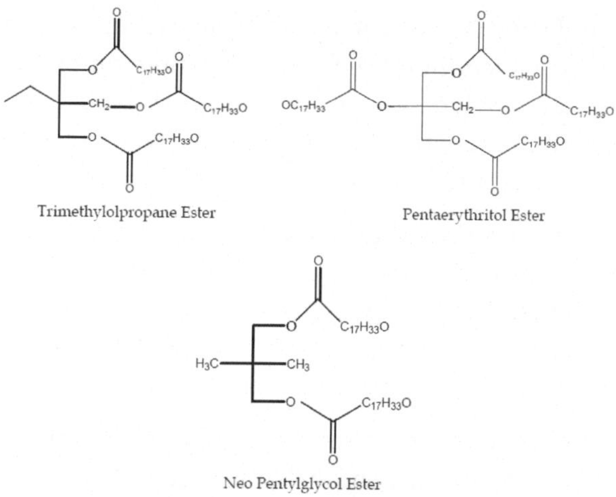

Fig. 7. Poliolesters molecular structures

The higher the molecule branching degree of this product the better the pour point, the higher the hydrolytic stability, the lower the VI. Regarding linearity, it is verified the opposite way. Regarding the double bonds, the higher the saturation, the better the oxidative stability, the worse the pour point (Wagner et al., 2001). Base oils from these superior alcohols, but with other vegetable oils, can be found in the market, with excellent performance.

To increase the transesterification reactions yield one must promote the reaction equilibrium shift towards the products. This can be reached by using a vacuum, which will remove the formed alcohol from the mixture.

Chemical or enzymatic catalysts may be used on the biolubricants esters synthesis. The chemical catalysis occurs in high temperatures (> 150oC), with the usage of homogeneous or heterogeneous chemical catalysts, with acid or alkaline nature (Abreu et al., 2004). The typical acid homogeneous catalysts are acid p-toluenesulfonic, phosphoric acid and sulfuric acid, while the alkaline are caustic soda, sodium ethoxide and sodium methoxide. The more popular heterogeneous catalysts are tin oxalate and cationic exchange resins. (Bondioli et al., 2003) performed the esterification reaction between caprilic acid and TMP, using tin oxide (SnO) as catalyst at 150°C. The yield was 99%, with the continuous removal of the produced water.

(Bondioli, 2004) reported the usage of strong acid ions exchange resins as catalysts in esterification and transesterification reactions. In the case of esterification reactions, the water plays a fundamental role on the catalyst performance. If on the one hand one must remove the produced water to increase the reaction yield, on the other hand the water has a positive effect on the dissociation of the strong acid groups of the resin. Thus, a completely dry resin does not present any catalytic activity, due to the impossibility of the sulfonic group dissociation.

Another limiting factor is the reactant diffusion inside a resin. Fatty materials possess high viscosity, which limits the catalysis using ion exchange resins. In the case of a required high catalytic efficiency, one must choose ion exchange resins with a limited crosslinking degree. Powder resins are more active than spherical ones on esterification reactions.

To esters synthesis, one must to use only acid-sulfonic ion exchange resins. Strong basic ion exchange resins may be attractive for transesterification reactions, however they have a limited stability when heated at temperatures higher than 40°C, and are neutralized by low concentrations of fatty acids. Another negative factor is the glycerin production during the reaction, which can make the resin waterproof.

In spite of these negative effects, ion exchange resins, when used as heterogeneous catalysts, present the following operational advantages:

- ❖ As solid acids or bases, in a batch process, they can easily be separated from the system at the reaction end;
- ❖ One may prepare the catalytic bed by packaging and produce a continuous process with higher productivity and catalytic efficiency;
- ❖ The possibility of regeneration decreases the process costs;
- ❖ Due to its molecular sieve action, there is a higher selectivity;
- ❖ These resins are less corrosive than the regular used acids and bases.

Biolubricants esters synthesis may be performed with efficiency using not only chemical catalysts but also biological ones (lipases). However, catalyst choice parameters must be based on the knowledge of each one's limitation. Thus, although the chemical via presents a main advantage because of the lower cost when compared to the enzymatic via, due to its higher availability in large amounts, it also presents some disadvantages, such as:

- ❖ Low catalyst selectivity, with several parallel reactions;
- ❖ Corrosion, mainly with sulfuric acid and sodium hydroxide as catalysts;
- ❖ Low conversion (40% in average), mainly with metal complex catalysts;
- ❖ Foam production (Basic catalysts);
- ❖ Almost any catalytic activity (H_2SO_4 and NaOH) with long chain alcohols;
- ❖ More severe operation conditions and higher energy consumption due to higher temperatures required.

Regarding the enzymatic catalysis, it occurs in milder temperatures (60°C), using lipases, triacyl ester hydrolases (glycerol ester hydrolases, E.C. 3.1.1.3). Normally, the lipases catalyze the glycerol ester hydrolysis in lipid/water interphases (Dossat et al., 2002). However, in aqua restrict systems, for example, solvents, lipases catalyze also the synthesis of such esters. Thus, they have been employed on the fat and oil modifications, in aqua restrict systems with or without the presence of organic solvents. Lipases from several microorganisms have been studied in the vegetable oil transesterification reactions, such as: *Candida rugosa, Chromobacterium viscosum, Rhizomucor miehei, Pseudomonas fluorescens* and *Candida antarctica*. The most used among these are *Rhizomucor miehei* (immobilized in macroporous anionic resin – Lipozyme) and *Candida rugosa*, in powder. In works made with sunflower oil, the *Candida rugosa* lipase usage showed a higher yield in the transesterification reaction, besides a lower cost than the *Rhizomucor miehei* lipase (Castro et al., 2004).

The transesterification reactions via enzymes may occur with or without the presence of organic solvents. Other interesting variable on this type of reactions is the added amount of alcohol. A large alcohol excess shifts the reaction equilibrium to the production of ester. However, literature data show that a very large excess (higher than 1:6, ester:alcohol) can cause inhibition of the enzymatic activity.

Another interesting characteristic regarding these reactions can be seen in transesterifications directly from the vegetable oils. These reactions have glycerin as subproduct, which, according to some authors, may be adsorbed on the enzyme surface, thus inactivating it (Dossat et al., 2002).

The enzymatic via shows some advantages, as well for example:

- High enzyme selectivity;
- High yields on the ester conversion;
- Milder reaction conditions, avoiding degradation of reactants and products;
- Lower energy consumption, due to low temperatures;
- Catalyst biodegradability;
- Easy recover of the enzymatic catalyst (Dossat et al., 2002).

A main disadvantage of this via is the high cost of the industrial scale process, due to the high cost of the enzymes. However, the development of more robust biocatalysts through molecular biology techniques or enzymes immobilization can make this process more industrially competitive in a few years.

The biolubricants esters synthesis can be carried out not only in batch reactors, but also in continuous reactors (fixed or fluidized bed). However, due to process simplicity, the batch is the majority choice. One illustrative example of a batch reactor is on figure 8.

(Lämsa, 1995) studied and developed new methods and processes regarding the esters production from vegetable oils, raw-materials for the biodegradable lubricants production, using not only chemical catalysts but also enzymatic catalysts. On the beginning it was synthesized 2-ethyl-1-hexyester of rapeseed oil, from 2-ethyl-1-hexanol and rapeseed oil, ranging catalysts (sodium hydroxide, potassium hydroxide, sodium methoxide, sodium ethoxide and sulfuric acid), molar ratio oil:alcohol (1:3 to 1:6), temperature (80 to 120°C) and pressure (2.0 to 10.6 MPa).

The established optimum conditions were: molar ratio (1:5), 0.5% alkaline catalyst (sodium methoxide), temperature range 80 to 105°C and pressure of 2.7 MPa. The obtained rapeseed yield was 97.6% in five hours of reaction.

The above described synthesis was also studied using *Candida rugosa* lipase as catalyst, with a yield of 87% in five hours of reaction. The best conditions were: molar ratio oil:alcohol (1:2.8), lipase concentration (3.4%), added water (1.0%) and temperature of 37°C.

(Lämsa, 1995) synthesized also a rapeseed methyl ester (biodiesel), reacting rapeseed with methanol (in excess) at 60°C, using 0.5% of alkaline catalyst. After four hours of reaction, the yield was 97%, with the separation of the formed glycerin and the distillation of the excess alcohol.

Fig. 8. Transesterification batch reactor

The same author still promoted the reaction between the rapeseed methyl ester and trimethylolpropane (TMP). This transesterification reaction followed a strategy of individual analyses of each variable behavior involved in the process. Firstly, it was studied the type and the amount of catalyst used, with the best results attributed to sodium methoxide (0.7%). Next, the molar ratio ester:TMP was evaluated, with the best value being 3.2:1 (small ester excess). Finally, the temperature and the pressure were studied, both of these variables have a strong effect on the yield. It was established the values of 85-110°C and 3.3 MPa for a yield of 98.9%, in 2.5 hours of reaction.

At last, the author performed the rapeseed methyl ester synthesis through enzymatic catalysis. The yields using lipases were high, but the reaction duration was extremely high (46 hours in average).

6. Biolubrificant properties

The main properties of a lubricant oil, which are basic requirements to the good performance of it, will be described as follows:

- Viscosity: the viscosity of lubricants is the most important property of these fluids, due to it being directly related to the film formation that protects the metal surfaces from several attacks. In essence, the fluid viscosity is its resistance to the flow, which is a function of the required force to occur slide between its molecule internal layers. For the biolubricants, there is not a pre-defined value, however, due to market reasons, the range 8 to 15 cSt at 100°C is the most required;

- Viscosity index (VI): it is an arbitrary dimensionless number used to characterize the range of the kinematic viscosity of a petroleum product with the temperature. A higher viscosity index means a low viscosity decrease when it increases the temperature of a product. Normally, the viscosity index value is determined through calculation (ASTM D2270 method), which takes in account the product viscosities at 40 and 100°C. Oils with VI values higher than 130 find a wide diversity of applications;

- Pour point: this essay was for a long period of time the only one used to evaluate the lubricants behavior at low temperatures. After pre-heating, the sample is cooled at a specified rate and observed in 3°C intervals to evaluate the flow characteristics. The lowest temperature where is observed movement in the oil is reported as the pour point. The lower the pour point, the better the base oil, having values lower than -36°C a wide market. Some pour point depressants may be used on the biolubricants formulations, but these are less efficient than when used with mineral oils;

- Corrosion: biolubricants, as mineral lubricants, must not be corrosives. Because of that, they must present 1B result (maximum) on the test ASTM D130, which consists on the observation of the corrosion in a copper plate after this plate is taken out from an oven, where it has been for 3 hours, immersed in the lubricant sample, at 150°C. The values 1A, 1B, etc., are attributed based on comparison with standards;

- Total acid number (TAN): this essay's goal is to measure the acidity of the lubricant, derived, in general, from the oxidation process, the fuel burning and some additives. In this essay, a sample, with known mass, is previously mixed with titration solvent and titrated in KOH in alcohol. It is determined the KOH mass by sample mass to the titration. It is desired values lower than 0.5 mgKOH/g, since higher TAN values contribute to increase the corrosion effects;

- Biodegradability: many vegetable oils and synthetic esters are inherently biodegradable. This means that they are not permanent and undergo physical and chemical changes as a result of its reaction with the biota, which leads to the removal of not favorable environmental characteristics. The negative characteristics are water immiscibility, eco toxicity, bioaccumulation in live organisms and biocide action against such organisms. For some applications, the lubricants must be readily biodegradable. The tests CEC L-33-T-82 and modified STURM are two of the most widely used to measure the lubricants biodegradability. To consider a lubricant as biodegradable, for example, it must present a result higher than 67% on the CEC test;

- Oxidative stability: most parts of the vegetable oils are unsaturated and trend to be less stable to oxidation than mineral oils. Low amounts of antioxidants (0.1-0.2%) are effective in mineral oil formulations. However, vegetable oils may require a large amount of such antioxidants (1-5%) to prevent its oxidative degradation. The most used essay to measure the oxidative stability of lubricants is the Rotary Pressure Vessel (RPVOT – ASTM D2272). A good lubricant must present an oxidation times higher than 180 minutes, on this method.

7. Conclusion

The biolubricants market has increased at an approximately 10% per year rate in the last ten years (Erhan et al., 2008). The driven forces of such increase are mainly the growing awareness regarding environmental friendly products and government incentives and regulations.

Even though, when compared to the mineral oil market, the biolubricants usage is very small, and, as mentioned before, concentrated in some countries of Europe and in the USA. In order to change the scenario, the biggest challenge to the industries is how to reduce the production costs of such products, therefore making its prices more attractive. The chemical process has low costs, but the yields are a little small. On the other hand, the enzymatic process, with high yields, possesses elevated costs. The newest technologies in lipases development and immobilization may contribute to decrease these costs and make these products cheaper.

Another important matter related to the biolubricants is the quality of their characteristics. On properties as viscosity, viscosity index and pour point, these products overcome the mineral oils based lubricants. But in terms of oxidative stability, efforts have been made to develop products with at least the same level of mineral oils. This can be achieved by chemical modification, acting on the biolubricant molecule, or by adding some special developed additives. The problem is that these additives must be biodegradable too, in order to not damage the biodegradability of the product as a whole. The additives and the lubricants industries have worked together towards the development of environmental friendly products.

The usage of each country's typical raw materials, like castor oil in Brazil, is used both for an economic reason and a social reason. In the Brazilian case, the small farmers of the poorest country regions are encouraged to plant castor, which is a very easily cultivated crop due to the Brazilian weather. They are able to sell these castor seeds for the oil and biodiesel producers, who can then produce biolubricants. This is a very interesting way to promote the social inclusion in underdeveloped countries. And another interesting feature of this crop is that there is not any food competition.

Finally, the biolubricants have a very important role in the future of mankind, because their potential to contribute to an environment free of pollution and with more equal opportunities for the entire World.

References

1. Abreu, F. R.; Lima, D. G.; Hamú, E. H.; Wolf C. & Suarez, P. A. Z. (2004). Utilization of Metal Complexes as Catalysts in the Transesterification of Brazilian Vegetable Oils with Different Alcohols. *Journal of Molecular Catalysis A: Chemical*, Vol. 209, pp. 29-33.
2. Azevedo, D. M. P. & Lima, E. F. (2001). *O Agronegócio da Mamona no Brasil*, Embrapa, (21st edition)., Brasília, Brazil.
3. Bartz, W. J. (1998). Lubricant and the Environment. *Tribology International*, Vol. 31, pp. 35-47. Birová, A.; Pavlovicová, A. & Cvengros, J. (2002). Lubricating Oils Base from Chemically Modified Vegetable Oils. *Journal of Synthetic Lubrication*, Vol. 18, No. 18-4, pp. 292- 299.
4. Bondioli, P.; Della Bella, L. & Manglaviti, A. (2003). Synthesis of Biolubricants with High Viscosity and High Oxidation Stability. *OCL*, Vol. 10, pp. 150-154.
5. Bondioli, P. (2004). The Preparation of Fatty Acid Esters by Means of Catalytic Reactions. *Topics in Catalysis*, Vol .27, No. 1-4 (Feb), pp. 77-81.
6. Castro, H. F.; Mendes, A. A.; Santos, J. C. & Aguiar, C. L. (2004). Modificação de Óleos e Gorduras por Biotransformação. *Química Nova*, Vol. 27, No. 1, pp. 146-156.
7. Dossat, V.; Combes, D. & Marty, A. (2002). Lipase-Catalysed Transesterification of High Oleic Sunflower Oil. *Enzyme and Microbial Technology*, Vol. 30, pp. 90-94.
8. Erhan, S. Z. & Asadauskas, S. (2000). Lubricant Basestocks from Vegetable Oils. *Industrial Crops and Products*, Vol. 11, pp. 277-282.
9. Erhan, S. Z., Sharma, B. K., Liu, Z., Adhvaryu A. (2008). Lubricant Base Stock Potential of Chemically Modified Vegetable Oils. *J. Agric. Food Chem.*, Vol. 56, pp. 8919-8925.

10. Kolwzan, B. & Gryglewicz, S. (2003). Synthesis and Biodegradability of Some Adipic and Sebacic Esters. *Journal of Synthetic Lubrication*, Vol. 20, No. 20-2, pp. 99-107.
11. Lal, K. & Carrick, V. (1993). Performance Testing of Lubricants Based on High Oleic Vegetable Oils. *Journal of Synthetic Lubrication*, No. 11-3, pp. 189-206.
12. Lämsa, M. (1995). *Environmentally Friendly Products Based on Vegetable Oils.* D.Sc. Thesis, Helsinki University of Technology, Helsinki, Finland.
13. Lastres, L. F. M. (2003). *Lubrificantes e Lubrificação em Motores de Combustão Interna.* Petrobras/CENPES/LPE, Rio de Janeiro, Brazil.
14. Murphy, W. R.; Blain, D. A. & Galiano-Roth, A. S. (2002). Benefits of Synthetic Lubricants in Industrial Applications. *J. Synthetic Lubrication*, Vol. 18, No. 18-4 (Jan), pp. 301-325.
15. Ravasio, N.; Zaccheria, F.; Gargano, M.; Recchia, S.; Fusi, A.; Poli, N. & Psaro, R. (2002). Environmental Friendly Lubricants Through Selective Hydrogenation of Rapeseed Oil over Supported Copper Catalysts. *App. Cat. A: Gen.*, Vol. 233, pp. 1-6.
16. Solomons, T. W. G. (1983). *Química Orgânica*, LTC, (1st edition), Rio de Janeiro, Brazil. Wagner, H.; Luther, R. & Mang, T. (2001). Lubricant Base Fluids Based on Renewable Raw
17. Materials. Their Catalytic Manufacture and Modification. *Applied Catalysis A: General*, Vol. 221, pp. 429-442.
18. Whitby, R. D. (1998). Synthetic and VHVI-Based Lubricants Applications, Markets and Price-Performance Competition. Course Notes, Rio de Janeiro, Brazil.
19. Whitby, R. D. (2005). Understanding the Global Lubricants Business – Regional Markets, Economic Issues and Profitability. Course Notes, Oxford, England.
20. Whitby, R. D. (2006). Bio-Lubricants: Applications and Prospects. In: *Proceedings of the 15th International Colloquium Tribology*, Vol. 1, pp. 150, Ostfildern, Germany, January, 2006.

8

Lubricating Greases Based on Fatty By-Products and Jojoba Constituents

1. Introduction

There has been a need since ancient times for lubricating greases. The Egyptians used mutton fat and beef tallow to reduce axle friction in chariots as far back as 1400 BC. More complex lubrications were tried on ancient axle hubs by mixing animal fat and lime, but these crude lubricants were in no way equivalent to the lubricating greases of modern times. Good lubricating greases were not available until the development of petroleum based oils in the late 1800's. Today, there are many different types of lubricating greases, but the basic structure of these greases is similar.

In modern industrial years, greases have been increasingly employed to cope with a variety of difficult lubrication problems, particularly those where the liquid lubricant is not feasible. Over the last several decades, greases making technology throughout the world, has undergone rapid change to meet the growing demands of the sophisticated industrial environment. With automation and mechanization of industry, modern greases, like all other lubricants, are designed to last longer, work better under extreme condition and generally expected to provide adequate protection against rust, water, and dust. So, greases are the important items for maintenance and smooth running of various machineries, automobiles, industrial equipments, instruments and other mechanical parts. Industrial development and advances in the field of greases have been geared to satisfy all these diverse expectations (Cann, 1997).

In general, lubricating greases contain a variety of chemical substances ranging from complicated mixtures of natural hydrocarbons in the base oils, well defined soaps and complex organic molecules as additives. Therefore, the more practical greases are lubricating oils which has been thickened in order to remain in contact with the moving surfaces, do not leak out under gravity or centrifugal action or be squeezed out under pressure. The majority of greases in the market are composed of mineral oil blended with soap thickeners. Additives enhance the performance and protect the greases and/or lubricated surfaces. Lubricating greases are used to meet various requirements in machine elements and components, including: valves, seals, gears, threaded connections, plain bearings, chains, contacts, ropes, rolling bearing and shaft/hub connections (Boner, 1954, 1976).

Developments in thickeners have been fundamental to the advances in grease technology. The contribution of thickeners has been so central to developments that many types of greases are often classified by the type of thickener used to give the required structured matrix and consistency. The two principal groups of thickeners are metal soaps and inorganic compounds. Soap-based greases are by far the most widespread lubricants.

In soap greases the metallic soap consists of a long-chain fatty acid neutralized by a metal such as lithium, sodium, calcium, aluminum, barium or strontium. A wide variety of fatty materials are used in the manufacture of base lubricating greases. In particular, lithium lubricating greases, first appeared during World War II, were made from lithium stearate pre-formed soap. Nowadays they are usually prepared by reacting lithium hydroxide, as a powder or dissolved in water, with 12-hydroxy stearic acid or its glycerides in mineral oils or synthetic oils Whether the free acid or its glycerides is preferred depends on the relationship between cost and performance (kinnear & Kranz, 1998; El-Adly, 2004a).

A comprehensive study of all aspects of grease technology with the corresponding literature references is beyond the scope of this short contribution. There are numerous textbooks available on this subject (Vinogradov, 1989; Klamann, 1984; Boner, 1976; Erlich, 1984; Lansdown, 1982).

Within the area of alternate sources of lubricants (El-Adly et al, 1999, 2004a, 2004b, 2005, 2009), a new frontier remains for researchers in the field of lubricating greases. Lithium greases have good multi-purpose properties, e.g. high dropping point, good water resistance and good shear stability. Alternative sources of fatty materials and additives involved in the preparation of such lithium greases will be found later in this chapter. The main objective is to explore the preparation, evaluation and development of lithium lubricating greases from low cost starting materials such as, bone fat, cottonseed soapstock and jojoba meal. The role of the jojoba oil and its meal as novel additives for such greases is also explored (El-Adly et al., 2004b).

2. Raw materials

The main components of lubricating greases, in general, are lubricating mineral oil, soaps and additives. The mineral oil consists of varying proportions of paraffinic, naphthenic and aromatic hydrocarbons, in addition to minor concentrations of non-hydrocarbon compounds. Soaps may be derived from animal or vegetable fats or fatty acids. Additives are added to lubricating greases, generally in small concentrations, to improve or enhance the desirable properties of the finished product. The use of these ingredients such as fats, fluids and additives, each of which consists of a number of chemical compounds, was originally dictated to a large extent by economic factors and availability. The raw materials mentioned in this chapter are, therefore, according to the following:

2.1 Lubricating fluid

Mineral oils are most often used as the base stock in grease formulation. About 99% of greases are made with mineral oils. Naphthenic oils are the most popular despite of their low viscosity index. They maintain the liquid phase at low temperatures and easily combine with soaps. Paraffinic oils are poorer solvents for many of the additives used in greases, and with some soaps they may generate at weaker gel structure. On the other hand, they are more stable than naphthenic oils, hence are less likely to react chemically during grease formulation.

Table 1. Physico-chemical properties of the lubricating fluids (Base oil B1&bright stock B2)

Characteristics	Base oil (B1)	Bright stock (B2)	Test Methods
Density, g/ml: at 15.56, °C	0.872	0.8975	ASTM D.1298
Refractive index, $n_D 20$	1.5723	1.5988	ASTM D.1218
ASTM-Color	1.0	1.0	ASTM D.1500
Kinematics viscosity, c St.			
at 40°C	50	78	
at 100°C	9	19	ASTM D.445
Viscosity index	233	225	ASTM D. 189
Dynamic Viscosity, @ 30 °C			
(20 rpm), cP	2100	2905	ASTM D. 189
Pour point, °C	-3	Zero	ASTM D.97
Total acid number, mg KOH/g@72 hr	0.12	0.2	ASTM D.664
Flash Point, °C	210	290	ASTM D.92
Molecular Weight	755	890	GPC*
Predominant, molecular weight	762	898	GPC*
Polydispersity	1.1023	1.253	GPC*
Structural group analysis hydrocarbon component, wt %			ASTM D-3238
$\%C_A$ (Aromatic Percentage)	19	20	Column chromatography
$\%C_P$ (Paraffinic Percentage)	61	68	
$\%C_N$ (Naphthinic Percentage)	20	12	
Mono-aromatic	14.9	13.2	
Di-aromatic	12.0	15.5	
Poly-aromatic	1.2	1.5	

GPC* Gel Permeation Chromatography

In this respect, two types of lube base oils are investigated as fluids part for preparing lithium lubricating greases: the first is a base mineral oil designated B1 and the second is a bright stock designated B2. The Physico-chemical properties of these oils were carried out using ASTM/ IP standard methods of analysis as shown in Table (1). Data in this table reveal that the bright stock could be classified as heavier oil than lube base oil. It may be pointed out, therefore, that the internal friction between oil layers in B2 is greater than in B1. This interpretation agrees with the data of gel permeation chromatography concerning molecular weights of B1 and B2. This is further supported by predominant molecular weights of B1 and B2 which are 762 and 898, respectively. In addition, the polydispersity (i.e., number of average molecular weight divided by mean molecular weight value, Mn/Mw) for bright stock is 1.2530 while it is 1.1023 for base mineral oil. This indicates that B1 and B2 have higher degree of similarity in hydrocarbon constituents (cross sectional areas of molecules are similar) and morphology of structure.

The rheological properties of the above mentioned oils were studied at different temperatures using Brookfield programmable Rheometer LV DV-III ULTRA. Different mathematical model (Herschel Bulkley, Bingham and Casson models) were applied to deduce the viscoelastic parameters. It was found that the fluids under investigation had a Newtonian behavior (El-Adly, 2009).

2.2 Fatty material

2.2.1 Cottonseed soapstock

Soapstock is formed by reacting crude vegetable oils with alkali to produce sodium soap as a by-product, which is separated from the oil by centrifuging. Typically, soapstock accounts for 5 to 10 wt. % of the crude oil. In general, soapstock from oilseed refining has been a source of fatty acids and glycerol. These processes are no longer cost effective. Consequently, in cottonseed oil extraction facilities, the treated soapstock is added to the animal meal to increase the energy content, reduce dust and improve pelleting of food products **(**Michael, 1996**)**. In general compositional information considering raw and acidulated cottonseed soapstock has been published **(**El-Shattory, 1979; Cherry & Berardi, 1983**)**.

2.2.2 Bone fat

The crude bone fat is produced by solvent extraction of crushed bone during the manufacture of animal charcoal. It is considered as by-product for this process. Also, it is extracted by wet rendering under atmospheric pressure from femur epiphyses of cattle, buffaloes and camels. The physical and chemical properties of the above mentioned bone fat was studied (El-Adly, 1999). It has low cost and possesses large-scale availability.

2.2.3 Physicochemical properties of the bone fat and soapstock

Data in Table (2) show the physicochemical properties of the bone fat and cottonseed soapstock carried out using ASTM/ IP standard methods of analysis.

Bone fat and cottonseed soapstock consist primarily of glycerides, that is, of various fatty acid radicals combined with glycerol. It is apparent from Table (2) that the saponification number for bone fat and soapstock are 180.0 and 198.0, respectively. These values were not only used as basis for figuring the amount of alkali required for a particular formulation, but also permitted speculation as to the identity of the fatty acids making up the fatty materials.

The results of gas liquid chromatography analysis of the esterified fatty acids in bone fat and hydrolyzed cottonseed soapstock are shown in Table (2). There is a wide variation in their fatty acids composition myristic, palmitic, stearic, oleic, linoleic and linolenic acid. Bone fat is composed of about 52% unsaturated fatty acids, mainly oleic acid, and 47% saturated fatty acids, being palmitic, stearic and myristic acid. However, soapstock contains more unsaturated fatty acids 71% and saturated 29%. This finding was supported by the iodine value measured for both fatty materials. The difference in their fatty constituents leads to the possibility of producing lithium lubricating grease.

Table 2. Physicochemical properties of bone fat and cottonseed soapstock

Property (in mole %)	Bone Fat	Soapstock	Test method
Saponification number	180	198	ASTM D-1962
Iodine value	45	60.0	ASTM D-2075
Titer, C°	35	45.0	ASTM D-1982
Palmitic acid	23.0	27.0	Gas chromatography
Myristic acid	9.0	trace	
Oleic acid	48.0	29.0	
Stearic acid	15.0	2.0	
Linoleic acid	4.0	42.0	
Linolenic acid	trace	trace	

2.3 Additives

The additives used in grease formulation are similar to those used in lubricating oils. Some of them modify the soaps, others improve the oil characteristics. The most common additives include anti-oxidants, rust and corrosion inhibitors, tackiness, and anti-wear and extreme pressure additives. Many studies reported detailed information about lubricating additives (Mang & Dresel, 2001; Shirahama, 1985). This chapter presents the utilization of jojoba oil and its meal as additives for the preparation of lithium lubricating greases.

2.3.1 Jojoba oil

Jojoba is known in botanical literatures as *Simmondsia chinenasis* (Link) of the family Buxaceaa and as *Simmondsia californica* Nutall. The first name is the correct one, although it perpetuates a geographical misnomer. In late 1970 sperm whale was included by the US Government in the list of endangered species and imports of oil, meal and other products derived from whales were banned. At that time, sperm oil consumption in the United States was about 40-50 million pounds per year, with half that figure used in lubricant applications. No single natural, or synthetic replacement with the unique qualities of sperm whale oil has yet been found, but enough experimental evidence has accumulated in the last years that jojoba oil is not only an excellent substitute of sperm oil but its potential industrial uses go beyond those of sperm oil (Wisniak, 1994). Sperm oil is widely used in lubricants because of the oiliness and metallic wetting properties, it imparts and its nondrying characteristics that prevent gumming and tackiness in end-use formulations. It is more important as a chemical intermediate since it is sulphonated, oxidized, sulfurized, sulfur-chlorinated and chlorinated to give industrial products that were used primarily as wetting agents and extreme pressure (EP) additives. The composition and physical properties of Jojoba are close enough to sperm oil to suggest the use of Jojoba oil as a substitute for most of the uses of sperm oil (Miwa & Rothfus, 1978). Sperm oil has been used as an extreme pressure and antiwear additive in lubricants for gears in differentials and transmissions, in hydraulic fluids that need a low coefficient of friction and in cutting and drawing oils. In some of these, sperm oil has been directly, but it is usually Sulfurized (sometimes epoxidized, chlorinated, or fluorinated). Gear lubricants (e.g., in automobile transmissions) commonly contain 5 to 25 percent of Sulfurized sperm oil (Peeler & Hartman, 1972).

Some of the first published results of sulfurized jojoba oil use a lubricant and extreme pressure (EP) additive were reported as patents (Flaxman, 1940; Wells, 1948). Wells pointed out several advantages of jojoba oil over sperm oil. Its slight odor is distinctly more pleasant than the fishy odor of sperm oil. Crude jojoba oil contains no glycerides so that the crude oil needs little or no treatment to prepare it for most industrial purposes.

In general, lubricant technology dealing with jojoba oil and its derivatives in the 70's concentrated on its replacement of sulfurized sperm oil products in such applications as industrial and automotive gear oils, hydraulic oils and metal working lubricants (Heilweil, 1988; Wills, 1985). In the 80's the lubrication industry has developed and research on jojoba has been shifting towards new derivatives with potential application to new technologies and newer areas of lubricant use. A monograph by Wisniak (1987) summarized the chemistry and technology of jojoba oil and jojoba meal.

2.3.1.1 Composition

The chemical composition of jojoba oil is unique in that it contains little or no glycerin and that most of its components fall in the chain-length range of C_{36}-C_{42}. Linearity and close- range composition are probably the two outstanding properties that give jojoba oil its unique characteristics. The oil is characterized of being a monoester of high molecular weight and straight chain fatty acids and fatty alcohols that has a double bond on each side of the ester. The molecular structure of the oil can be represented by the following general formula:

$$CH_3\text{-}(\text{-}CH_2\text{-})_7\text{-}CH{=}CH\text{-}(CH_2\text{-})_m\text{-}COO\text{-}(\text{-}CH\text{-})_n\text{-}CH{=}CH\text{-}(CH_2\text{-})_7\text{-}CH_3$$

Where, m and n are between 8 to 12 (Miwa 1971, 1980; Spencer et al, 1977; Greene & Foster, 1933).

They were the first to report that jojoba nuts contain about 46-50% of liquid oil which resembles sperm whale oil in its analytical characteristics. Qualitative tests suggested that the oil might consist mainly of fatty acid esters of decyl alcohol. Shortly thereafter, detailed analysis of the chemical constituents was reported (Greene & Foster, 1933). The main components were eicosenoic and docosenoic acids and eicosanol and docosanol. Because of the problems of the high resistance of the oil to saponification, the difficulties in isolating pure fractions and the lack of convenient and reliable quantitative analytical techniques the characterization of jojoba oil was developed by Miwa (1971 & 1980).

Jojoba oil is unusually stable towards oxidation especially at high temperature. Kono et al, (1981) mentioned that the oxidative stability of jojoba oil was due, at least in part, to the presence of tocopherol and other natural antioxidants. Also, some of the antioxidants separated and identified by molecular distillation of the oil and analysis of the distillate by gas chromatography/ mass spectrometry. The, ά , γ and δ isomers of tocopherol are present, in varying quantities depending on the origin of the oil, γ-isomer being most abundant.

2.3.1.2 Physical properties

Jojoba oil is chemically purer than most natural substances. It is soluble in common organic solvents such as benzene, petroleum ether, chloroform, carbon tetrachloride, and carbon disulfide, but it is immiscible with ethanol, methanol, acetic acid, and acetone (Miwa & Hagemann, 1978). It is usually a low-acidity, light-golden fluid that requires little or no refining. It is non-volatile and free from rancidity. Even after repeated heating to temperatures above 285°C for 4 days it is essentially unchanged (Daugherty et al., 1953). Its boiling point (at a pressure of 757 mmHg, under nitrogen) rises to 418 °C but drops rapidly to a steady 398 °C (Miwa 1973; Wisniak, 1987). Neutralization of the oil is not usually required and bleaching to a water-clear fluid can be done with common commercial techniques. Some properties of the oil are listed in Table 3 (El-Adly et al, 2009).

Data in Table (3) reveal that the possibilities for economic development of the oil and its suitability to produce lubricants and lubricant additives for use in the preparation of lubricating greases. This view is in agreement with a study on using of jojoba oil as oxidation, thermal and mechanical stabilities to improve the properties of lithium lubricating grease (Ismail, 2008).

Table 3. Physico-chemical properties of Jojoba oil (El-Adly et al, 2009)

Characteristics	Jojoba oil	Test Method
Density, g/ml @ 25/25, °C	0.863	ASTM D-1298
Refractive index, $n_D 20$	1.4652	ASTM D-1218
Kinematics viscosity, c St.		
at 40°C	26	ASTM D-445
at 100°C	7.5	ASTM D-445
Viscosity index	257	ASTM D- 189
Dynamic Viscosity, @ 30 °C (rpm 6), cP	58.4	ASTM D-97
TAN, mg KOH/g	2.0	ASTM D-664
Flash Point, °C	310	ASTM D-92
Iodine Value	80	ASTM D-2075
Average Molecular Weight	604	GPC
Surface tensions mN/m		24
Oxidation stability test (min)	23	IP 229

2.3.2 Jojoba meal

Table 4. Amino acid composition (%) of deoiled meal of two varieties of jojoba meal (Verbiscar and Banigan, 1978)

Amino acid	Apache 377	SCJP 977
Lysine	1.05	1.11
Histidine	0.486	0.493
Arginine	1.56	1.81
Aspartic acid	2.18	3.11
Threonine	1.14	1.22
Serine	1.04	1.11
Glutamic acid	2.40	2.79
Proline	0.958	1.1.
Glycine	1.50	1.41
Alanine	0.832	0.953
Valine	1.10	1.19
Methionine	0.186	0.210
Isoleucine	0.777	0.866
Leucine	1.46	1.57
Tyrosine	1.04	1.05
Phenylalanine	0.919	1.07
Cystine+ cystine	0.791	0.519
Tryptophan	0.492	0.559

A byproduct of jojoba seeds is the meal remaining after the oil has been pressed and extracted. This material constitutes about 50% of the seed and contains 25-

30 % crude protein. Table (4), presents the amino acid composition (%by wieght) of deoiled meal of two varieties of jojoba meal (Verbiscar &Banigan, 1978). Basic information on the composition of jojoba meal, polyphenolic compounds, carbohydrate contents, and Simmondsin compounds have been reported (Verbiscar et al., 1978; Cardeso et al., 1980; Wisniak, 1994). On the other hand, the possibility of using the meal as fuel has already been considered (Kuester, 1984 & Kuester et al., 1985). El-Adly et.al (2 004b) reported the novel application of jojoba meal as additive for sodium lubricating grease.

Table (5) presents the anion and cation concentrations in jojoba meal determined through the sulphuric acid wet ashing. The anion concentrations are measured using ion chromatography (IC) model DIONEX LC20 equipped with electrochemical detector model DIONEX ED50, while the cation concentrations determined by inductively coupled plasma/ atomic emission (ICP/AE) spectrometer model flame Modula spectra.

Table 5. Anions and cations contents of the jojoba meal

Cations	Concentration ppm	Concentration ppm Anions	Concentration, ppm
Calcium	1178	Phosphate	12718
Lithiuum	1.73	Chloride	1286
Potassium	7304	Sulphate	8600
Sodium	566	fluoride	135
Magnesium	2079		
Alumminum	33.4		
Iron	124		
Copper	13.9		
Manganese	20.1		
Barium	1.51		
Zinc	29.8		
Cobalt	3.56		
Nickel	0.34		
Strontium	3.99		

Table (5) also reveals that the main anions in jojoba meal are phosphate (12718 ppm) and chloride (1286 ppm) but the main cations are magnesium, calcium, potassium and sodium. This indicates the possibility of using and optimizing the organometalic compounds in jojoba meal as additives for the lubricating greases.

3. Grease preparation and evaluation

3.1 Lithium greases preparation

Lithium base lubricating greases can be prepared either by batch or continuous processes. Such products can be manufactured from either preformed soap or soap prepared in *situ*. From the standpoint of economy and versatility, the latter method is preferable and is therefore used by most manufacturers. The exception to this last statement is in the case of synthetic lubricating fluids. Preformed soaps are desirable in such case because some of these fluids such as diesters will hydrolyze in the presence of alkalies and heat (Boner, 1954, 1976). The lithium lubricating greases mentioned in this chapter were prepared using batch processing. The studied greases were prepared in two steps according to the following:

- Saponification process was performed on a mixture of fatty materials and fluids by alkaline slurry within the temperature range 190 to 195oC. The autoclave was charged, while stirring, with a mixture of 25% wt of light mineral oil and 14% wt of fatty materials (bone fat and soapstock). The autoclave was closed and heating started. Then about 3% wt lithium hydroxide/oil slurry is gradually pumped into the autoclave. The temperature of the reaction mixture must be raised to 190-195oC and held at this temperature for approximately 60 min. to ensure complete saponification. After completion of the saponification step, jojoba oil and or jojoba meal in different concentrations was added. A sample was then taken to examine its alkalinity/acidity. Corrections were made by adding fatty materials or Lithium hydroxide oil slurry as required reaching a neutral product i.e. complete saponification
- Cooling process was performed after the completion of the saponification reaction. The reaction mixture was cooled gradually while adding the rest of the base lube oil to attain the required grease consistency.

The obtained greases were tested and classified according to the standards methods, National Lubricating Greases Institute (NLGI) and the Egyptian Standards (ES). Also, the physico-chemical characteristics of all the prepared greases under investigation were determined using standard methods of analysis. These include penetration, dropping point, apparent viscosity, oxidation stability, total acid number, oil separation and four balls. In general, test methods are used to judge the single or combined and more or less complex properties of the greases. The last summary containing detailed descriptions of ASTM and DIN methods was reported (Schultze, 1962); but the elemental analysis of the greases is nowadays performed by spectroscopic methods, e.g. X-ray fluorescence spectrometry, inductively coupled plasma atomic emission, or atomic absorption spectrometry, with attention being directed mostly to methods of preparation (Robison et al 1993; Kieke,1998). Also, Thermogravimetry and differential scanning calorimetry tools are used to evaluate of base oil, grease and antioxidants (Pohlen, 1998; Gatto &Grina, 1999).

3.2 Effect of the fatty materials and fluid part concentrations on the prepared greases

The physical and chemical behaviors of greases are largely controlled by the consistency or hardness. The consistency of grease is its resistance to deformation by an applied force.

Table 6. Effect of the fatty material and fluid concentrations on characterization of prepared greases

Ingredient \ Symbol	G1A	G1B	G1C	G1D	G1E	G1F	G1G	Test method
Base oil, Wt %	79.0	79.0	80.0	-	-	-	30	
Brightstock, Wt %	-	-	-	80.0	80.0	80.0	50	
Soap stock, Wt %	18	-	8.5	17.0	-	8.5	8.5	
Bone fat, Wt %	-	18.0	8.5	-	17.0	8.5	8.5	
LiOH, Wt %	3.0	3.0	3.0	3.0	2.8-3	2.8-3	2.8-3	
Penetration								ASTM D-217
Unworked	300	300	300	290	290	290	285	
worked	310	310	310	300	300	300	290	
Dropping point, °C	170	173	174	174	175	177	178	ASTM D-566
Copper Corrosion 3h/100°C	Ia	Ia	Ia	Ia	Ia	Ia	Ia	ASTM D-4048
Oxidation Stability 99± 96h, pressure drop, psi	4.2	4.1	4.5	4.0	4.0	4.1	4.0	ASTM D-942
Alkalinity, Wt%	0.3	0.4	0.4	0.5	0.5	0.5	0.5	ASTM D-664
TAN, mg KOH/gm @ 72h	0.34	0.34	0.33	0.33	0.32	0.30	0.28	ASTM D-664
Oil Separation, Wt%	2.5	2.5	2.3	2.3	2.2	2.2	2	ASTM D-1724
Code grease								
NLGI	2	2	2	2	2	2	2	
Egyptian standard	LB	LB	LB	LB	LB	LB	LB	
Apparent Viscosity, cP, @ 90 °C	39600	39650	39680	39700	39710	39750	39891	ASTM D-189
Yield stress, D/cm2	60.2	61.3	62.1	62.9	63.6	64.3	65.0	
Four ball weld load, Kg	160	162	165	166	168	169	170	ASTM D-2596

Also, it is defined in terms of grease penetration depth by a standard cone under prescribed conditions of time and temperature (ASTM D-217, ASTM D-1403). In order to standardize grease hardness measurements, the National Lubricating Grease Institute (NLGI) has separated grease into nine classification, ranging from the softest, NLGI 000, to the hardest, NLGI 6. On the other hand, the drop point is the temperature at which grease shows a change from a semi-solid to a liquid state under the prescribed conditions. The drop point is the maximum useful operating temperature of the grease. It can be determined in an apparatus in which the sample of grease is heated until a drop of liquid is formed and detaches from the grease (ASTM D-266, ASTM D-2265).

In order to evaluate the effect of fatty materials type and fluid on the prepared lithium grease properties, grease blends G_{1A}, G_{1B}, G_{1C}, G_{1D}, G_{1E}, G_{1F} and G_{1G} have been prepared and formulated according to the percent ingredient listed in Table (6).

Data in Table (6) indicate the effect of different ratios from soapstock, bone fat, base oil and bright stock on the properties of the prepared lithium lubricating greases. It is evident from these results that the dropping point of lithium grease blend made from bone fat or soapstock alone is lower than that of lithium grease containing a mix from each both fatty materials and fluids. This clearly indicates that the most powerful thickener in the saponification process is the equimolar ratio from bone fat and soapstock. In other words, both fatty materials have synergistic effect during the saponification reaction. The mechanical efficiency of the formulated greases is according to the following order $G_{1G} > G_{1F} > G_{1E} > G_{1D} > G_{1C} > G_{1B} > G_{1A}$. On the other hand, the above mentioned test showed that the difference of penetration values between unworked and worked (60 strokes) greases follows an opposite order. Based on this finding, it is concluded that the most efficient lube oil in saponification is the light base oil (B1). This is attributed to the fact that lighter oil B1 is easily dispersed in fatty materials during saponification step at temperature 190oC and form stable soap texture. After completion of saponification, the bright stock (B2) is suitable in the cooling step which leads to heavier consistency and provides varying resistance to deformation. This reflects the role of the effect of mineral oil viscosity and fatty materials on the properties of the prepared grease.

It is apparent from the data in Table (6) that the oil separation, oxidation stability, total acid number and mechanical stability for the prepared grease G_{1G} are 2.0, 3.0, 0.68 and 5.0 respectively. This indicates that the best formula is G_{1G} compared with G_{1A}, G_{1B}, G_{1C}, $G_{1D,}$ G_{1E}, and G_{1F}. Based on the above mention results and correlating these results with the apparent viscosity dropping point and penetration, clearly indicates that the suitable and selected formula for the lithium lubricating grease is G_{1G}.

3.3 Effect of the jojoba oil additive on properties of the selected prepared grease

To evaluate the role of jojoba oil as additive for the Selected Prepared Grease G_{1G}, different concentrations from jojoba oil were tested. In this respect, three concentrations of jojoba oil of 1wt%, 3 wt% and 5wt% were added to the selected grease G1G yielding G_{2A}, G_{2B} and G_{2C}, respectively, as shown in Table (6). Worth mentioning here, Jojoba oil ratio was added to the prepared greases after the completion of saponification process. Data in Table (7) show that the results of the penetration and dropping point tests for lithium grease prepared G_{2A}, G_{2B} and G_{2C} produced from different ratio of jojoba oil. These results show that the difference of penetration values between unworked and worked (60 double strokes) lithium lubricating greases are in the order $G_{2C} < G_{2B} < G_{2A}$. This means that the resistance to texture deformation decreases with increase of jojoba oil ratio in the prepared grease. It may be indicated also that on increasing the ratio jojoba oil additive to the prepared greases would increase binding and compatibility of the grease ingredient. As a result, the dropping point values for prepared greases G_{2A}, G_{2B} and G_{2C} increased to 178, 180 and 183°C, respectively.

Table (7) shows, in general, the positive effect of all concentrations of jojoba oil additive on the proprieties of G_{2A}, G_{2B} and G_{2C}. In this respect, the 5%wt of additive of jojoba oil showed a marked improvements effect. Such improvements may be attributed to the unique properties of jojoba oil, e.g. high viscosity index 257, surface tension 45 mN/m and its chemical structure (Wisniak, 1987). Based on these properties and correlation with the dropping point, penetration, oil separation, oxidation stability, dynamic viscosity, consistency index and yield stress data, its clear that the suitable and selective grease formula is G_{2C}.

Table 7. Effect of addition of Jojoba oil on properties of the selected prepared grease G_{1G}

Symbol / Ingredient & property	G2A	G2B	G2C	Test method
G_1g, wt%	99	97	95	
Jojoba oil, wt%	1	3	5	
Penetration at 25°C				
Un worked	284	278	277	
worked	289	282	280	ASTM D-217
Dropping point, °C	180	182	187	ASTM D-566
Oxidation Stability 99±96h,				
pressure, drop, psi	3.5	3.2	3.0	ASTM D-942
Alkalinity, Wt%	0.16	0.14	0.14	ASTM D-664
Total acid number, mg				
KOH/g, @72h	0.20	0.18	0.16	ASTM D-664
Oil separation, Wt%	1.8	1.8	1.7	ASTM D-1724
Copper Corrosion				
3h/100°C	Ia	Ia	Ia	ASTM D-4048
Code Grease				
NLGI	2	2	2	
Egyptian Standard	LB	LB	LB	
Apparent Viscosity, cP, @				
90 °C	39891	41090	41294	ASTM D-189
Yield stress, D/cm2	75.6	78.1	80.6	
Four ball weld load, Kg	188	190	195	ASTM D -2596

3.4 Effect of the jojoba meal additive

Because greases are colloidal systems, they are sensitive to small amounts of additives. To study the effect of jojoba meal additive on the properties of the selected grease G_{2C}, five grades of lithium lubricating greases containing different concentrations of jojoba meal additive were prepared. These concentrations included 1 wt%,, 2 wt%,, 3 wt%, 4 wt% and 5 wt% yielding G_{3A}, G_{3B}, G_{3C}, G_{3D} and G_{3E} greases, respectively.

These greases have been prepared and formulated according to the percent ingredient listed in Table (8).

Table 8. Effect of addition of jojoba meal on properties of the selected prepared grease G_{2C}

Symbol Ingredient& property	G3A	G3B	G3C	G3D	G3E	Test method
G_{2C}, Wt %	99	98	97	96	95	
Jojoba meal, Wt %	1	2	3	4	5	
Penetration at 25°C	282	280	278	275	275	ASTM
Un worked worked	287	285	280	277	277	D-217
						ASTM
Dropping point, °C	188	190	192	195	198	D-566
						ASTM
Oxidation Stability 99± 96h, pressure, drop, psi	2.5	2.3	2.0	1.5	1.5	D-942
						ASTM
Intensity of (C=O) group @ 72h,	1.2	1.0	1.0	0.995	0.937	D-942
						ASTM
Intensity of (OH) group@ 72h	0.821	0.7921	0.7501	0.7023	0.6813	D-942
						ASTM
Alkalinity, Wt%	0.12	0.13	.14	0.15	0.15	D-664
						ASTM
Total acid number, mg KOH/g @ 72 h	0.15	0.15	0.14	0.12	0.12	D-664
						ASTM
Oil separation, Wt%	1.8	1.8	1.7	1.7	1.6	D-1724
						ASTM
Copper Corrosion 3h/100°C	Ia	Ia	Ia	Ia	Ia	D-4048
Code grease NLGI	2	2	2	2	2	
Egyptian Standard	LB	LB	LB	LB	LB	
						ASTM
Apparent Viscosity, cP, @ 90 °C	41820	42032	42232	42611	42652	D-189
Yield stress, D/cm2	80.6	82.5	85.0	86.4	86.6	
						ASTM
Four ball weld load ,Kg	235	240	245	250	250	D-2596

Data in this table reveal that all concentrations of the JM exhibit marked improvements in all properties of the investigated greases compared with the corresponding grease G_{2C} without jojoba meal. In addition, the difference of penetration values between unworked and worked for greases G_{3A-3E} decreased markedly by increasing jojoba meal content in the range of 1wt to 3wt%. Further increase of the jojoba meal concentration up to 4 and 5% by wt shows almost no difference. Parallel data are obtained concerning dropping point, dynamic viscosity, oil separation and total acid number of greases G_{3A-3E}. Such improving effect, as mentioned above, could be attributed to the high polarity of jojoba meal constitutes, which result in increasing both the compatibility and electrostatic forces among the ingredients of the prepared greases under investigation. Based on the improvement in the dynamic viscosity, consistency, dropping point and oil separation of the addition jojoba meal to the selected grease G_{2C} (Table 8), a suggested mechanism for this improvement is illustrated in the Schemes 1& 2. This suggested mechanism explains the ability of jojoba meal ingredients (amino-acids and polyphenolic compounds) to act as complexing agents leading to grease G_{3D} which is considered the best among all the investigated greases. This agrees well with previous reported results in this connection (El-Adly et al, 2009).

The aforementioned studies on the effects of fatty materials, jojoba oil and meal reveal that the selective greases are G_{1G}, G_{2C} and G_{3D}, respectively.

3.5 Evaluation of the selected greases (G_{1G}, G_{2C} and G_{3D})

3.5.1 Rheological behavior

Lubricating grease, according to rheological definition, is a lubricant which under certain loads and within its range of temperature application, exhibits the properties of a solid body, undergoes plastic strain and starts to flow like a liquid should the load reach the critical point, and regains solid body like properties after the removal of stress (Sinitsyn, 1974).

Rheology is the cornerstone of any quantitative analysis of processes involving complex materials. Because grease has rather complex rheological (Wassermann, 1991) properties it has been described as both solid and liquid or as viscoelastic plastic solids. It is not thick oil but thickened oil. The grease matrix is held together by internal binding forces giving the grease a solid character by resisting positional change. This rigidity is commonly referred to as consistency. When the external stress exceed the threshold level of sheer (stress or strain)- the yield value-the solid goes through a transitional state of plastic strain before turning into a flowing liquid. Consistency can be seen the most important property of a lubricating grease, the vital difference between grease and oil. Under the force of gravity, grease is normally subjected to shear stresses below the yield and will therefore remain in place a solid body. At higher level of shear, however, the grease will flow. Therefore, it is the utmost important to be able to determine the exact level of yield (Gow, 1997).

The rheological measurement of the selected greases is tested using Brookfield Programmable Rheometer HADV-III ULTRA in conjunction with software RHEOCALC. V.2. All Rheometer functions (rotational speed, instrument % torque scale, time interval, set temperature) are controlled by a computer. The temperature is controlled by connection with bath controller HT-107 and measured by the attached temperature probe. In this respect, the rheological behavior of the selected greases G_{1G}, G_{2C} and G_{3D} are determined at 90 °C and 120 °C. Figures 1 and 2 afford nearly linear plots having different yield values. Also, they indicate that the flow behavior of greases at all temperatures obey plastic flow. This is due to operative forces among lithium soap, lubricating fluid, jojoba oil and its meal. Also, the variety in fatty acids (soapstock and bone fat compositions) lead to the soap particles will arrange themselves to form soap crystallites, which looks a fiber in the grease. These soap fibers are disposed in a random manner within a given volume. This packing will automatically ensure many fiber contacts, and as a result, an oil-retentive pore network is formed, which is usually known as the gel network. When a stress is applied to this network, a sufficient number of contact junctions will rupture to make flow possible. The resistance value associated with the rupture is known as yield stress. Therefore yield stress can be defined as the stress value required to make a grease flow (Barnes, 1999).

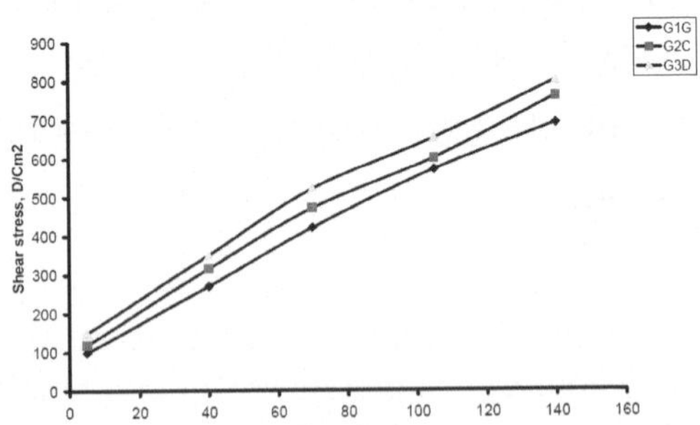

Fig. 1. Variation of shear stress with shear rate for G_{1G}, G_{2C} and G_{3D} at 90°C

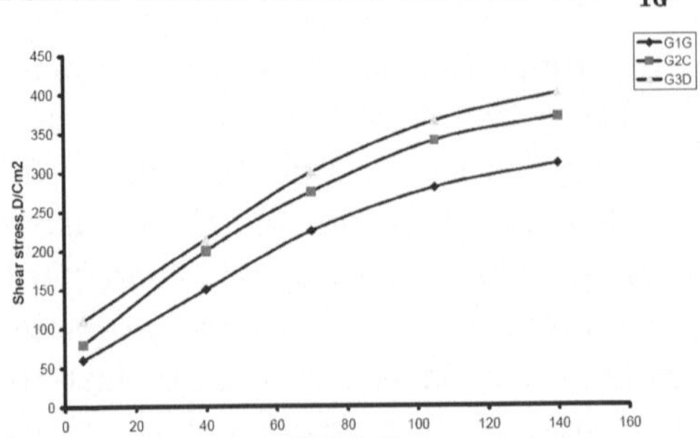

Fig. 2. Variation of shear stress with shear rate for G_{1G}, G_{2C} and G_{3D} at 120°C

In this respect, Rheological data apparent viscosity and yield stress (Tables 6, 7 & 8), for the selected greases show improvement and reinforcement in the order $G_{3D} > G_{2C} > G_{1G}$. This is attributed to the ability of jojoba meal to enhance the resistance to flow for G_{3D}, due to the action of the jojoba meal containing amino acids which act as chelating compounds, columbic interactions and hydrogen bonding, with Li-soap Scheme (1& 2). Also, according to the basic information on the composition of the jojoba meal (Verbiscar, et al., 1978; Cardeso, et al., 1980; Wisniak, 1994), amino acids, wax ester, fatty materials, polyphenolic compounds and fatty alcohols in jojoba meal could be acting as natural emulsifiers leading to increase in the compatibility among the grease ingredients. There is evidence that soap and additive have significant effects on the rheological behavior.

The flow and viscoelastic properties of a lubricating grease formed from a thickener composed of lithium hydroxystearate and a high boiling point mineral oil are investigated as a function of thickener concentration (Luckham & Tadros, 2004).

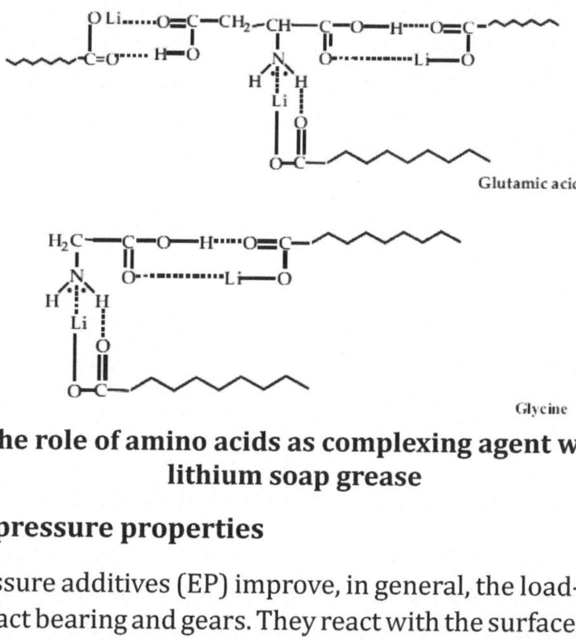

Scheme 1. The role of amino acids as complexing agent with texture of lithium soap grease

3.5.2 Extreme-pressure properties

Extreme pressure additives (EP) improve, in general, the load-carrying ability in most rolling contact bearing and gears. They react with the surface to form protective films which prevent metal to metal contact and the consequent scoring or welding of the surfaces. The EP additives are intended to improve the performance of grease. In this respect, the selected greases are usually tested in a four ball machine where a rotating ball slides over three stationary balls using ASTM-D 2596 procedure. The weld load data for the selected greases G_{1G}, G_{2C} and $\mathbf{G_{3D}}$ are 170, 195 and 250 Kg, respectively. These results indicate that the selected grease containing jojoba oil and jojoba meal G_{3D} exhibit remarkable improvement in extreme pressure properties compared with grease without additives G_{1G} and grease G_{2C} with jojoba oil alone. This may be attributed to the synergistic effect of the complex combination among Li-soap, amino acids, and polyphenolic compounds scheme (1 &2), in addition to the role of anion (PO43-, SO42-, Cl- and F-) and cation (Li+, Na+, K+, Ca2+, Mg2+, Al3+, Fe2+, Cu2+, Ba2+, Sr2+, Mn2+, Zn2+, Co2+ and Ni2+) in jojoba meal. These chemical elements are in such a form, that under pressure between metal surfaces they react with the metal to produce a coating film which will either sustain the load or prevent welding of the two metals together. This view introduces the key reasons for the improvements of the load- carrying properties and agrees well with the data previously reported by El-Adly et al (2004).

On other hand, it has been found that some thickening agents used in grease formulation inhibit the action of EP additives (Silver & Stanly 1974). The additives most commonly used as anti-seize and anti-scuffing compounds are graphite and molybdenum disulphide.

3.5.3 Oxidation stability

The oxidation stability of grease (ASTM D-942) is the ability of the lubricant to resist oxidation. It is also used to evaluate grease stability during its storage. The base oil in grease will oxidize in the same way as lubricating oil of a similar type. The thickener will also oxidize but is usually less prone to oxidation than the base oil. So, anti-oxidant additive must be selected to match the individual grease. Their primary function is to protect the grease during storage and extend the service life, especially at high temperatures.

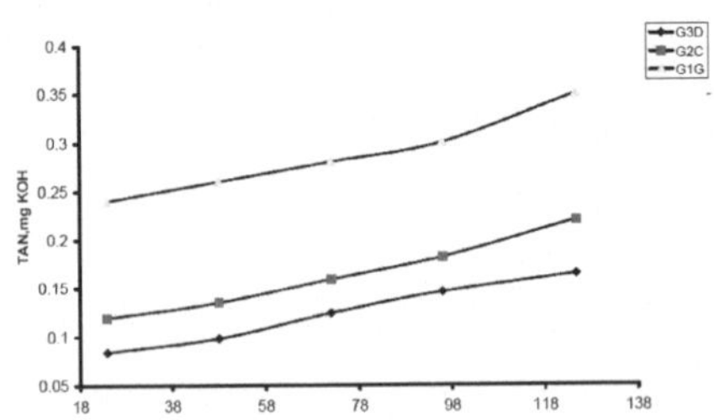

Fig. 3. Effect of deterioration time on Total Acid Number for selected greases

Oxidative deterioration for the selected greases G_{1G}, G_{2C} and G_{3D} are determined by the total acid number at oxidative times ranging from zero to 120 hours Figures (3). In addition, pressure drop, in psi. at 96 hour for greases G_{1G}, G_{2C} and G_{3D} are 4.0, 3.0 and 1.5 psi respectively. These results give an overview on the efficiency of the jojoba meal and jojoba oil in controlling the oxidation reactions compared with the grease without additive G_{1G}. Jojoba oil in conjunction with jojoba meal additive proves to be successful in controlling and inhibiting the oxidation of the selected grease G_{3D}. Inhibition of oxidation can be accomplished in two main ways: firstly by removal of peroxy radicals, thus breaking the oxidation chain, secondly, by obviating or discouraging free radical formation. A suggested mechanism for this inhibition is illustrated in the Schemes (2 & 3). The efficiency of jojoba meal ingredients as antioxidants is here postulated due to the presence of phenolic groups and hyper conjugated effect. Accordingly, Simmondsin derivatives and polyphenolic compounds which are considered the main component of jojoba meal include in their composition electron rich centers, which act as antioxidants by destroying the peroxides without producing radicals or reactive oxygenated products.

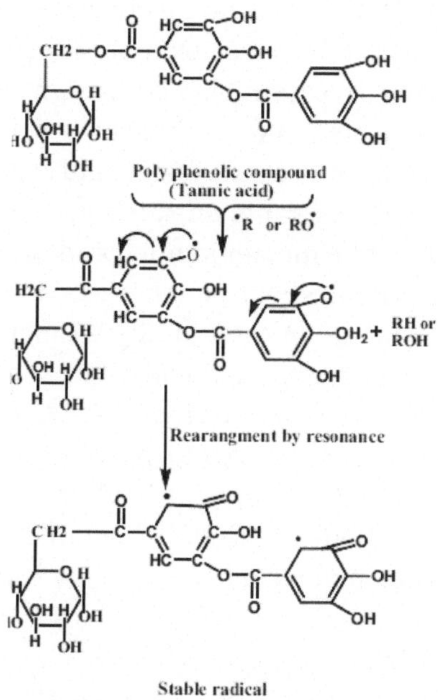

Scheme 2. The role of Polyphenolic compounds as antioxidant for prepared lithium grease

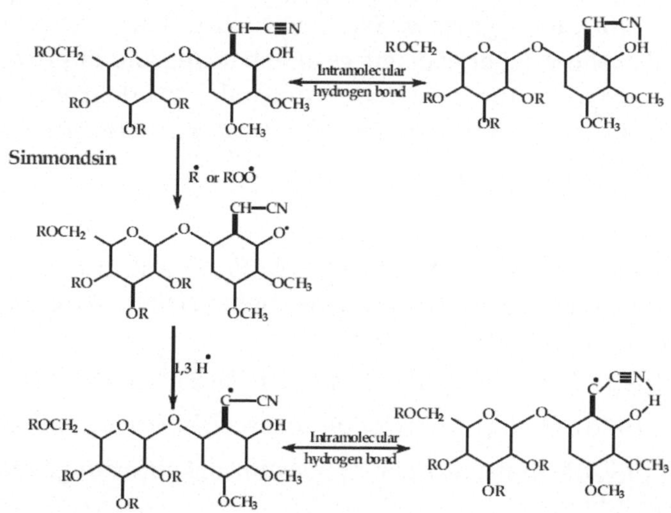

Scheme 3. The role of the Simmondsin as antioxidant for prepared lithium grease

4. Future research

Base oils used to formulate greases are normally petroleum or synthetic oils. Due to growing environmental awareness and stringent regulations on the petroleum products uses, research and development in the area of eco-friendly grease is now gaining importance. Since biodegradable synthetic ester lubricant is higher in cost, vegetable oils are drawing attention economically as biodegradable alternates for synthetic esters. Looking forward into the next decade, the need for more advanced science in grease technology is essential. The design of special components is becoming increasingly complicated and machines are becoming much smaller and lighter in weight and are required to run faster and withstand heavier loads. To be able to develop the optimal lubricants for these new conditions, the mechanism behind grease lubrication must be further studied and understood. There will be an increased specialization in both products and markets and the survival of individual lubricants companies will depend on their ability to adapt to changing conditions. Not only machines but also new materials will affect the development of greases. Biogreases (El-Adly et al 2010) and nanogrease have better lubricating properties such as, wear protection, corrosion resistance, friction reduction, heat removal, etc. In this respect, anti-friction, anti- wear and load-carrying environment friendly additives are prepared from non-traditional vegetable oils and alkyl phenols of agricultural, forest and wasteland origin (Anand, et al, 2007).

5. Conclusion

Lubricating grease is an exceptionally complex product incorporating a high degree of technology in all the related sciences. The by-products, soapstock, bone fat, jojoba meal, produced from processing crude vegetable oils are valuable compounds for lubricating greases. Such byproducts have varieties of chemical compounds which show synergistic effect in enhancing and improving the grease properties. Advantages of these byproducts include also their low cost and large scale availability. Research in this area plays a great role in the economic, scientific and environmental fields.

References

1. Anand, O. N; Vijay, k.; Singh, A.K. & Bisht, R.P. (2007). Anti-friction, Anti-Wear and Load- Carrying Characteristics of Environment Friendly Additive Formulation, *Lubrication Science* Vol.19, pp. 159-167.
2. Barnes, J.(1999). *Non-Newtonian Fluid Mech.* Vol.81, pp. 133-178.
3. Boner, C. J. (1976). Modern Lubrication Greases, *Scientific Publications (GB) Ltd.*
4. Boner, C.J. (1954). Manufacture and Application of Lubricating Greases, *New York Reinhold Publishing*.
5. Cann, P.M. (1997). Grease Lubrication Films in Rolling Contacts, *Eurogrease* Nov-Dec 1997, pp. 6-22.
6. Cardoso, F. A. & Price, R. L. (1980). Extraction, Charachterization and Functinal Properties of Jojoba Proteines. In: M. Puebla (Ed.) Proceedinf of *Forth International Conference on Jojoba and its Uses*, Hermosillo, pp 305-316.
7. Cherry, J. P, & Berardi, I.C. (1983). Cottonseed, *Handbook of Processing and utilization in Agriculture*, Vol.II, edited by I.A.wolff, CRC press Inc, Boca Raton

8. Daugherty, P.M.; Sineath, H.H. & Wastler, T.A. (1953). Industrial Raw Material of Plant Origin, IV.A Survey of *Simmondsia Chinensis, Bull.Eng. Exp. Sta., Georgia Inst.Technol., 15(13).*
9. El-Adly R. A. (1999). Producing Multigrade Lubricating Greases from Animal and Vegetable Fat By-products. *J. Synthetic Lubrication*. Vol.16, No.4, pp. 323-332.
10. El-Adly R. A.; El-Sayed S. M. & Ismail M. M. (2005). Studies on The Synthesis and Utilization of Some Schiff's Bases: 1. Schiff's Bases as Antioxidants for Lubricating Greases. *J. Synthetic Lubrication* Vol.22, pp. 211-223.
11. El-Adly, R.A & Enas A. Ismail. (2009). Study on Rheological Behavior of Lithium Lubricating Grease Based on Jojoba Derivatives. *11thLubricating Grease Conference,* Mussoorie, India. February 19-21 2009 (NLGI India Chapter)
12. El-Adly, R.A.; El-Sayed, S.M. & Moustafa, Y.M. (2004). A Novel Application of Jojoba Meal as Additives for Sodium Lubricating Grease, *The 7th International Conference on Petroleum & the Environment,* Egyptian petroleum Research Institute In Cooperation with EURO-Arab Cooperation Center & International Scientists Association, Cairo, Egypt. March 27-29 2004.
13. El-Adly,R.A. (2004). A Comparative Study on the Preparation of Some Lithium Greases from Virgin and Recycled Oils, *Egypt J. Petrol* Vol.13, No, 1. pp. 95-103.
14. El-Adly, R.A.; Enas, A.Ismail. & Modather, F. Houssien. (2010). A Study on Preparation and Evaluation of Biogreases Based on Jojoba Oil and Its Derivates , *The 13th International Conference on Petroleum & the Environment,* Egyptian petroleum Research Institute In Cooperation with EURO-Arab Cooperation Center & International Scientists Association, Egypt, March 7-9 2010.
15. El-Shattory, Y. (1979). Statistical Studies on Physical and Chemical Characteristics, Phospholipids and Fatty Acid Constitution of Different Processed Cottonseed Soapstock, *Rev. Fr. Corps Gras.* Vol, 26, pp.187-190.
16. Erlich, M. (ed). (1984). NLGI Lubricating Grease Guide, *NLGI, Kansas City.*
17. Flaxman,M.T.(1940). Sulfurized Lubricating Oil, *U.S.Patant* 2,212,899.
18. Greene, R.A. & Foster, E. D. (1933). The Liquid Wax of Seeds of *Simmondsia Californica*, Bot. Gaz., Vol.94, pp. 826-828.
19. Heilweil, I.J, (1988). Review of Lubricant Properties of Jojoba Oil and Its Derivatives, *Am.Oil Chem. Soc.*Vol, 9, pp 246-260.
20. Ismail, I.A. (2008). A Study on the Utilization of Jojoba Oil and Meal as Additives for Lubricating Oils and Greases, Ph.D. Thesis Ain Shams University, Egypt.
21. Gatto, V. J. & Grina, M.A. (1999). Effect of Base Oil Type, Oxidation Test Conditions and Phenolic Antioxidant Structure on the Detection and Magnitude of Hindered Phenol/ Diphenylamine Synergism, Lubrication Engineering, Vol.55, pp.11-20.
22. Gow, G. (1997). Lubricating Greases, in Chemistry and Technology of Lubricants, 2nd edn, (Eds R.M. Mortiers, S. T. Orszulik), Blackie Academic and Professional, London, pp 307-319.

23. Kieke, M. L. (1998). Microwave Assisted Digestion of Zinc, Phosphorus and Molybdenum in Analysis of Lubricating greases, *NLGI Spoksman* Vol.62, pp. 29-35.
24. Kinnear, S. & Kranz, K. (1998). An Economic Evaluation of 12- Hydroxyl Stearic Acid and Hydrogenated Castor Oil as Raw Materials for Lithium Soap Lubricating Grease, *NLGI Spokesman*, Vol.62, No.5 pp.13-19.
25. Klamann, D. (1984). Lubrications and Related Products: Synthesis, Proprties, Applications, International Standards, *Verlag Chemie, Weinheim.*
26. Kono, Y.; Tomita, K.; Katsura, H. & Ohta, S. (1981). Antioxidant in Jojoba Crude Oil, In: Puebla, (Editor), Proceedings of the Fourth International Conference on Jojoba, Hermosillo, pp 239-256.
27. Kuester, J. L. (1984). *Energy Biomass Wastes* vol.8,pp 1435
28. Kuester, J. L.; Fernandez Carmo,T.C. & Heath, G. (1985). *Fundam. Thermochem .Biomass Convers*: 875.
29. Lansdown, A. R. (1982). Lubrication, A Practical Guide to Lubricant Selection, Pergamon Press, Oxford
30. Luckham, P. F.& Tadros,Th.F. (2004). Steady Flow and Viscoelastic Properties of Lithium Grease Containing Various Thickener Concentration, *Journal of colloid and Interface Science*, vol.274, pp 285-293
31. Mang, T. & Dresel, W. (2001). Lubricants and Lubrication, *WILEY-VCH,* ISBN 3-527-295-36- 4, New York
32. Michael, K. Dowd. (1996). Compositional Characterization of Cottonseed Soapstock, *J.Am.Oil.Chem.Soc,* Vol 73, No.1o, pp 1287-1295.
33. Miwa, T.K, (1980). Chemical Structure and Propreties of Jojoba Oil, In: M. Puebla (Editor), Proceeding of *the Fourth International Conference on Jojoba*, Hermosillo, pp pp 227- 235.
34. Miwa, T.K. & Hagemann, J.W. (1978). Physical and Chemical Properties of Jojoba Liquid and Solid Waxes, In: Proceedings of *the Second International Conference on jojoba and Iits Uses,* Bnsenada, pp 245-252, 1976.
35. Miwa, T.K. & Rothfus, J.A. (1978). In-depth Comparison of Sulfurized Jojoba and Sperm Whale Oils a Extreme Pressure Extreme Temperature Lubricants, In: D.M. Yermanos (Editor), Proceeding of *the Third International Conference on Jojoba and Its Uses*, Riverside, Calif., pp 243-267
36. Miwa, T.K. (1971). Jojoba Oil wax Esters and Derived Fatty Acids and Alcohols, Gas Chromatographic Analysis , *J.Am.Oil Chem.*, Vol.48, pp 299-264
37. Miwa, T.K. (1973). Chemical Aspects of Jojoba Oil, a Unique Liquid Wax from Desert Shrub *Simmondsia californica*, Cosmet. Perfum, Vol.88, pp 39-41
38. Peeler, R.F.& Hartman, L.M,(1972). Evaluation of Sulfurized Sperm Oil Replacements, *NLGI Spokesman,* vol.37, No.17
39. Pohlen, M. J. (1998). DSC- A Valuable Tool for the Grease Laboratory, NLGI Spokesman, Vol.62, pp 11-16.

40. Robison, P. D.; Salmon, S.G.; Siber, J. R. & Williams, M.C. (1993). Elemintal Analysis of Greases, NLGI Spokesman, Vol.56, pp 157-160.

41. Schultze, G. R. (1962). Wesen and Eufbau Derschmierfette in Zerbe, C. Mineralole and verwandte Produkte, *2nd edn, Springer* Berlin, pp .405-432.

42. Shirahama.(1985). The Effects of Temperature and Additive Interaction on Valve Train Wear, Proc. *JSLE.Int. Trib.Conf.* Tokyo,Japan, pp 331-336 8-10 July

43. Silver B.H.& Stanley R.I. (1974). Effect of The Thickener on The Efficiency of Load Carrying Additives in Greases, *Tribology International*, Vol.7, pp 113-118.

44. Sinitsyn, V. V. (1974). The Choice and Application of Plastic Greases, *Khimiya, Moscow.*

45. Spencer, G.F.; Plattner, R.D.& Miwa, T. K, (1977). Jojoba Oil Analysis by High Pressure Liquid Chromatography/Mass Spectrometry, *J. Am. Oil Chem. Soc.*, vol.54, pp 187- 189.

46. Verbiscer,A. J.; Banigen,T. F. ;Weper, C. W. ; Reid, B. L.; Tlei, J. E.& Nelson, E. A. (1978). Detoxification and Analysis of Jojoba Meal. In: D. M. Yermoanos (Ed.) Proceeding of the *Third International Conference on Jojoba and Its Uses*, Riverside, Calif. pp 185- 197.

47. Vinogradov, G.V. (1989). Rheological and Thermophysical Properties of Grease, Gordon and Breach Science Publications, London.

48. Wells, F. B. (1948). Process of Making Sulfurized Jojoba oil *U.S. Patent* 2,450,403.

49. Wills, J. G. (1985). Jojoba, New Crop for Arid Lands, New Material for Industry, National Research Council, National Academy Press, Washington, No,6, pp,130-150.

50. Wisniak, J. (1987). The Chemistry and Technology of Jojoba Oil, *American Oil Chemists Society,* Champaign, Illinois.

51. Wisniak, J. (1994). Potential Uses of Jojoba Oil and Meal-a Review, *Industrial Crops and Products* Vol.3 pp, 43-68.

52. Wassermann, G. From Heraklit to Blair,W. S. (1991). *Rheology,* Vol.91 pp 32-38.

9

Characterization of Lubricant on Ophthalmic Lenses

1. Introduction

When people started wearing eye-glasses from the 13th century until the middle of 20th century, the glass was the only material used for ophthalmic lenses. However, plastic lenses were rapidly developed and began to be widely used when PPG Industries, Inc. developed CR-39® in 1940; CR-39®, i.e., allyl diglycol carbonate (ADC), is a thermosetting resin that can be used as a lens material with a refractive index of 1.5. The features of this material are as follows: (1) it is a lightweight material (its specific gravity is half of that of glass), (2) it has strong impact resistance (i.e., it is shatter proof, which guarantees high safety), (3) it is stainable (i.e., has high fashionability), and (4) it can be used in a variety of frames (i.e., it has high fashionability or high workability). The quest for thinner lenses led to an increase in the refractive index of lenses, and current lenses have a super-high refractive index of 1.74 or 1.76.

The biggest drawback of plastic lenses was that they could be "easily scratched," but they were improved sufficiently for practical use, by using a hard coating (HC), i.e., an overcoat formed on the plastic substrate. Subsequently, anti-reflection (AR) coating films were added to increase the clearness of the lens, to reduce the reflection from the ophthalmic lens as viewed by another person, and even to enhance measures for preventing scratches. In recent years, further value-adds have been made to plastic lenses, with the use of lubricants in the top layers for increasing durability, preventing contamination due to scratches on spectacle lenses, and facilitating "easy removal" of dirt.

Research on lubricants used for the improvement of tribology characteristics has progressed rapidly; it has been supported from the end of the 1980s by the development of surface analysis methods (Kimachi et al., 1987; Mate et al., 1989;

Novotny et al., 1989; Newman et al., 1990; Mate et al., 1991; Toney et al., 1991; Novotny et al., 1994; Sakane et al., 1999; Tani, 1999; Tadokoro et al., 2001; Tadokoro et al., 2003) and by the technology for high-density magnetic disc recording used in personal computers. The main lubricant selected was perfluoropolyether (PFPE), because it possesses thermal stability, oxidation stability, low vapor pressure, low surface tension, and good boundary lubricity. It was effective in reducing the frictional wear of the surfaces of the magnetic disc and magnetic head, and thus, hundreds of thousands of stable data read-and-write operations could be conducted. The main parameters that determine lubricant properties are the structure, thickness, and state of the lubricant, and various methods were used to investigate them.

On the other hand, the purpose of using a lubricant for ophthalmic lenses is to improve a scratch resistance, to prevent contamination, and to facilitate "easy removal" of dirt; the tribology characteristics of such a lubricant are similar to those of the lubricant used on magnetic discs, and has possibilities of application. There are two differences between lubricants used for ophthalmic lenses and those used for magnetic discs: (1) the film thickness of the lubricant used for magnetic discs does not need to be reduced, because the recording density achieved by using the lubricant for the magnetic disc increases exponentially when the gap between the magnetic disc surface and magnetic head is reduced as much as possible (to approximately 1 nm), and (2) the lubricant for ophthalmic lenses needs to be solid, but magnetic discs can be solid or liquid if stiction, in which a magnetic head sticks to the surface of a magnetic disc does not occur. However, in the case of ophthalmic lenses, dirt, dust, and fingerprints frequently block the view of the user, and the user cleans the lenses with water or rubs them with a soft cloth or paper; therefore, liquid lubricants can cause adhesion problems and does not last for a long time. In reality, conference presentations and papers are limited to information provided by the authors (Tadokoro et al, 2009; Tadokoro et al, 2010; Tadokoro et al, 2011). This chapter discusses tribology, with a focus on the characterization of lubricants, and presents analysis and evaluation results based on the film thickness, structure, distribution, and abrasion resistance of lubricants reported by the authors.

2. Scratches and dirt

Figure 1 shows optical microscopic pictures of ophthalmic lens returned by a consumer who complained about the quality. The different colors in the picture demonstrate the peeling of the AR coating films along the scratch, and thus, the small scratches become visible. Details on how and when the lenses were used are unknown, but it must be understood that scratches actually occur and this problem must be taken into account; this picture shows the importance of surface reforming based on the use of lubricants. While scratch-free lenses cannot be made only by modifying lubricants, the lubricant is one of the most important factors that affect the formation of scratches. Figure 2 shows the results of an abrasion test conducted by scrubbing a lens 20 times with 20 kg steel wool for different lubricants. The results show that the formation of scratches can be controlled by changing the structure or the distribution state of the lubricant. Finally as an example of the comparison of dirt adhesion, figure 3 shows the adhesion of cedar pollen on the lens. In Japan, hay fever, a seasonal allergy caused by cedar pollen, is very common (30% of the citizens have this

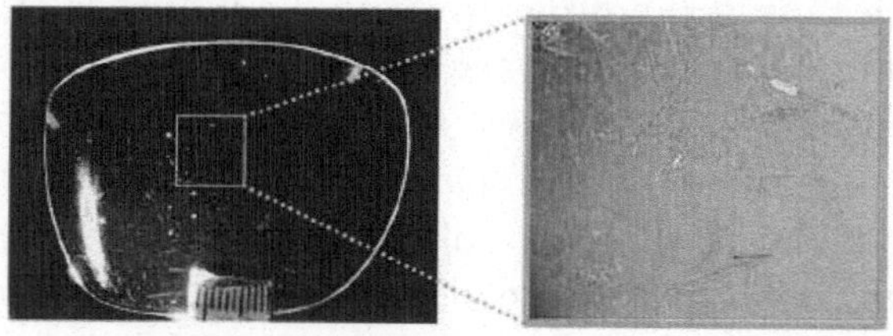

Fig. 1. Damaged ophthalmic lens and scratches

allergy). The results in figure 3 show that changing the surface condition reduces the amount of pollen adhered to the ophthalmic lens brought indoors. As in the example of scratches, the results show the possibility that the surface condition can be controlled to change the amount of dirt that adheres to the lenses.

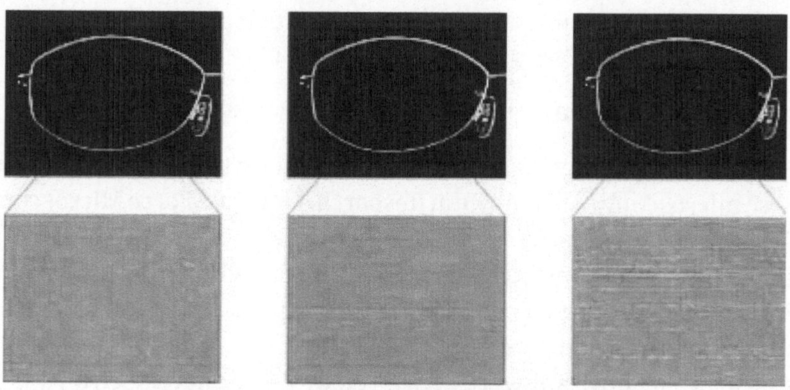

Fig. 2. Scratch test results for 3types lubricants: the lens was scrubbed 20 times with 2 kg steel wool

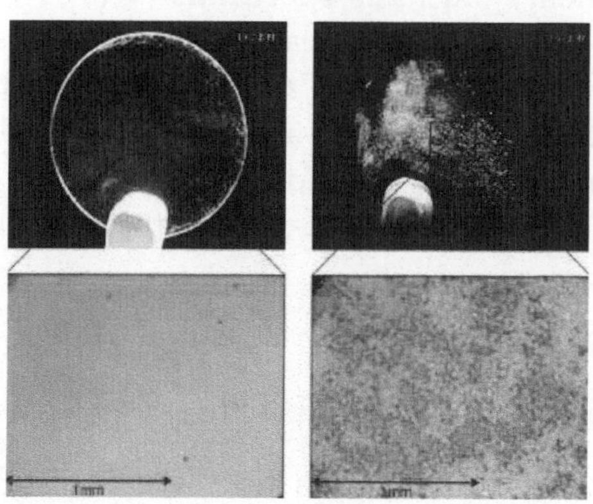

Fig. 3. Comparison between surface condition and cedar pollen adheres to the lens

2.1 Experimental

2.1.1 Sample preparation

Commercial ophthalmic lenses of allyl diglycole carbonate (ADC, CR-39®) were used in this study. In addition, the detailed estimations of lubricants were carried out directly on silicon wafer in order to avoid the influence of surface curvature, roughness, or amorphous states of actual ophthalmic lenses. The structures of the ophthalmic lenses were as follows: a sol-gel based underlayer on the plastic lens substrate was deposited by dip coating or spin coating methods. The HC material was made using a silica sol and 3-glycidoxypropyltrimethoxysilane. The thickness of HC was approximately 3500 nm. AR coating layers, composed of a sandwich structure between low-index material (SiO_2) and high-index material (Ta_2O_5), were deposited by vacuum deposition methods after the HC underlayer was cleaned by ultrasonic washing with detergent and de-ionized water. The total film thickness was approximately 620nm. The PFPE lubricants, which were also commercial products, were deposited over the AR coating layers by the vacuum deposition methods. The main structure of lubricants A, B, C, G, and H has ($-CF_2-CF_2-O-$)m-(CF_2-O-)n, the main structure of lubricants D and F has ($-CF(CF_3)-CF_2-O-$) m', the main structure of lubricant E has ($-CF_2-CF_2-CF_2-O-$)m''.

2.1.2 Analysis and evaluation methods

The surface morphology and the lubricant film distribution were examined by atomic force microscopy (AFM; Asylum Research, Molecule Force Microscope System MFP-3D). The film thickness, morphology of the cross section, and elemental analysis were used by transmission electron microscopy (TEM-EDS; JEOL, JEM-200FX-2). For the TEM observation, a Cr protective layer was deposited onto the lubricants layer in order to identify a top surface of the lubricants films. The film thickness and the coverage ratio of the lubricant were measured by X-ray photoelectron spectroscopy (XPS; Physical Electronics, PHI ESCA5400MC). Structure analysis was conducted by time-of-flight secondary ion mass spectrometry (TOF- SIMS; ULVAC-PHI, PHI TRIFT-3 or PHI TRIFT-4) and XPS. The wear properties of lubricants were evaluated by contact angle measurement (Kyowa Interface Science Co.,Ltd.; Contact angle meter, model CA-D) and by the use of an abrasion tester (Shinto Scientific Co., Ltd.; Heidon Tribogear, Type 30S). The abrasion test was rubbed in the Dusper K3(Ozu corp.) to have wrapped around the eraser under the condition of 2 kg weight and 600 strokes.

2.2 Results and discussion

2.2.1 Cross-sectional structure, film thickness and coverage of lubricants

Figure 4 shows an example of TEM photograph of lubricant B on a silicon wafer. Figure 5 and figure 6 show an EDS analysis area of TEM photograph and an EDS spectrum of lubricant B. Table 1 summarized the lubricant film thickness and coverage ratio by XPS and TEM. The thickness of the lubricant layer was estimated to be 2.6 nm. And also, we recognized fluorine element in this area by TEM-EDS. These data indicate that both the film thicknesses and the coverage ratios were almost identical across all films. Here, we directly measured the film thickness by TEM. Despite the fact that the lubricant layer was comprised of organic materials, the existence of the lubricant film was directly observed and the film thickness was successfully measured by TEM. Generally, the issue of TEM measurement is sample damage by electron beam. For the reason of successful measurement by TEM, it seems that the lubricant damage of ophthalmic lens is stronger than that of the magnetic disk for electron beam.

It is well-known that the film thickness is proportional to a logarithmic function of the intensity ratio of photoelectrons. According to Seah and Dench (1979), they reported the escape depth of electrons of organic materials with electron kinetic energy by the following equation; they provide a set of relations for different classes of material over the energy range 1 eV – 6keV (Briggs & Seah, 1990).

$$\lambda_m = 49/E_k2 + 0.11\bullet E_k0.5 \quad (1)$$

$$\lambda_{lub\ F} = \lambda_m / \rho \quad (2)$$

where $\lambda_{lub\ F}$ is the escape depth of F_{1s} photoelectron of lubricants, λ_m is the escape depth of monolayers for organic materials, E_k is electron kinetic energy, and ρ is the density of material.

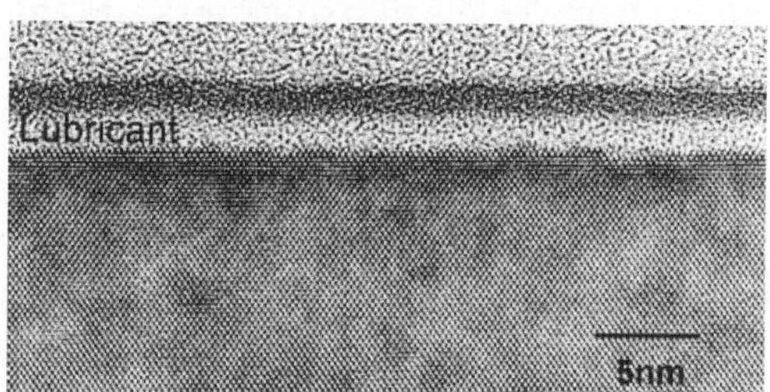

Fig. 4. TEM cross-sectional photograph (glue/Cr layer/lubricant/Si wafer) of lubricant B

Fig. 5. TEM photograph of lubricant B on a silicon wafer. (Blue area shows the EDS analysis area)

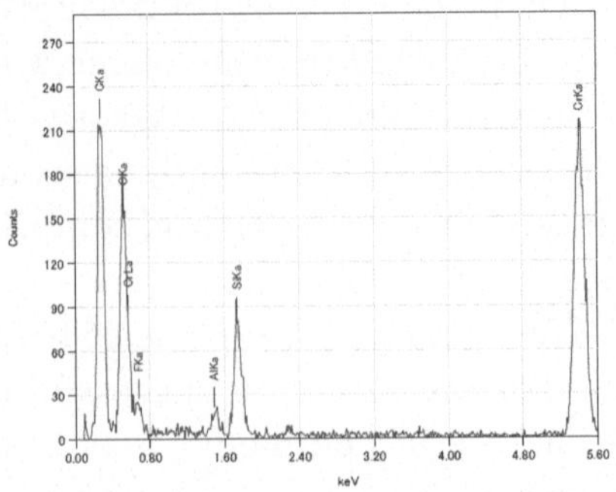

Fig. 6. TEM-EDS spectrum for lubricant B on a silicon wafer

Table 1. Film thickness and coverage ratio of lubricant by XPS and TEM

	Lub. film thickness (nm)	Lub. film coverage by XPS (%)	Lub. film coverage by TEM (%)
Sample A	1.5-1.7	98 over	100
Sample B	2.3-2.7	98 over	100
Sample C	2.3-2.7	98 over	100
Sample D	2.1-2.5	98 over	-----
Sample E	1.7-2.2	98 over	-----

The lubricants film thickness of XPS was calculated by the following equation (3). Table-2 summarizes the parameters used. We experimentally calculated the A factor by using equation (3) from TEM's film thickness and the intensity ration of F_{1s} and Si_{2p} photoelectron (the experimental A factor is 0.116).

$$T = \lambda_{lub\ F} \cdot \sin\vartheta \cdot \ln [A \cdot (I_{lub\ F} / I_{Si}) + 1] \quad (3)$$

where T is the film thickness of lubricants , ϑ is the detection angle of XPS measurement, $I_{lub\ F}$ is the intensity of F_{1s} photoelectrons, I_{Si} is the intensity of Si_{2p} photoelectrons, A is the correction factor (calculated value: 0.116, i.e., lubricants films thickness by TEM).

According to Kimachi et al. (1987), they have derived an expression for the coverage ratio of lubricants on magnetic disks using an island model. In the present study, we propose a modified equation (4) for the coverage of our lubricants using the F_{1s} and the Si_{2p} photoelectrons.

$$A \cdot (I_{lub\ F}/I_{Si}) = \{r \cdot [1-\exp(-T/(\lambda_{lub\ F} \cdot \sin\vartheta))]\}/\{(1-r)+r \cdot \exp(-T/(\lambda_{Si} \cdot \sin\vartheta))\} \quad (4)$$

where r is the coverage ratio from 0 to 1.

Figure 7 shows an example of the relationship between the logarithmic function of the intensity ratio of photoelectron and the coverage ratio. Table 1 already summarized the lubricant film thickness and coverage ratio by XPS and TEM. The coverage ratio of lubricants by XPS is estimated to be over 98%. However, the coverage ratio of TEM seems to be covered a fully 100 % on Si wafer. In case of an actual XPS measurement, a coverage ratio of 100% is unlikely to occur due to the influence of surface roughness, the density of actual lubricants films, and the photoelectron signal of Si_{2p}. Therefore, it seems that the lubricant layer completely covers on the Si wafer when the coverage ratio is approximately 100%. By using this XPS technique, we can easily monitor the lubricant thickness and coverage ratio on a production line for quality control.

Table 2. The escape depth and parameters used

	B.E (eV)	$\lambda_{lub\ F}$ (nm)	ρ (kg/m3)
Lub. F_{1s}	689	1.45	1.8x103

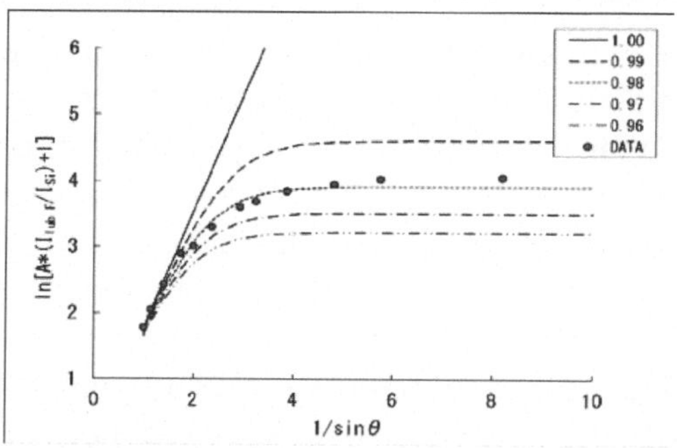

Fig. 7. Coverage calculation results of sample B by XPS measurement

2.2.2 The distribution state of lubricants

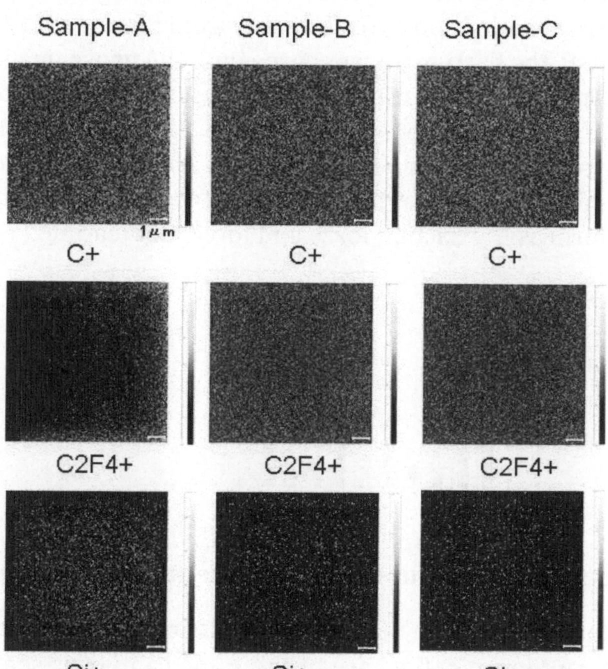

Fig. 8. TOF-SIMS image (C+, C2F4+, and Si+ fragment ions) for each sample

Figure 8 illustrates the lubricant distribution of samples A, B, and C by TOF-SIMS analysis. The image was obtained by detecting the positive ion fragments of C+, C2F4+, and Si+. The ion signal intensity is displayed on a scale of relative brightness; bright areas indicate high intensity of each type of fragment ion. Figure 9 shows the comparison of lubricants fragment ion for samples A, B and C. From fugure-9, we recognized that these samples have same main structure of (-CF_2-CF_2-O-)m-(CF_2-O-) n. The lubricant distribution determined by this analysis was consistent with the actual lubricant distribution. The behavior of the lubricant distribution obtained is attributable to suggest chemical structure and mechanical property of lubricant. Therefore, in terms of elemental fragment ions, the distribution of the lubricant appears to be homogenous at the 10mm scale from figure 8.

Figure 10 illustrates the lubricant distribution of samples A, B, and C by AFM topographic image and friction force image at the 10 mm scale. Figure 11 shows a frequency analysis of phase separation for sample A and sample B. A red histogram shows the whole area, a blue area shows the phase separation A of lubricants, and a green area shows phase separation B of lubricants. Area distribution of sample 2 has approximately two times larger than that of sample 1. Figure 12 shows the lubricant image of sample B by using phase image and force modulation image. The components between the in-phase (input-i: elasticity) and the quadrature (input-q: viscosity) divided phase image are shown in figure 13. From the TOF-SIMS fragment image in figure 8, we recognized the homogeneity of lubricant distribution for sample A, sample B and sample C. However, we found that the uniformity or heterogeneity of an image depended upon the sample and the scale, except for topographic images by AFM measurement added some functionality from figures 10, 12, and 13. Here, in the case of sample B, the friction images agree with the phase images and the phase images agree to the force modulation images. Thus, the friction force image reveals the distribution of friction behavior on the surface. Also, the force modulation image indicates the distribution of hardness; the darker areas correspond to softer areas. Thus, the phase image suggests friction or hardness behavior because it assumes the same image form as the friction force and force modulation.

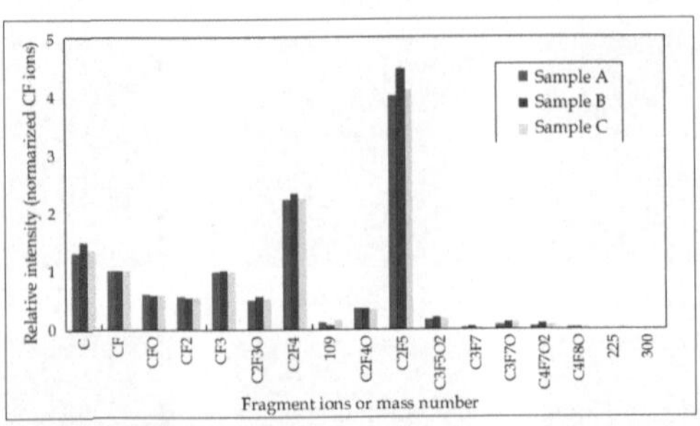

Fig. 9. The comparison of lubricants fragment ion for sample A, B and C

By friction force microscopy (FFM), the twisting angle is proportional to the tip height of the cantilever in the case of the same cantilever shape and the same material (Matsuyama, 1997).

$$\vartheta_o = \mu \cdot F_L \cdot (h_t + t/2) \cdot L/(r \cdot G \cdot w \cdot t3) \quad (5)$$

where ϑ_o: twisting angle, L: length of cantilever, r: correction factor (calculated value 0.3 to ~0.4), G: shear modulus, w: width of cantilever, t: thickness of cantilever, μ: friction coefficient, F_L: load force, h_t: height of cantilever.

In previous work (Tadokoro et al., 2001), we observed the morphology of lubricants on the magnetic disk surface by FFM. The images of lubricants obtained by a high-response cantilever of tip height 8.4 mm were clearer than those by a standard cantilever of tip height 3 mm in the same load force. The sensitivity of the high-response cantilever was about 2 to 3 times greater than that of the standard cantilever when compared in the same sample area. These observations seemed to experimentally support the theoretical predictions, and the effects of load force for the standard cantilever agree with the theoretical equation. However, FFM has two disadvantages. If the area is too small (i.e., <1 mm) and is low-friction material, the friction force signal is drastically reduced. Moreover, there might be damage to the lens surface because the friction force image is made by contact. On the other hand, the disadvantage of force modulation methods is that the tip can change shape and is a possible source of contamination because it is always pushed into the sample (indentation). Therefore, we believe that it is more convenient to use phase images than friction force images or force modulation images for determining the island structures of shapes with similar surface morphologies.

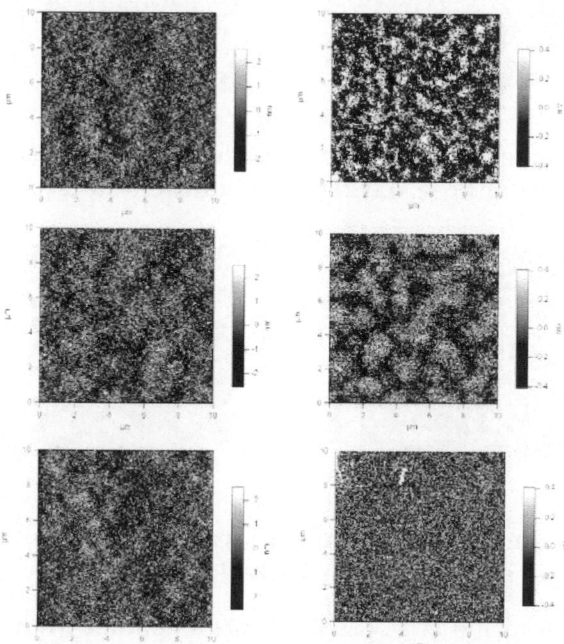

Fig. 10. Topographic image (left side), FFM image (right side; bright area indicates higher friction, darker area indicates lower friction); upper image is sample A, middle image is sample B, lower image is sample C

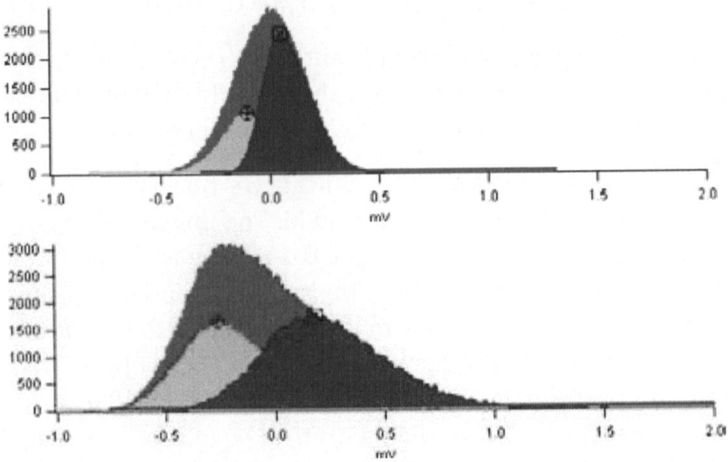

Fig. 11. Frequency analysis of phase separation by FFM (top distribution: sample A, bottom distribution: sample B), it shows red histogram for whole area, blue area for lubricant phase separation A, and green area for lubricant phase separation B

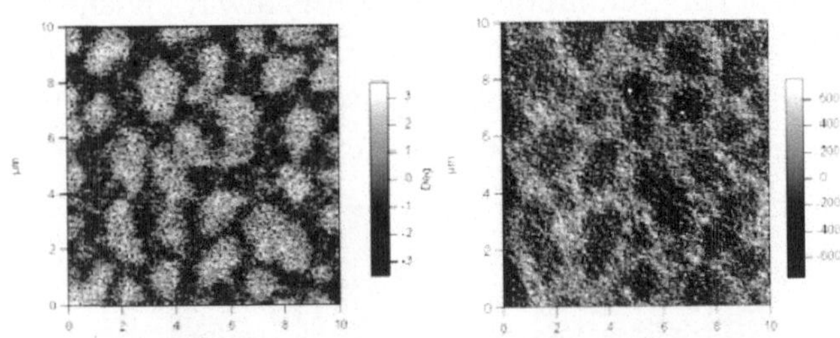

Fig. 12. Phase image (left side), force modulation image (right side; bright area indicates harder area, darker area indicates softer area) of sample B

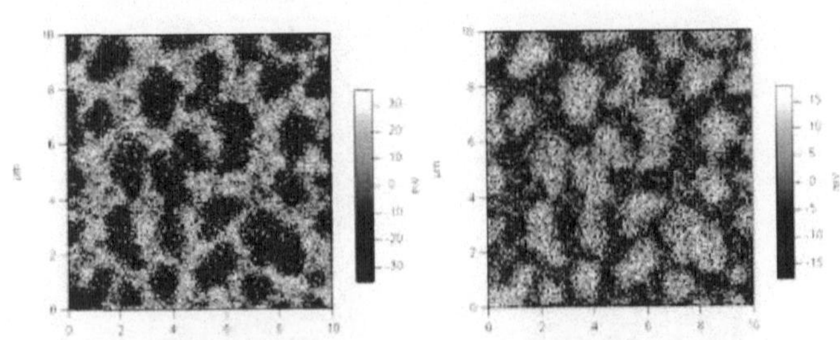

Fig. 13. In-phase image (input-i: left side) and quadrature image (input-q: right side) of sample B divided by phase image

According to Cleveland et al. (1998), if the amplitude of the cantilever is held constant, the sine of the phase angle of the driven vibration is then proportional to changes in the tip- sample energy dissipation. This means that images of the cantilever phase in tapping- mode AFM are closely related to maps of dissipation.

Our phase images suggest that the bright area corresponds to a higher phase because a phase image is taken in repulsive mode. The bright area is more energy-dissipated than the dark area, which means the bright area is softer or more adhesive. Because the phase image was divided by the components of the in-phase (input-i) and the quadrature (input-q), the relation of the in- phase (input-i) and the quadrature (input-q) is converse. It seems that an in-phase image (input-i) has the same tendency as the force modulation image: its darker area corresponds to a softer area. In general, the relation between an in-phase image (input-i) and a quadrature image (input-q) is the relation between elasticity and viscosity. Our observations seem to experimentally support this relation. Figure 13 demonstrates that the bright area of the in-phase image has lower energy dissipation than the darker area, which means the bright area is harder or less adhesive. On the other hand, the darker area in figure 12 (the force modulation image) corresponds to a softer area. If ophthalmic lens surface is sticky, a lot of contaminants can easily attach to the lens surface. Fortunately, the lubricant material is fluorocarbon, which has low surface energy. Thus, the contaminant is easily removed from the lens surface wiping the surface with a cloth. From these results, in the case of sample B, it appears that these island structures are mixtures of soft regions and hard regions at the 10 mm scale.

Figure 14 illustrates the lubricant distribution of sample D by AFM topographic image and phase image at the 10 mm scale. Figure 15 shows the lubricant image of sample C by topographic image, phase image, in-phase image (input-i), and quadrature image (input-q) at the 1 mm scale. Figure 16 shows the lubricant image of sample C by topographic image, phase image, in-phase image (input-i), and quadrature image (input-q) at the 500 nm scale. The topographic image, phase image, in-phase (input-i), and quadrature image (input-q) of sample D at the 1 mm scale are shown in figure 17. Finally, the topographic image, phase image, in-phase (input-i), and quadrature image (input-q) of sample E at the 1 mm scale are shown in figure 18.

In the case of samples C, D, and E at the 10 mm scale, island structures cannot be observed by phase image, although it seems that the lubricant is homogeneous in these areas. However, samples C, D, and E reveal some island structures at smaller scales (i.e., 500 nm scale and 1 mm scale). We earlier discussed the relation between friction force image, force modulation image, and phase image. Nevertheless, the signal mark depends upon the measurement mode; these images reveal island structures in cases of similar morphology.

In the case of sample C, it seems that the grain is too small and some clusters gather with different dissipation energies. The topographic image of sample D reveals unevenness of grain, but the phase image clearly shows the grain boundary. This suggests that the grain boundary in sample D is accumulated lubricants rather than grain. On the other hand, sample E has grain but the grain boundary in the phase image is not clearly apparent. It seems that the lubricant in the grain boundary is in accord with the lubricant on the grain, and the lubricant of sample E is more homogenous than that of sample C or D.

In some ophthalmic lenses, island structures can be observed on the lens surface at the 10 mm scale, whereas in others it is necessary to use the 1 mm or 500 nm scale. From these lubricant images we have determined that the morphologies of the lubricants of commercial phthalmic lenses vary widely and thus perform differently in terms of wear property and dirt protection. Therefore, the methods described here are useful and suitable for investigation of lubricants on ophthalmic lens surfaces.

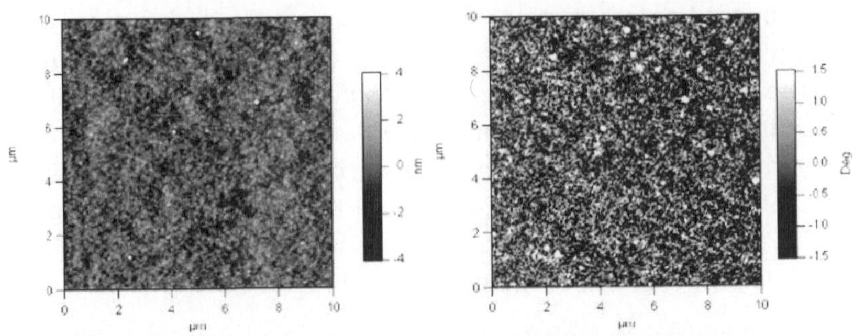

Fig. 14. Topographic image (left side), phase image (right side) of sample D

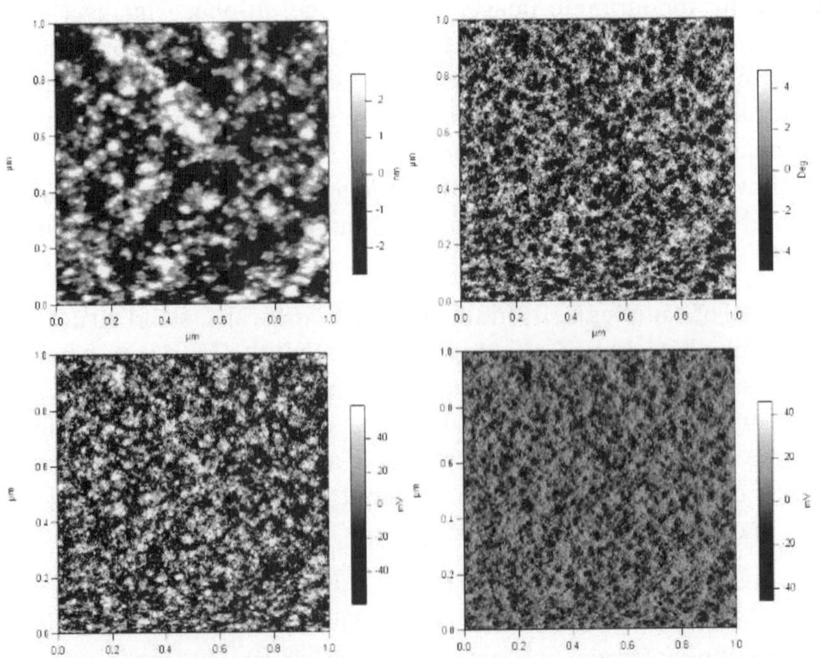

Fig. 15. Topographic image (upper left), phase image (upper right), input-i image (lower left,) and input-q image (lower right) of sample C at the 1 mm scale

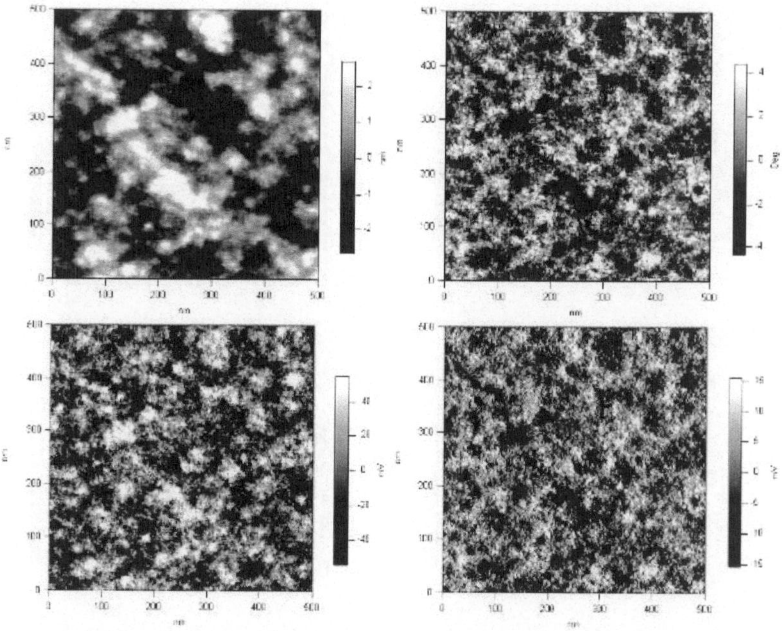

Fig. 16. Topographic image (upper left), phase image (upper right), input-i image (lower left), and input-q image (lower right) of sample C at the 500 nm scale

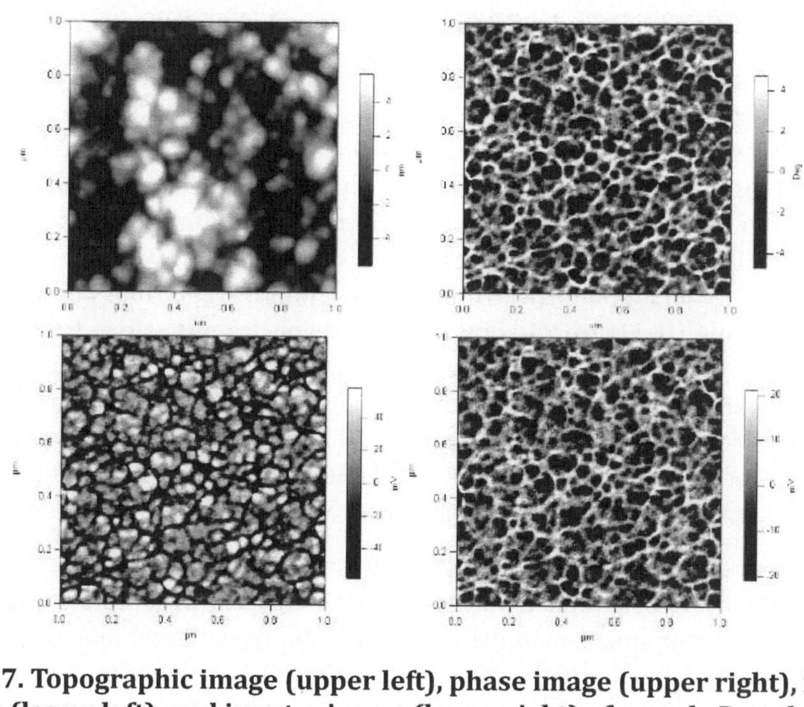

Fig. 17. Topographic image (upper left), phase image (upper right), input-i image (lower left), and input-q image (lower right) of sample D at the 1 mm scale

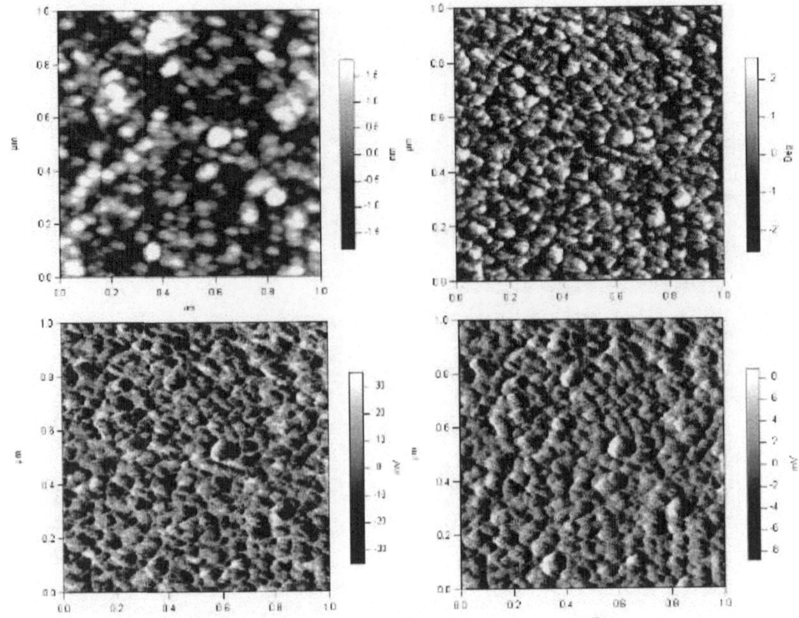

Fig. 18. Topographic image (upper left), phase image (upper right), input-i image (lower left), and input-q image (lower right) of sample E at the 1 mm scale

2.2.3 X-ray damage of lubricants and chimerical structures

Figure 19 shows the X-ray damage ratio of F_{1s} spectra for sample F, G, and H as a function of X-ray exposure time under the condition of X-ray power 300W and Mg-Kα source by XPS. Figures 20 - 22 show the changing chemical structure of C_{1s} for samples F-G as a function of exposure time (initial structure shown for reference, structure after 30 min, and structure after 60 min), as determined by XPS. Figure 23 shows the initial structure and of the mass spectra of positive fragment ions, as obtained by TOF-SIMS (upper spectrum: sample F, middle spectrum: sample G, lower spectrum: sample H). Figure 24 shows the mass spectra of positive fragment ions after 60 min X-ray exposure by XPS (upper spectrum: sample F, middle spectrum: sample G, lower spectrum: sample H). Figure 25 shows the mass spectra of negative fragment ions for sample F, as obtained by TOF-SIMS (upper spectrum: initial, lower spectrum: after 60 min, obtained by XPS). Table 3 summarized the film thickness and coverage ratio of lubricant before and after XPS damage.

From figure 19, we found that the X-ray damage in the case of sample F is greater than that in the case of sample G and sample H. In the case of sample G and sample H, the lubricant component of fluorine remained on the surface; fluorine was kept on approximately 80% on the surface after 60 min of exposure to X-rays. On the other hand, the lubricant component of sample F decreased by approximately 40% after exposure for 60 min.

On the basis of the initial structures shown in figure 23 and figure 25, it is concluded that the main structure of sample F has a side chain structure (-CF (CF_3)-CF_2-O-)m', similar to that in Fombline Y or Krytox. This periodic relation of 166 amu (C_3F_6O) continues up till mass numbers of approximately 5000 amu. In the case of magnetic disks, the high molecular structure of the lubricants was realized and maintained by dip coating or spin coating.

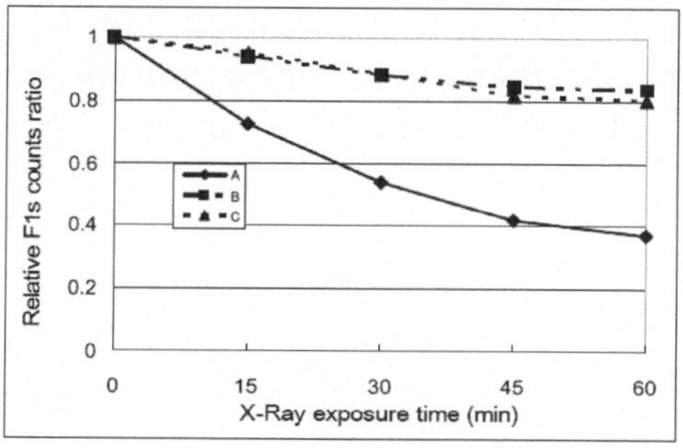

Fig. 19. Relationship between F1s intensity and X-Ray exposure time during XPS

However, the ophthalmic lens of lubricants was deposited by lamp heating methods into vacuum. Nevertheless, some main structure of lubricants was contained high-polymeric structures. On the other hand, the main structures of sample G and sample H has a straight chain structure without the side chain structures (-CF_2-CF_2-O-)m-(CF_2-O-)n, similar to the main structure of Fombline Z. From figure 20, 24 and 25, we found that the main chemical structure of lubricants for sample F is decreasing and destroying as a function of exposure time by XPS.

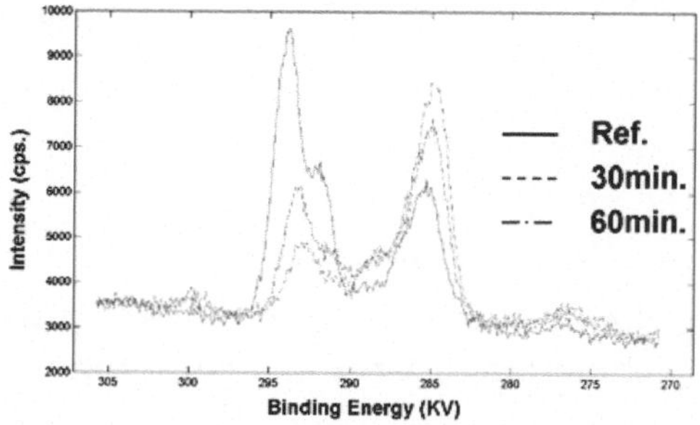

Fig. 20. Changing chemical structure of C_{1s} spectrum for sample F as a function of X-ray exposure time by XPS

These observations suggest that the straight chain structure of (-CF_2-CF_2-O-)m-(CF_2-O-)n is more robust to X-ray damage during XPS than the side chain structure (-CF (CF_3)-CF_2-O-)m'. We attribute this difference in the strength of the structures to the presence or absence of the chemical structure of the side chain. TEM or XPS measurement reveals that the film thickness of the lubricants is 2–3 nm. According to Tani (1999), he found double steps on the lubricant film with 2.9 nm thickness that was almost completely cover the surface by the mean molecular radius of gyration with coil of lubricant molecular. Therefore, it seems that the 2-3 coils of lubricant molecular have been stacked on the surface of the ophthalmic lens.

In the case of sample F, the molecular interaction in the side chain structure of CF_3 is weaker than that in the straight chain structure of CF_2 because in CF_3, three-dimensional structures overlap and this leads to repulsion between fluorine atoms. Therefore, the damage due to exposure to X-rays during XPS in the case of sample F is more than that in the case of sample G or that in the case of sample H. It is predicted that the trend observed in the adhesion properties of lubricants will be the same as that observed in the case of these damages.

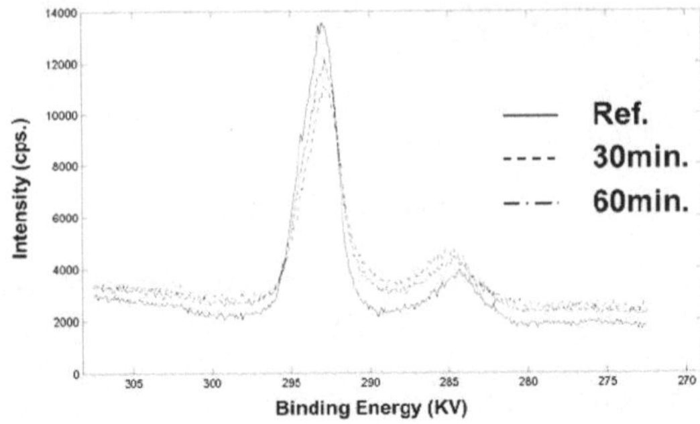

Fig. 21. Changing chemical structure of C_{1s} spectrum for sample G as a function of X-ray exposure time by XPS

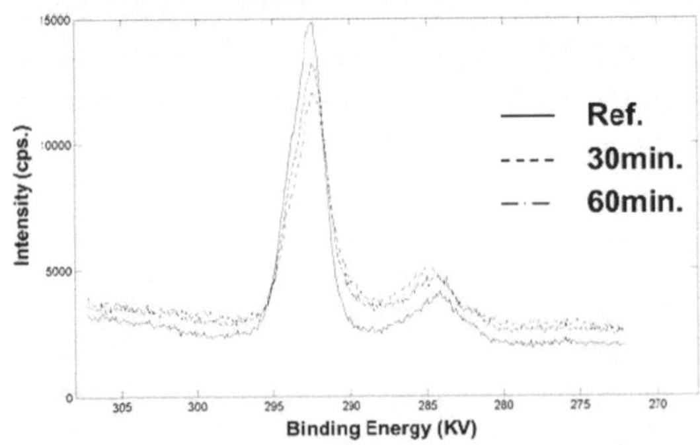

Fig. 22. Changing chemical structure of C_{1s} spectrum for sample H as a function of X-ray exposure time

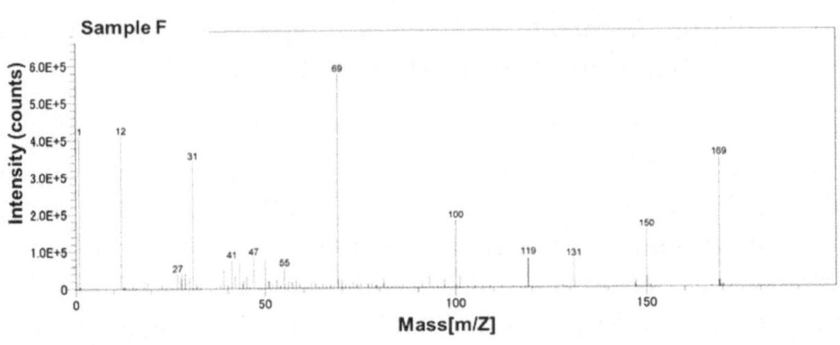

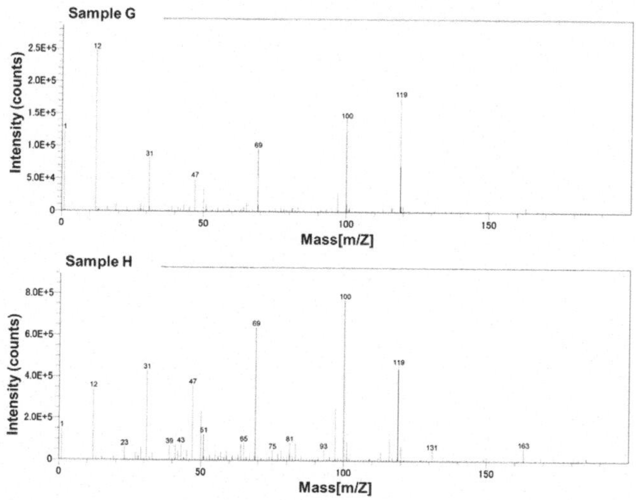

Fig. 23. Initial structure of the mass spectra of positive fragment ions, as determined by TOF- SIMS (upper spectrum: sample F, middle spectrum: sample G, lower spectrum: sample H)

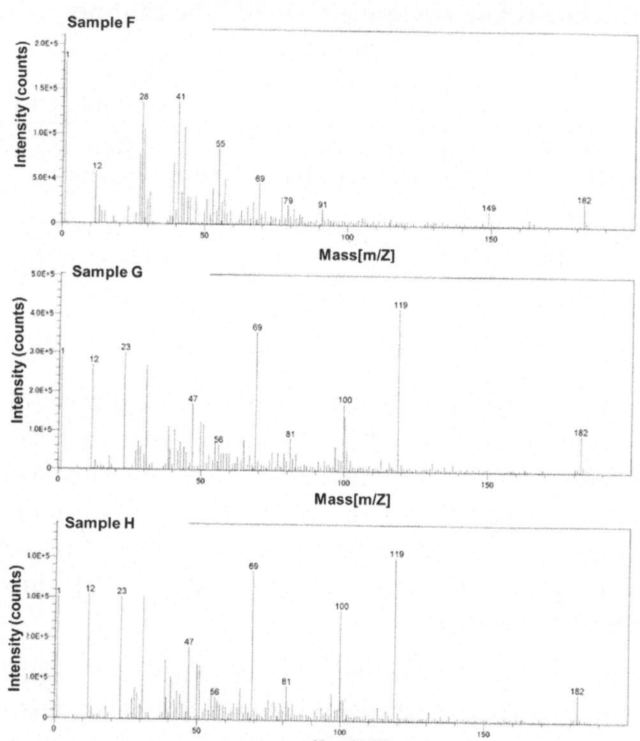

Fig. 24. The mass spectra of positive fragment ions after 60 min X-ray exposure by XPS, as determined by TOF-SIMS (upper spectrum: sample F, middle spectrum: sample G, lower spectrum: sample H)

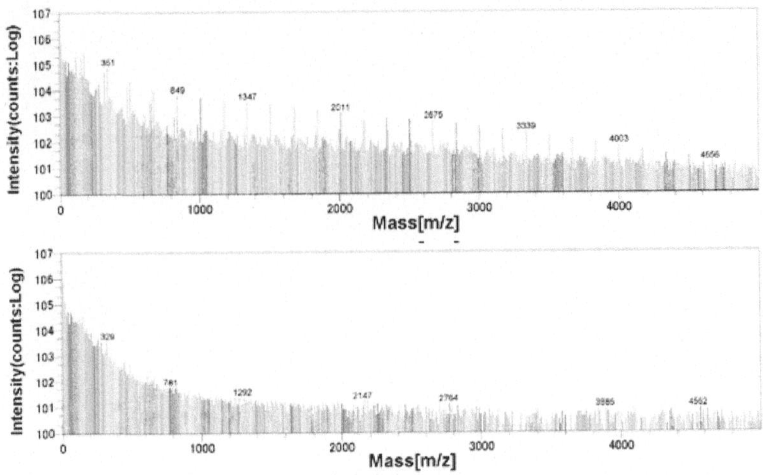

Fig. 25. Mass spectra of negative fragment ions for sample A, as determined by TOF-SIMS (upper spectrum: initial, lower spectrum: after 60 min X-ray exposure by XPS)

Table 3. Film thickness and coverage ratio of lubricant before and after XPS damage

	Initial		**After 60min X-ray explosured**	
	Lub. film thickness (nm)	**Lub. film coverage (%)**	**Lub. film thickness (nm)**	**Lub. film coverage (%)**
Sample F	2.4-2.9	98 over	0.9-1.3	88-91
Sample G	2.3-2.7	98 over	1.8-2.2	94-95
Sample H	2.3-2.7	98 over	1.7-2.1	94-95

2.2.4. Abrasion test

The water contact angle for sample F, sample G, and sample H before and after the abrasion test is listed in table 4. The XPS spectrum for each sample before and after abrasion test is shown in figures 26 – 28. Figures 29 – 31 show the topographic image and the phase image for each sample before and after abrasion test (image on the upper left image: initial topographic image, upper right image: initial phase image, lower left image: topographic image after abrasion test, lower right image: phase image after abrasion test).

The results in table 4 indicate that the water contact angles in the case of sample G and sample H decreased slightly after the abrasion test was performed. In contrast, the water contact angle of sample F decreased drastically from 116° to 89° after the sample was scratched by a 2 kg weight over 600 strokes. In the case of sample F, it seems that the water repellant of lubricant was declined because it was decreased the lubricants quantity of sample F by abrasion test. A phase image that was obtained by AFM revealed the distribution of unevenness (roughness), the viscosity, elasticity, friction force, adhesion, and soft-hardness from the energy dissipation of interaction between tip and sample. In a previous study, we showed that the energy dissipation in the areas corresponding to bright areas in the phase image is greater than that in the areas corresponding to dark areas in the image. This result, along with a comparison of the phase image and force modulation image, reveals that the bright area is softer or more adhesive than the dark area. The initial phase images for each sample comprise a mixture of small soft areas and small hard areas (or small adhesive areas and small non-adhesive areas). In the case of sample F, a scratch is observed along the scan area in the image obtained after the abrasion test. Just like, the lubricants were removed by rubbing. Therefore, the water contact angle decreased when the lubricants were removed. On the other hand, in the case of sample G and sample H, we observed that the cluster of lubricants was larger than the initial cluster. Further, there is no scratch in the image obtained after the abrasion test. We guess that lubricants repeated the attaching and moving, the mixtures of soft regions and hard regions were grown by rubbing process. Thus, there is no significant change in the water contact angle. These results indicate that the trend in lubricant damage during XPS agrees with the trend in durability during the abrasion test. Therefore, we found that we can select suitable lubricants for an ophthalmic lens by XPS measurement.

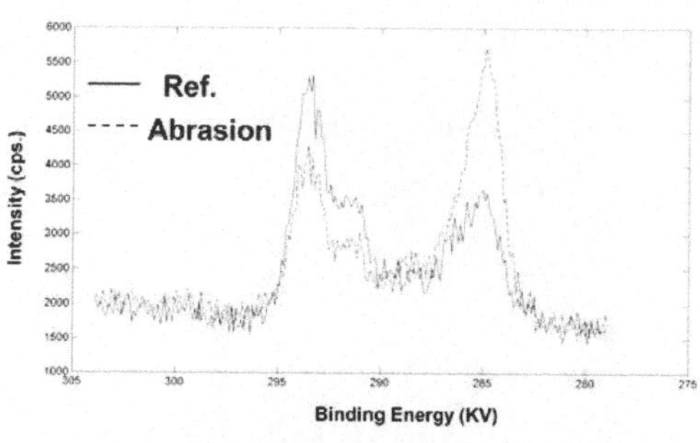

Fig. 26. Changing chemical structure of C_{1s} spectrum for sample F before and after the abrasion test

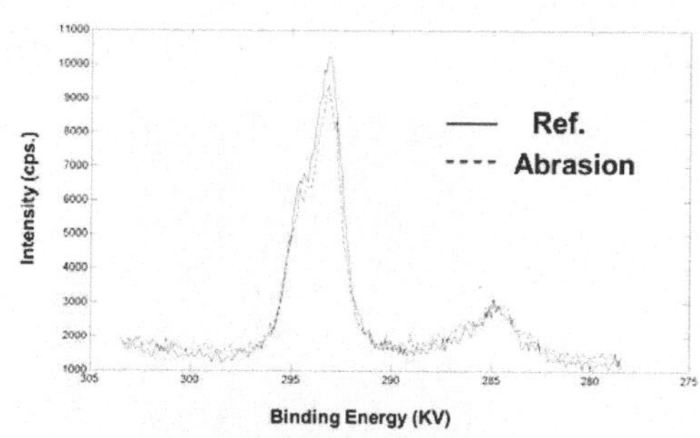

Fig. 27. Changing chemical structure of C_{1s} spectrum for sample G before and after the abrasion test

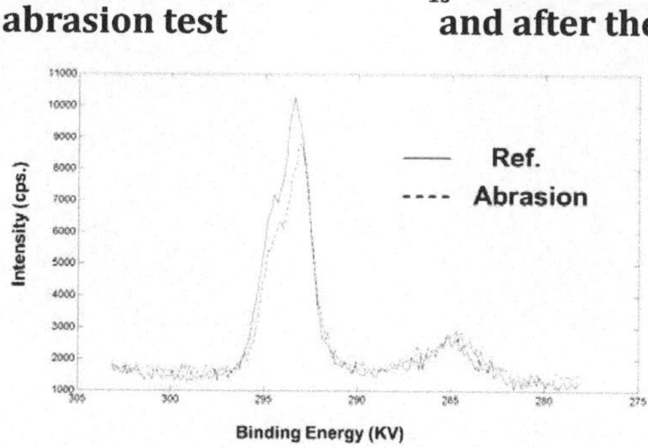

Fig. 28. Changing chemical structure of C_{1s} spectrum for sample H before and after the abrasion test

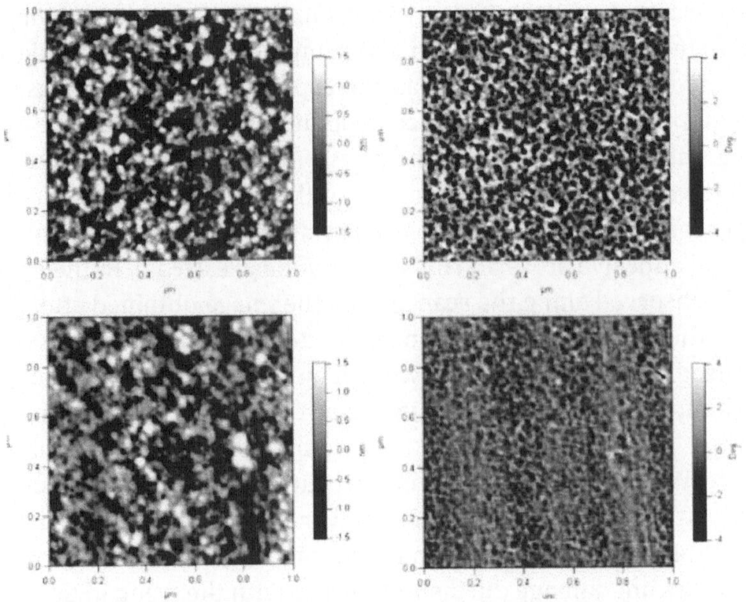

Fig. 29. Topographic image and phase image obtained for sample F (upper left image: initial topographic image, upper right image: initial phase image, lower left image: topographic image after abrasion test, lower right image: phase image after abrasion test)

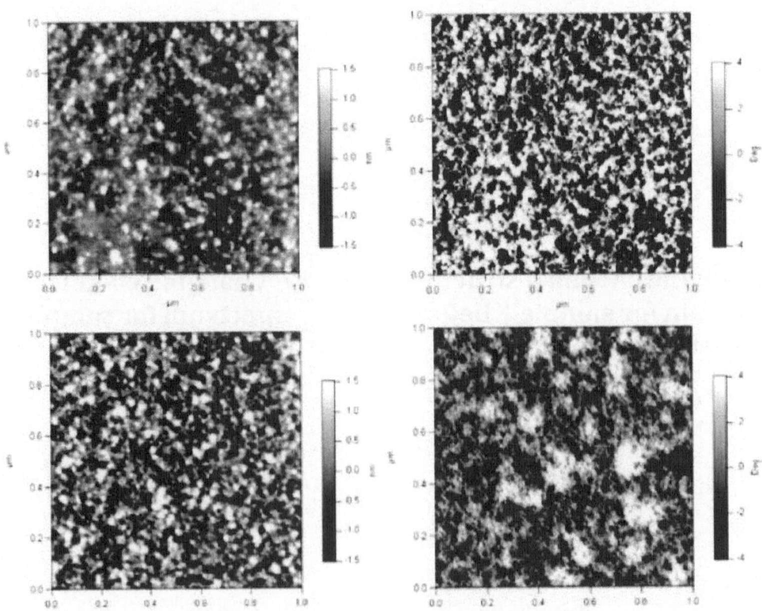

Fig. 30. Topographic image and phase image obtained for sample G (upper left image: initial topographic image, upper right image: initial phase image, lower left image: topographic image after abrasion test, lower right image: phase image after abrasion test)

Table 4. Film thickness and water contact angle before and after the abrasion test

	Initial		After abration test	
	Lub. film thickness (nm)	Contact angle	Lub. film thickness (nm)	Contact angle
Sample F	2.4-2.9	116°	1.1-1.5	89°
Sample G	2.3-2.7	110°	2.1-2.5	107°
Sample H	2.3-2.7	111°	1.9-2.5	108°

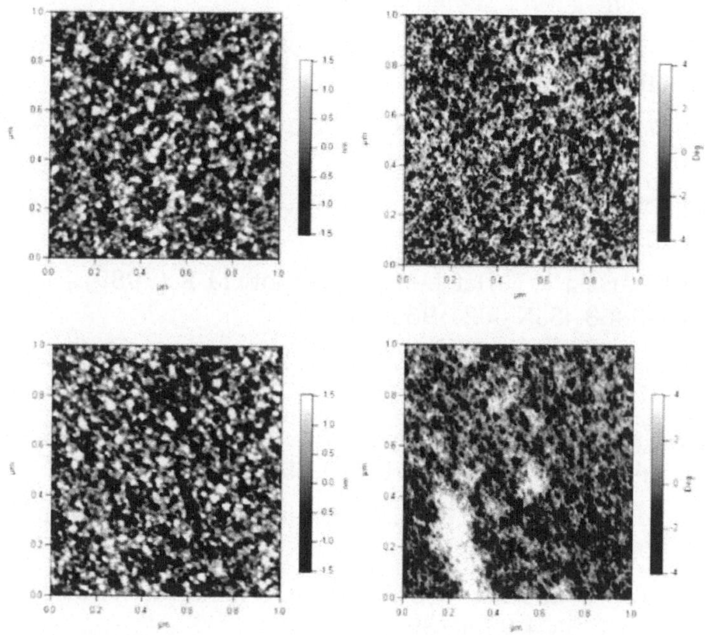

Fig. 31. Topographic image and phase image obtained for sample H (upper left image: initial topographic image, upper right image: initial phase image, lower left image: topographic image after abrasion test, lower right image: phase image after abrasion test)

3. Conclusion

We evaluated various methods for the analysis of lubricants on ophthalmic lenses. The lubricant film thickness can be directly determined by TEM measurement. The coverage ratio, the X-ray damage and the chemical structure can be investigated by XPS analysis. And also, TOF-SIMS analysis was used the investigation of X-ray damage and the chemical structure. In particular, AFM with an additional functional mode is a highly effective method for examining the morphology of lubricants; while determining the island structures of shapes with similar surface morphologies, it is more convenient to use phase images than friction force images and force modulation image. This information can be used to improve the tribological performance of ophthalmic lenses surface in order to meet customer demand.

References

1. Briggs, D. & Seah, M. P. (1990). Practical surface analysis 2'nd edition, John Wiley & Sons Ltd., pp209, ISBN 0471920819
2. Cleveland, J. P., Anczykowski, B., Schmid, A. E., & Elings, V. B. (1998). Applied Physics Letters, Vol. 72, No. 20, pp. 2613-2615, ISSN 0003-6951
3. Kimachi, Y., Yoshimura, F., Hoshino, M., & Terada, A. (1987), IEEE. Trans. mag., Vol.23, pp. 2392- 2394, ISSN 0018-9464
4. Matsuyama, K. (1997). J. of Japanese society of tribologists, Vol.42 pp. 823-828, ISSN 0915- 1168
5. Mate, C. M., Lorenz, M. R. & Novotny, V. J. (1989). J. Chem. Phys., Vol.90, pp. 7550-7555, ISSN 0021-9606
6. Mate, C. M. & Novotny, V. J. (1991). J. Chem. Phys., Vol.94, pp. 8420-8427, ISSN 0021-9606
7. Newman, J. G. & Viswanathan, K. V. (1990). J. Vac. Sci. Technol. A8, pp. 2388-2392, ISSN 0734-2101
8. Novotny, V. J., Hussla, I., Turlet, J.-M., & Philopott, M. R. (1989). J. Chem. Phys., Vol. 90, pp. 5861-5868, ISSN 0021-9606
9. Novotny, V. J., Pan, X., & Bhatia, C. S. (1994). J. Vac. Sci. Technol. A12, pp. 2879-2886, ISSN 0734-2101
10. Sakane, Y. & Nakao, M. (1999). Magnetics Conference, INTERMAG 99, IEEE. Trans. mag., vol.35, pp. 2394-2396, ISBN 0-7803-5555-5
11. Seah, M. P. and Dench, W. A. (1979). Surface and interface analysis, Vol.1, pp. 2-11, ISSN 1096-9918
12. Tadokoro, N. & Osakabe, K. (2001). Proc. Int. Tribol. Conference Nagasaaki 2000, pp. 2191- 2196, ISBN 4-9900139-6-4
13. Tadokoro, N., Yuki M., & Osakabe, K. (2003). Applied surface science, 203-204, pp. 72-77, ISSN 0169-4332
14. Tadokoro, N., Khraikratoke, S., Jamnongpian, P., Maeda, A., Komine, Y., Pavarinpong, N., Suyjantuk, S., & Iwata, N. (2009). Proc. World. Tribol. Congress 2009, pp. 749, ISBN 978-4-9900139-9-8
15. Tadokoro, N., Pannakarn, S., Khraikratoke S., Kamura, H., & Iwata, N. (2010). Proc. the 8th ICCG8, pp. 343-348, ISBN 978-3-00-031387-5
16. Tadokoro, N., Pannakarn, S., Wisuthtatip, J., Kunchoo, S., Parnich, V., Takashiba, K., Shimizu, K., and Higuchi, H. (2011). J. of Surface analysis, Vol.13 pp. 190-193, ISSN 1341-1756
17. Tani, H. (1999). Magnetics Conference, INTERMAG 99, IEEE. Trans. mag., vol.35, pp.2397- 2399, ISBN 0-7803-5555-5
18. Toney, M. F., Mate C. M., & Pocker, D. (1991). IEEE. Trans. mag., Vol.34, pp. 1774-1776, ISSN 0018-9464

10

Lubricating Oil Additives

1. Introduction

1.1 Lubrication (Rizvi, 2009)

The principle of supporting a sliding load on a friction reducing film is known as lubrication (Ludema, 1996). The substance of which the film is composed is a lubricant, and to apply it is to lubricate. These are not new concepts, nor, in their essence, particularly involved ones. Farmers lubricated the axles of their ox carts with animal fat centuries ago. But modern machinery has become many times more complicated since the days of the ox cart, and the demands placed upon the lubricant have become proportionally more exacting. Though the basic principle still prevails the prevention of metal-to-metal contact by means of an intervening layer of fluid or fluid-like material.

1.2 Lubricants (Rizvi, 2009; Ludema, 1996; and Leslie, 2003)

All liquids will provide lubrication of a sort, but some do it a great deal better than others. The difference between one lubricating material and another is often the difference between successful operation of a machine and failure.

Modern equipment must be lubricated in order to prolong its lifetime. A lubricant performs a number of critical functions. These include lubrication, cooling, cleaning and suspending, and protecting metal surfaces against corrosive damage. Lubricant comprises a base fluid and an additive package. The primary function of the base fluid is to lubricate and act as a carrier of additives. The function of additives is either to enhance an already-existing property of the base fluid or to add a new property. The examples of already-existing properties include viscosity, viscosity index, pour point, and oxidation resistance. The examples of new properties include cleaning and suspending ability, antiwear performance, and corrosion control.

Engine oil at the dawn of the automotive era was not highly specialized or standardized, and exceedingly frequent oil changes were required.

Engine oil lubricants make up nearly one half of the lubricant market and therefore attract a lot of interest. The principal function of the engine oil lubricant is to extend the life of moving parts operating under many different conditions of speed, temperature, and pressure. At low temperatures the lubricant is expected to flow sufficiently in order that moving parts are not starved of oil. At higher temperatures they are expected to keep the moving parts apart to minimize wear. The lubricant does this by reducing friction and removing heat from moving parts. Contaminants pose an additional problem, as they accumulate in the engine during operation. The contaminants may be wear debris, sludges, soot particles, acids, or peroxides. An important function of the lubricant is to prevent these contaminants from doing any damage.

The lube oil base stock is the building block with respect to which appropriate additives are selected and properly blended to achieve a delicate balance in performance characteristics of the finished lubricant. Various base stock manufacturing processes can all produce base stocks with the necessary characteristics to formulate finished lubricants with the desirable performance levels. The key to achieving the highest levels of performance in finished lubricants is in the understanding of the interactions of base stocks and additives and matching those to requirements of machinery and operating conditions to which they can be subjected.

1.3 Additives

Additives, (Rizvi, 2009, Ludema, 1996, and Leslie, 2003), are chemical compounds added to lubricating oils to impart specific properties to the finished oils. Some additives impart new and useful properties to the lubricant; some enhance properties already present, while some act to reduce the rate at which undesirable changes take place in the product during its service life. Additives, in improving the performance characteristics of lubricating oils, have aided significantly in the development of improved prime movers and industrial machinery.

Modern passenger car engines, automatic transmissions, hypoid gears, railroad and marine diesel engines, high speed gas and steam turbines, and industrial processing machinery, as well as many other types of equipment, would have been greatly retarded in their development were it not for additives and the performance benefits they provide.

Additives for lubricating oils were used first during the 1920s, and their use has since increased tremendously. Today, practically all types of lubricating oil contain at least one additive, and some oils contain additives of several different types. The amount of additive used varies from a few hundredths of a percent to 30% or more.

Over a period of many years, oil additives were identified that solved a variety of engine problems: corrosion inhibition, ability to keep particles such as soot dispersed, ability to prohibit acidic combustion products from plating out as varnish on engine surfaces, and ability to minimize wear by laying down a chemical film on heavily loaded surfaces. In addition, engine oil became specialized so that requirements for diesel engine oils began to diverge from requirements for gasoline engines, since enhanced dispersive capability was needed to keep soot from clumping in the oil of diesel engines.

The more commonly used additives are discussed in the following sections. Although some are multifunctional, as in the case of certain viscosity index improvers that also function as pour point depressants or dispersants or antiwear agents that also function as oxidation inhibitors, they are discussed in terms of their primary function only.

1.3.1 Friction Modifiers (FM) (Ludema, 1996)

These are additives that usually reduce friction (Battez et al., 2010 & Mel'nikov, 1997). The mechanism of their performance is similar to that of the rust and corrosion inhibitors in that they form durable low resistance lubricant films via adsorption on surfaces and via association with the oil, Figure (1.1).

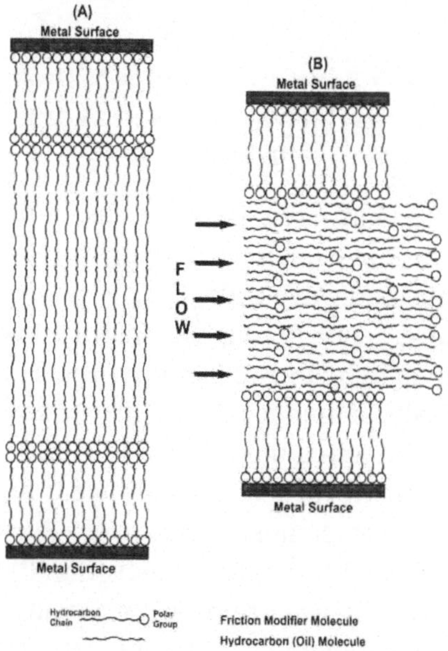

Fig. 1.1 Adsorption of friction modifiers on metal (A) Steady state (B) Under shear

Common materials that are used for this purpose include long-chain fatty acids, their derivatives, and the molybdenum compounds. In addition to reducing friction, the friction modifiers also reduce wear, especially at low temperatures where the anti-wear agents are inactive, and they improve fuel efficiency.

1.3.2 Anti-wear agents (A.W.) and extreme-pressure (E.P.) additives

Anti-wear (AW) (Rizvi, 2009, Ludema, 1996, Leslie, 2003, and Masabumi, 2008), agents have a lower activation temperature than the extreme-pressure (EP) agents. The latter are also referred to as anti-seize and anti-scuffing additives. Organosulfur and organo-phosphorus compounds, Figure (1.2), such as organic polysulfides, phosphates, dithiophosphates, and dithiocarbamates are the most commonly used AW and EP, Rizvi, 2009, Ludema, 1996, Leslie, 2003, agents.

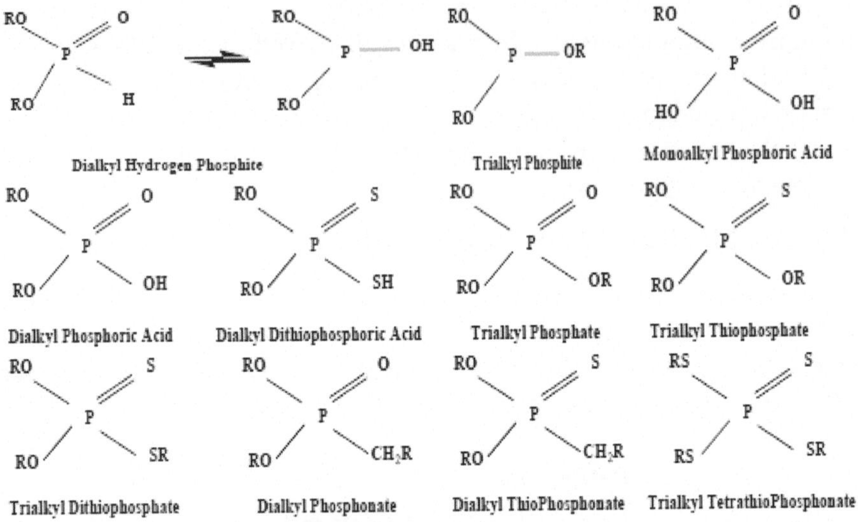

Fig. 1.2 Common phosphorus derivatives used as antiwear agents / extreme-pressure

As the power of engines has risen, the need for additives to prevent wear has become more important. Initially engines were lightly loaded and could withstand the loading on the bearings and valve train. Corrosive protection of bearing metals was one of the early requirements for engine oils. Fortunately, the additives used to protect bearings usually had mild antiwear properties. These antiwear agents were compounds such as lead salts of long- chain carboxylic acids and were often used in combination with sulfur-containing materials. Oil-soluble sulfur-phosphorous and chlorinated compounds also worked well as antiwear agents. However, the most important advance in antiwear chemistry was made during the 1930s and 1940s with the discovery of zinc dialkyldithiophosphates (ZDDP) (Masabumi, et. al., 2008). These compounds were initially used to prevent bearing corrosion but were later found to have exceptional antioxidant and antiwear properties. The antioxidant mechanism of the ZDDP was the key to its ability to reduce bearing corrosion. Since the ZDDP suppresses the formation of peroxides, it prevents the corrosion of Cu/ Pb bearings by organic acids. Antiwear and extreme-pressure additives function by thermally decomposing to yield compounds that react with the metal surface. These surface-active compounds form a thin layer that preferentially shears under boundary lubrication conditions.

After the discovery of ZDDP, Figure (1.3) it rapidly became the most widespread antiwear additive used in lubricants. As a result, many interesting studies have been undertaken on ZDDP with many mechanisms proposed for the antiwear and antioxidant action (Masabumi, et. al., 2008).

Extreme pressure additives form extremely durable protective films by thermo-chemically reacting with the metal surfaces. This film can withstand extreme temperatures and mechanical pressures and minimizes direct contact between surfaces, thereby protecting them from scoring and seizing.

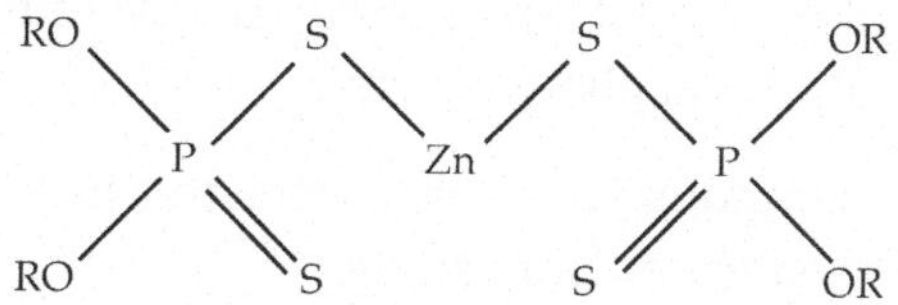

Fig. 1.3 Zinc dithiophosphate as antiwear additives / extreme pressure

1.3.3 Antioxidant additives (AO)

One of the most important aspects of lubricating oils is that the oxidation stability be maximized. Exposure of hydrocarbons to oxygen and heat will accelerate the oxidation process. The internal combustion engine is an excellent chemical reactor for catalyzing the process of oxidation. Also, the engine's metal parts, such as copper and iron, act as effective oxidation catalysts. Thus, engine oils are probably more susceptible to oxidation than any other lubricant application.

Oxidation mechanism of lubricating oils

The lubricating oils consist of hydrocarbons with (C_{20} - C_{70}) carbon atoms. At higher temperature these hydrocarbons are oxidized to form fatty acids, fatty alcohols, fatty aldehydes and ketones, fatty esters and fatty peroxides as shown in the following mechanism, Figure (1.4). All these compounds form the solid asphaltic materials. For this reason, the addition of antioxidants is necessary to all lubricating oils to prevent the formation of such compounds.

Initiation

$$RH \xrightarrow{+O_2} R\cdot$$

Chain propagation

$$R\cdot + O_2 \longrightarrow ROO\cdot$$

$$ROO\cdot + RH \longrightarrow ROOH + R\cdot$$

Chain branching

$$ROOH \longrightarrow RO\cdot + \cdot OH$$

$$RO\cdot + RH \longrightarrow ROH + R\cdot$$

$$\cdot OH + RH \longrightarrow H_2O + R\cdot$$

Termination

$$2R\cdot \longrightarrow R\text{-}R$$

Fig. 1.4 Oxidation mechanism of lubricating oils

Inhibition effects of antioxidants on lubricating oil oxidations

It is known that high temperature, high pressure, high friction, and high metal concentration in motors, lead to oxidation of lubricating oil it is necessary to improve oil stabilities against oxidation. Oxidation generally increase oil viscosity and results in formation of the following compounds:

❖ Resins, which are oxygen-containing compounds, soluble in oil and can, lead to lacquer formation.

❖ Lacquers, slightly colored, relatively plastic, and can form deposits on various engine parts (particularly on piston skirts).

❖ Insoluble asphaltic compounds, when associated in the oil with combustion residues and condensed water from sludge.

❖ Acidic compounds and hydroperoxides, which may promote corrosion, particularly of hard alloy bearings.

Thus, addition of antioxidant additives to lubricating oils prevents the formation of all resins, lacquers and acidic compounds.

There is no relationship between the two rates of increasing the viscosity and acidity in the oxidation process. The rate of viscosity increased in direct proportion with the rate of decomposition of an organic peroxides. By studying the antioxidant additives mechanism in turbine aviation oils; it was showen that these antioxidants reacted with oxygenic free radical compounds to form the antioxidant N - oxide derivatives and thus decrease the quantity acids, alcohols, esters in the media. However, any lubricating oil exposed to air and heat will eventually oxidize. Antioxidants are the key additive that protects the lubricant from oxidative degradation, allowing the fluid to meet the demanding requirements for use in engines and industrial applications.

Antioxidant additives mechanism

To define the oxidation stability for lubricating oils, it is necessary to check the rate of acidity and viscosity increase with oxidation time during the oxidation process. Lubricating oils are susceptible to degradation by oxygen. The oil oxidation (Rizvi, 2009, Ludema, 1996, and Leslie, 2003) process is the major cause of oil thickening. This manifests itself as sludge and varnish formation on engine parts, leading to increased engine wear, poor lubrication, and reduced fuel economy. Antioxidants are essential additives incorporated into lubricant formulations to minimize and delay the onset of lubricant oxidative degradation.

The rate of acidity and viscosity was increased (in the oxidation process for oils) due to the continuous repetition of the oxidation process, where a chain reaction occurs. The oxidation process can be considered to progress in the following manner:

$$RH \rightarrow R^{\bullet} + H^{\bullet} \quad (1.1)$$

$$R^{\bullet} + O_2 \rightarrow ROO^{\bullet} \quad (1.2)$$

$$ROO^{\bullet} + RH \rightarrow ROOH + R^{\bullet} \quad (1.3)$$

Decomposition of the hydroperoxide molecule caused by the so-called branching reaction leads to the formation of oxygen bearing compounds. Their oxidation products form high – molecular – weight –oil –insoluble polymers that settle as deposit causing an increase of oil viscosity. Decomposition of the hydroperoxide as follow:

$$ROOH \rightarrow RO\bullet + \bullet OH \quad (1.4)$$

During oxidation process, in a median period oil viscosity is increased although its acidity is remaining constant because there are alcohol's of unsaturated hydrocarbons, produced from decomposition of hydroperoxides process, so it neutralizes the effect of acidity formed from another oxygen bearing compound such as: aldehydes, ketones and acids.

The proceeding lubricant degradation mechanism makes clear several possible counter measures to control lubricant degradation. Blocking the energy source is one path but is effective only for lubricants used in low-shear and temperature situations. However, more practical for most lubricant applications are the trapping of catalytic impurities and the destruction of hydrocarbon radicals, alkyl peroxy radicals, and hydroperoxides. This can be achieved through the use of radical scavengers, peroxide decomposers, and metal deactivators.

The radical scavengers are known as *primary antioxidants*. They donate hydrogen atoms that react with alkyl radicals and/or alkyl peroxy radicals, interrupting the radical chain mechanism of the auto-oxidation process. The primary antioxidant then becomes a stable radical, the alkyl radical becomes a hydrocarbon, and the alkyl peroxy radical becomes an alkyl hydroperoxide. Hindered phenolics and aromatic amines are the two chemical classes of primary antioxidants for lubricants. The transfer of a hydrogen from the oxygen or nitrogen atom to the radical forms quinones or quinine imines that do not maintain the radical chain mechanism.

The peroxide decomposers are known as *secondary antioxidants* (Rizvi, 2009, Ludema, 1996, and Leslie, 2003). Sulfur and/or phosphorus compounds reduce the alkyl hydroperoxides in the radical chain to alcohols while being oxidized in a sacrificial manner. Zinc dialkyldithiophosphate, phosphites, and thio-ethers are examples of different chemical classes of secondary antioxidants.

There are two types of metal deactivators: chelating agents and film forming agents . The chelating agents will form a stable complex with metal ions, reducing the catalytic activity of the metal ions. Thus, the deactivators can show an antioxidant effect. Film-forming agents act two ways. First, they coat the metal surface, thus preventing metal ions from entering the oil. Second, they minimize corrosive attack of the metal surface by physically restricting access of the corrosive species to the metal surface.

Several effective antioxidants classes have been developed over the years and have seen use in engine oils, automatic transmission fluids, gear oils, turbine oils, compressor oils, greases, hydraulic fluids, and metal-working fluids. The main classes of oil-soluble organic and organo-metallic antioxidants are the following types:

1. Sulfur compounds
2. Phosphorus compounds
3. Sulfur-phosphorus compounds
4. Aromatic amine compounds
5. Hindered phenolic compounds
6. Organo-alkaline earth salt compounds
7. Organo-zinc compounds
8. Organo-copper compounds
9. Organo-molybdenum compounds

1.3.4 Anti-foam (A.F.) agents

The foaming of lubricants, (Rizvi, 2009, Ludema, 1996, and Leslie, 2003), is a very undesirable effect that can cause enhanced oxidation by the intensive mixture with air, cavitation damage as well as insufficient oil transport in circulation systems that can even lead to lack of lubrication. Beside negative mechanical influences the foaming tendency depends very much on the lubricant itself and is influenced by the surface tension of the base oil and especially by the presence of surface-active substances such as detergents, corrosion inhibitors and other ionic compounds.

In many applications, there may be considerable tendency to agitate the oil and cause foaming, while in other cases even small amounts of foam can be extremely troublesome. In these cases, a defoamant may be added to the oil. It is thought that the defoamant droplets attach themselves to the air bubbles and can either spread or form unstable bridges between bubbles, which then coalesce into larger bubbles, which in turn rise more readily to the surface of the foam layer where they collapse, thus releasing the air.

1.3.5 Rust and corrosion inhibitors

Rust inhibitors, (Rizvi, 2009, Ludema, 1996, and Leslie, 2003), are usually compounds having a high polar attraction toward metal surfaces. By physical or chemical interaction at the metal surface, they form a tenacious, continuous film that prevents water from reaching the metal surface. Typical materials used for this purpose are amine succinates and alkaline earth sulfonates.

Rust inhibitors can be used in most types of lubricating oil, but the selection must be made carefully to avoid problems such as corrosion of nonferrous metals or the formation of troublesome emulsions with water. Because rust inhibitors are adsorbed on metal surfaces, an oil can be depleted of rust inhibitor in time.

A number of kinds of corrosion can occur in systems served by lubricating oils. Probably the two most important types are corrosion by organic acids that develop in the oil itself and corrosion by contaminants that are picked up and carried by the oil.

Corrosion by organic acids can occur, for example, in the bearing inserts used in internal combustion engines. Some of the metals used in these inserts, such as the lead in copper- lead or lead-bronze, are readily attacked by organic acids in oil, The corrosion inhibitors form a protective film on the bearing surfaces that prevents the corrosive materials from reaching or attacking the metal. The film may be either adsorbed on the metal or chemically bonded to it. It has been found that the inclusion of highly alkaline materials in the oil will help to neutralize these strong acids as they are formed, greatly reducing this corrosion and corrosive wear.

1.3.6 Detergent and dispersant (D / D) additives

Modern equipment must be lubricated in order to prolong its lifetime. One of the most critical properties of the automotive lubricants, especially engine oils, is their ability to suspend undesirable products from thermal and oxidative degradation of the lubricant. Such products form when the byproducts of fuel combustion, such as hydroperoxides and free radicals, go past piston rings into the lubricant and, being reactive species, initiate lubricant oxidation. The resulting oxidation products are thermally labile and decompose to highly polar materials with a tendency to separate from the bulk lubricant and form surface deposits and clog small openings.

Oxidation inhibitors, detergents (Rizvi, 2009, Ludema, 1996, Leslie, 2003, and Ming et. al., 2009), and dispersants (Alun, 2010) make up the general class of additives called *stabilizers and deposit control agents*. These additives are designed to control deposit formation, either by inhibiting the oxidative breakdown of the lubricant or by suspending the harmful products already formed in the bulk lubricant. Oxidation inhibitors intercept the oxidation mechanism, and dispersants and detergents perform the suspending part (Kyunghyun, 2010). Detergents are metal salts of organic acids that frequently contain associated excess base, usually in the form of carbonate. Dispersants are metal-free and are of higher molecular weights than detergents. The two types of additives work in conjunction with each other.

The final products of combustion and lubricant decomposition include organic and inorganic acids, aldehydes, ketones, and other oxygenated materials.

The acids have the propensity to attack metal surfaces and cause corrosive wear. Detergents, especially basic detergents, contain reserve base that will neutralize the acids to form salts. While this decreases the corrosive tendency of the acids, the solubility of the salts in the bulk lubricant is still low. The organic portion of the detergent, commonly called "soap", has the ability to associate with the salts to keep them suspended in the bulk lubricant. However, in this regard, detergents are not as effective as dispersants because of their lower molecular weight. The soap in detergents and the dispersants also have the ability to suspend non- acidic oxygenated products, such as alcohols, aldehydes, and resinous oxygenates. The mechanism by which this occurs is depicted in Figure (1.5).

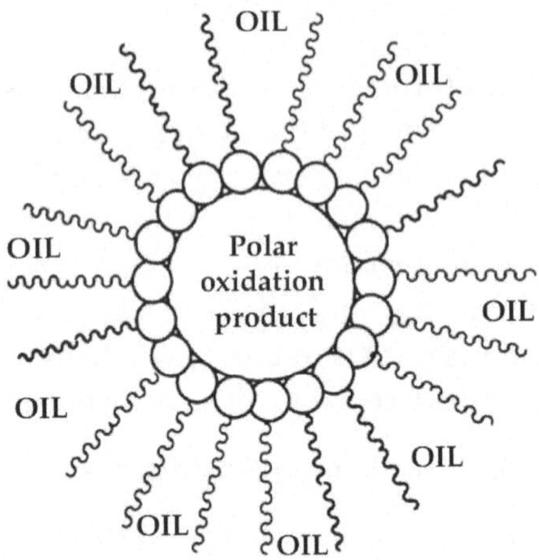

Fig. 1.5 Oil Suspension of polar oxidation products

Dispersants and detergents together make up the bulk, about 45–50%, of the total volume of the lubricant additives manufactured. This is a consequence of their major use in engine oils, transmission fluids, and tractor hydraulic fluids, all of which are high-volume lubricants.

As mentioned, detergents neutralize oxidation-derived acids as well as help suspend polar oxidation products in the bulk lubricant. Because of this, these additives control rust, corrosion, and resinous buildup in the engine. Like most additives detergents contain a surface-active polar functionality and an oleophilic hydrocarbon group, with an appropriate number of carbon atoms to ensure good oil solubility. Sulfonate, phenate, and carboxylate are the common polar groups present in detergent molecules. However, additives containing salicylate and thiophosphonate functional groups are also sometimes used, Figure (1.6).

As mentioned, common metals that can be used to make neutral or basic detergents include sodium, potassium, magnesium, calcium, and barium. Calcium and magnesium find most extensive use as lubricant additives, with a preference for calcium due to its lower cost. The use of barium-derived detergents is being curbed due to concerns for barium's toxicity.

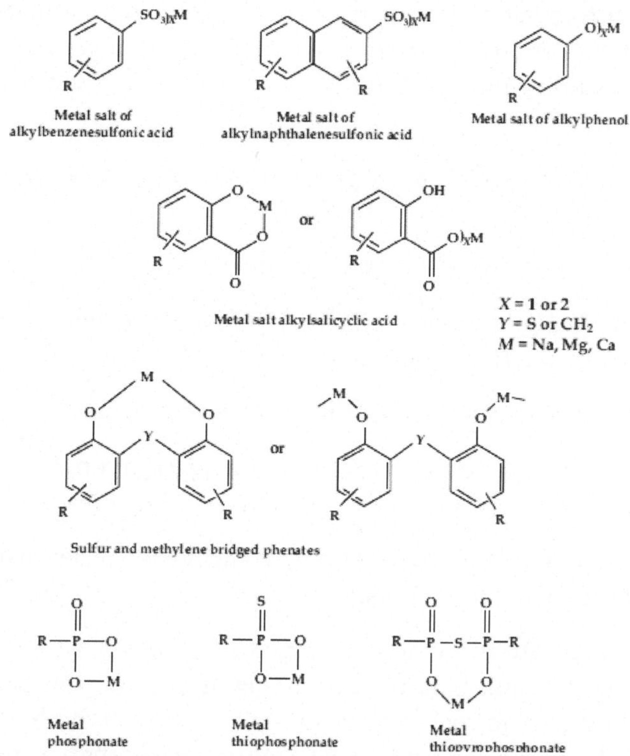

Fig. 1.6 Idealized structures of neutral salts (soaps)

Technically, one can use metal oxides, hydroxides, and carbonates to manufacture neutral (non-overbased) detergents; for non-overbased detergents, oxides and hydroxides are the preferred bases. For sodium, calcium, and barium detergents, sodium hydroxide, calcium hydroxide, and barium hydroxide are often used. For magnesium detergents, however, magnesium oxide is the preferred base. Dispersants differ from detergents in three significant ways:

1. Dispersants are metal-free, but detergents contain metals, such as magnesium, calcium, and sometimes barium. This means that on combustion detergents will lead to ash formation and dispersants will not.

2. Dispersants have little or no acid-neutralizing ability, but detergents do. This is because dispersants have either no basicity, as is the case in ester dispersants, or low basicity, as is the case in imide / amide dispersants. The basicity of the imide/amide dispersants is due to the presence of the amine functionality. Amines are weak bases and therefore possess minimal acid-neutralizing ability. Conversely, detergents, especially basic detergents, contain reserve metal bases as metal hydroxides and metal carbonates. These are strong bases, with the ability to neutralize combustion and oxidation-derived inorganic acids, such as sulfuric acid and nitric acid, and oxidation-derived organic acids.

3. Dispersants are much higher in molecular weight, approximately 4–15 times higher, than the organic portion (soap) of the detergent. Because of this, dispersants are more effective in fulfilling the suspending and cleaning functions than detergents.

The dispersants suspend deposit precursors in oil in a variety of ways. These comprise:

- ❖ Including the undesirable polar species into micelles.
- ❖ Associating with colloidal particles, thereby preventing them from agglomerating and falling out of solution.
- ❖ Suspending aggregates in the bulk lubricant, if they form.
- ❖ Modifying soot particles so as to prevent their aggregation. The aggregation will lead to oil thickening, a typical problem in heavy-duty diesel engine oils.
- ❖ Lowering the surface / interfacial energy of the polar species in order to prevent their adherence to metal surfaces.

At the low-temperature regions, such as the piston skirt, the deposits are not heavy and form only a thin film. For diesel engine pistons, this type of deposit is referred to as "lacquer"; for gasoline engine pistons, this type of deposit is called "varnish". The difference between lacquer and varnish is that lacquer is lubricant-derived and varnish is largely fuel- derived. In addition, the two differ in their solubility characteristics. That is, lacquer is water-soluble and varnish is acetone soluble. Lacquer usually occurs on piston skirts, on cylinder walls, and in the combustion chamber. Varnish occurs on valve lifters, piston rings, piston skirts, valve covers, and positive crankcase ventilation (PCV) valves.

Another component of the combustion effluent that must be considered is soot. Soot not only contributes toward some types of deposits, such as carbon and sludge, but it also leads to a viscosity increase. These factors can cause poor lubricant circulation and lubricating film formation, both of which will result in wear and catastrophic failure.

Deposit control by dispersants

Fuel and lubricant oxidation and degradation products, such as soot, resin, varnish, lacquer, and carbon, are of low lubricant (hydrocarbon) solubility, with a propensity to separate on surfaces. The separation tendency of these materials is a consequence of their particle size. Small particles are more likely to stay in oil than large particles. Therefore, resin and soot particles, which are the two essential components of all deposit-forming species, must grow in size via agglomeration prior to separation. Alternatively, soot particles are caught in the sticky resin, which is shown in parts A and B of Figure (1.7). Dispersants interfere in agglomeration by associating with individual resin and soot particles. The particles with associated dispersant molecules are unable to coalesce because of either steric factors or electrostatic factors. Dispersants consist of a polar group, usually oxygen- or nitrogen-based, and a large non polar group. The polar group associates with the polar particles, and the non polar group keeps such particles suspended in the bulk lubricant. This is shown in parts C and D of Figure (1.7). Neutral detergents, or soaps, operate by an analogous mechanism.

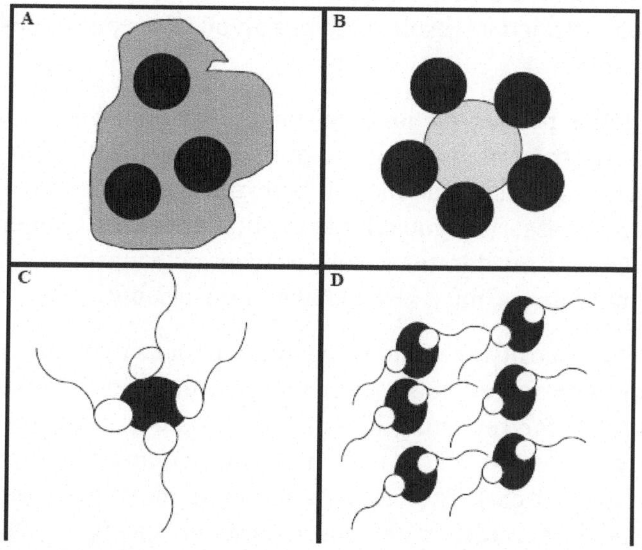

Fig. 1.7 Mechanism of soot-resin-additive interaction

Dispersant structure

A dispersant molecule consists of three distinct structural features:

A *hydrocarbon group*, a *polar group*, and a connecting group or a *link* (see Figure 1.8). The *hydrocarbon group* is polymeric in nature and, depending on its molecular weight; dispersants can be classified into *polymeric dispersants* and *dispersant polymers*. Polymeric dispersants are of lower molecular weight than dispersant polymers. The molecular weight of polymeric dispersants ranges between 3000 and 7000 as compared to dispersant polymers, which have a molecular weight of 25,000 and higher. While a variety of olefins, such as polyisobutylene, polypropylene, polyalphaolefins, and mixtures thereof, can be used to make polymeric dispersants, the polyisobutylene derived dispersants are the most common.

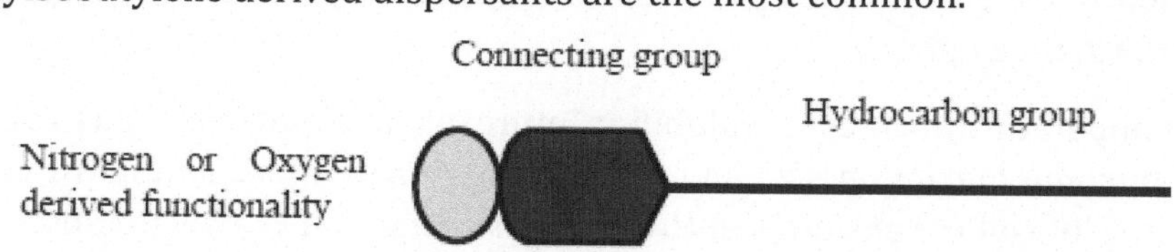

Fig. 1.8 Graphic representation of a dispersant molecule

1.3.7 Viscosity index improvers

Probably the most important single property of a lubricating oil is its viscosity. A factor in the formation of lubricating films under both thick and thin film conditions, viscosity (Rizvi, 2009, Ludema, 1996, Leslie, 2003 and Margareth, et. al., 2010), affects heat generation in bearings, cylinders, and gears; it governs the sealing effect of the oil and the rate of consumption or loss; and it determines the ease with which machines may be started under cold conditions. For any piece of equipment, the first essential for satisfactory results is to use an oil of proper viscosity to meet the operating conditions.

In selecting the proper oil for a given application, viscosity is a primary consideration. It must be high enough to provide proper lubricating films but not so high that friction losses in the oil will be excessive. Since viscosity varies with temperature, it is necessary to consider the actual operating temperature of the oil in the machine. Other considerations, such as whether a machine must be started at low ambient temperatures, must also be taken into account.

The kinematic viscosity of a fluid is the quotient of its dynamic viscosity divided by its density, both measured at the same temperature and in consistent units. The most common units for reporting kinematic viscosities now are the stokes (St) or centistokes (cSt; 1 cSt = 0.01 St), or in SI units, square millimeters per second (mm2/s; 1 mm2/s = 1 cSt). The viscosity of any fluid changes with temperature, increasing as the temperature is decreased, and decreasing as the temperature is increased. Thus, it is necessary to have some method of determining the viscosities of lubricating oils at temperatures other than those at which they are measured. This is usually accomplished by measuring the viscosity at two temperatures, then plotting these points on special viscosity–temperature charts developed by ASTM. The two temperatures most used for reporting viscosities are 40ºC (104ºF) and 100ºC (212ºF).

VI improvers are long chain, high molecular weight polymers that function by causing the relative viscosity of an oil to increase more at high temperatures than at low temperatures. Generally this result is due to a change in the polymer's physical configuration with increasing temperature of the mixture. It is postulated that in cold oil the molecules of the polymer adopt a coiled form so that their effect on viscosity is minimized. In hot oil, the molecules tend to straighten out, and the interaction between these long molecules and the oil produces a proportionally greater thickening effect.

As temperature increases, solubility improves, and polymer coils eventually expand to some maximum size and in so doing donate more and more viscosity. The process of coil expansion is entirely reversible as coil contraction occurs with decreasing temperature (see Figure 1.9).

Different oils have different rates of change of viscosity with temperature. For example, a distillate oil from a naphthenic base crude would show a greater rate of change of viscosity

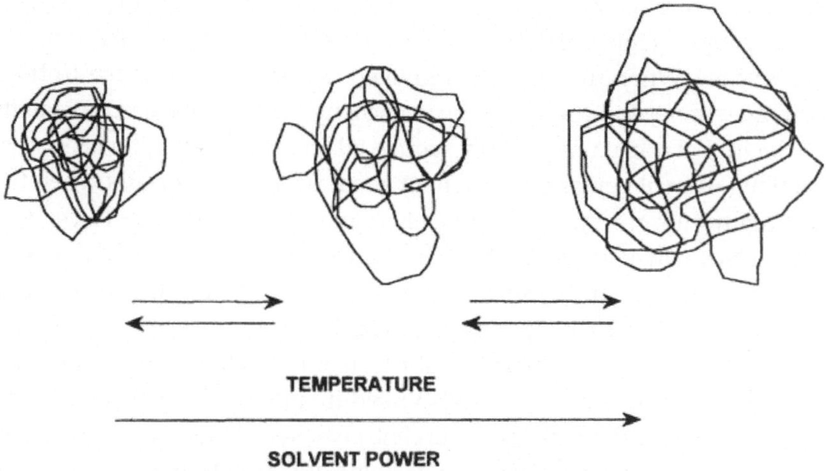

Fig. 1.9 Polymer coil expansion

with temperature than would a distillate oil from a paraffin crude. The viscosity index is a method of applying a numerical value to this rate of change, based on comparison with the relative rates of change of two arbitrarily selected types of oil that differ widely in this characteristic. A high *VI* indicates a relatively low rate of change of viscosity with temperature; a low *VI* indicates a relatively high rate of change of viscosity with temperature. For example, consider a high *VI* oil and a low *VI* oil having the same viscosity at, say, room temperature: as the temperature increased, the high *VI* oil would thin out less and, therefore, would have a higher viscosity than the low *VI* oil at higher temperatures. The *VI* of an oil is calculated from viscosities determined at two temperatures by means of tables published by ASTM. Tables based on viscosities determined at both 104ºF and 212ºF, and 40ºC and 100ºC are available. Finished mineral-based lubricating oils made by conventional methods range in *VI* from somewhat below 0 to slightly above 100. Mineral oil base stocks refined through special hydroprocessing techniques can have *VIs* well above 100.

Additives called *VI* improvers can be blended into oils to increase *VIs*; however, *VI* improvers are not always stable in lubricating environments exposed to shear or thermal stressing. Accordingly, these additives must be used with due care to assure adequate viscosity over the anticipated service interval for the application for which they are intended.

Among the principal *VI* improvers are methacrylate polymers and copolymers, acrylate polymers, olefin polymers and copolymers, and styrene butadiene copolymers, Figure (1.10). The degree of *VI* improvement from these materials is a function of the molecular weight distribution of the polymer.

The long molecules in *VI* improvers are subject to degradation due to mechanical shearing in service. Shear breakdown occurs by two mechanisms. Temporary shear breakdown occurs under certain conditions of moderate shear stress and results in a temporary loss of viscosity. Apparently, under these conditions the long molecules of the *VI* improver align themselves in the direction of the stress, thus reducing resistance to flow. When the stress is removed, the molecules return to their usual random arrangement and the temporary viscosity loss is recovered. This effect can be beneficial in that it can temporarily reduce oil friction to permit easier starting, as in the cranking of a cold engine. Permanent shear breakdown occurs when the shear stresses actually rupture the long molecules, converting them into lower molecular weight materials, which are less effective *VI* improvers. This results in a permanent viscosity loss, which can be significant. It is generally the limiting factor controlling the maximum amount of *VI* improver that can be used in a particular oil blend. *VI* improvers are used in engine oils, automatic transmission fluids, multipurpose tractor fluids, and hydraulic fluids. They are also used in automotive gear lubricants. Their use permits the formulation of products that provide satisfactory lubrication over a much wider temperature range than is possible with straight mineral oils alone.

(a) Polymethacrylates (PMA)

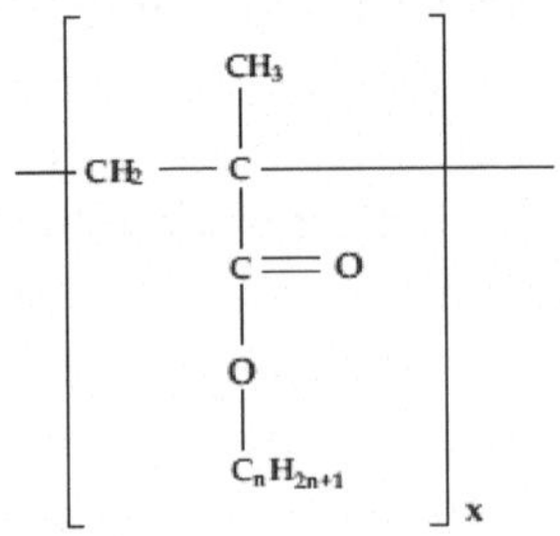

These are polymerised esters of methacrylic acid. They normally exhibit pour-point depressing activity. Dispersant properties can be obtained by incorporating polar groups in the molecular structure.

(b) Polyisobutenes (PIB)

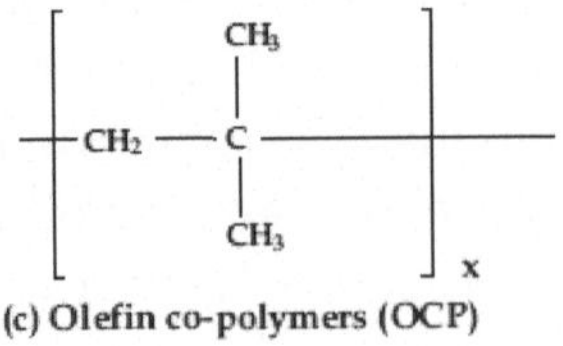

These are non-dispersant polymers and they have no effect on the pour point of formulated lubricants. They have limited use in modern formulations.

(c) Olefin co-polymers (OCP)

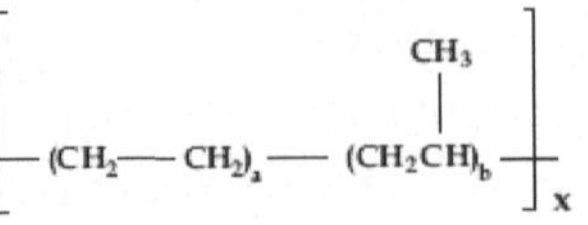

These are usually co-polymers of ethylene and propylene. Dispersant properties can be obtained incorporating polar groups in the moleculare structure.

(d) Styrene/diene co-polymers

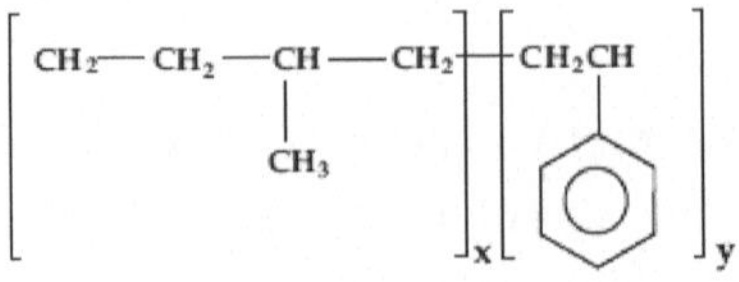

The molecular weight distribution is optimised to give good shear stability in crankcase applications. Its uniquely effective thickening power in solution gives an overall thickening efficiency that is superior to other polymers of equivalent shear stability.

Fig. 1.10 Viscosity index improvers

1.3.8 Pour point depressants

The pour point, (Rizvi, 2009, Ludema, 1996, and Leslie, 2003), *PP* of a lubricating oil is the lowest temperature at which it will pour or flow when it is chilled without disturbance under prescribed conditions. Most mineral oils contain some dissolved wax and, as an oil is chilled, this wax begins to separate as crystal that interlock to form a rigid structure that traps the oil in small pockets in the structure.

When this wax crystal structure becomes sufficiently complete, the oil will no longer flow under the conditions of the test. Since, however, mechanical agitation can break up the wax structure; it is possible to have an oil flow at temperatures considerably below its pour point. Cooling rates also affect wax crystallization; it is possible to cool an oil rapidly to a temperature below its pour point and still have it flow.

While the pour point of most oils is related to the crystallization of wax, certain oils, which are essentially wax free, have viscosity-limited pour points. In these oils the viscosity becomes progressively higher as the temperature is lowered until at some temperature no flow can be observed. The pour points of such oils cannot be lowered with pour point depressants, *PPDs*, since these agents act by interfering with the growth and interlocking of the wax crystal structure.

Certain high molecular weight polymers function by inhibiting the formation of a wax crystal structure that would prevent oil flow at low temperatures, Figure (1.11).

Crystal Morphology Without Pour Point Depressant

Crystal Morphology With Pour Point Depressant

Fig. 1.11 The mechanism of the pour point depressant performance

Two general types of pour point depressant are used:

1. Alkylaromatic polymers adsorb on the wax crystals as they form, preventing them from growing and adhering to each other.
2. Polymethacrylates co-crystallize with wax to prevent crystal growth.

The additives do not entirely prevent wax crystal growth, but rather lower the temperature at which a rigid structure is formed. Oils used under low-temperature conditions must have low pour points.

Oils must have pour points (1) below the minimum operating temperature of the system and (2) below the minimum surrounding temperature to which the oil will be exposed.

While removal of the residue waxes from the oil is somewhat expensive, pour point depressants are an economical alternative to reduce the pour point of lubricants. The most common pour point depressants are the same additives used for viscosity index improvement. The mechanism through which these molecules reduce pour point is still poorly understood and somewhat controversial.

It has been suggested that these molecules adsorb into the wax crystals, (Chen et. al., 2010, and Bharambe, 2010) and redirect their growth, forming smaller and more isotropic crystals that interfere less with oil flow.

Depending on the type of oil, pour point depression of up to 50ºF (10ºC) can be achieved by these additives, although a lowering of the pour point by about (20Fº – 30Fº) (-6.67Cº to - 1.1Cº) is more common.

There is a range of pour point depressant additives of different chemical species (102 - 105).

Polymethacrylates

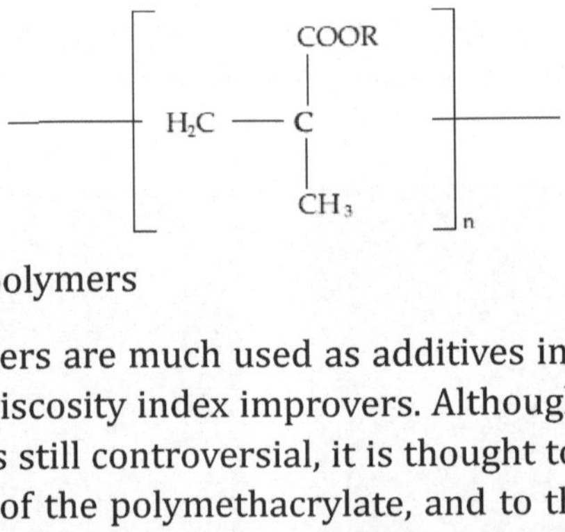

Polymethacrylates polymers

Methacrylate polymers are much used as additives in lubricating oils, as pour point depressants and viscosity index improvers. Although the mechanism of such pour point depression is still controversial, it is thought to be related to the length of the alkyl side chains of the polymethacrylate, and to the nature of the base oil. R in the ester has a major effect on the product, and is usually represented by a normal paraffinic chain of at least 12 carbon atoms. This ensures oil solubility. The molecular weight of the polymer is also very important. Typically these materials are between 7000 and 10,000 number average molecular weights. Commercial materials normally contain mixed alkyl chains, which can be branched.

Polyacrylates

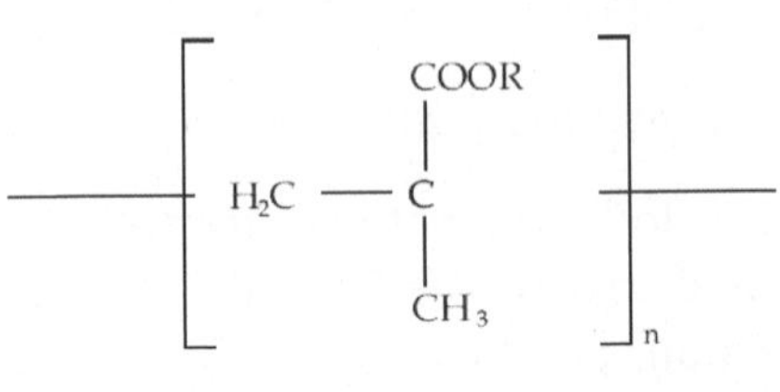

These are very similar in behavior to the polymethacrylates.

Di (tetra paraffin phenol) phthalate

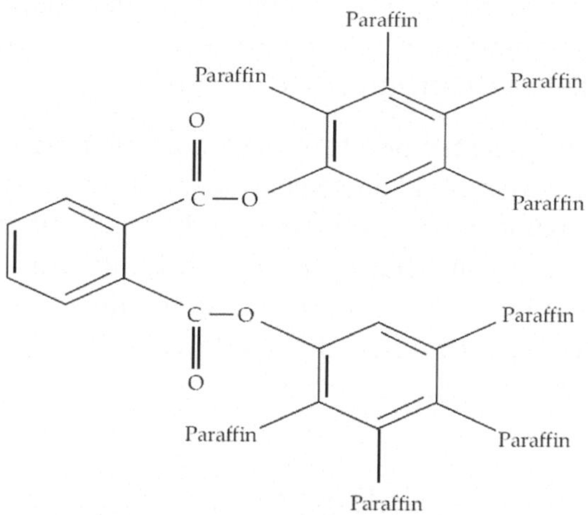

Condensation products of tetra paraffin phenol

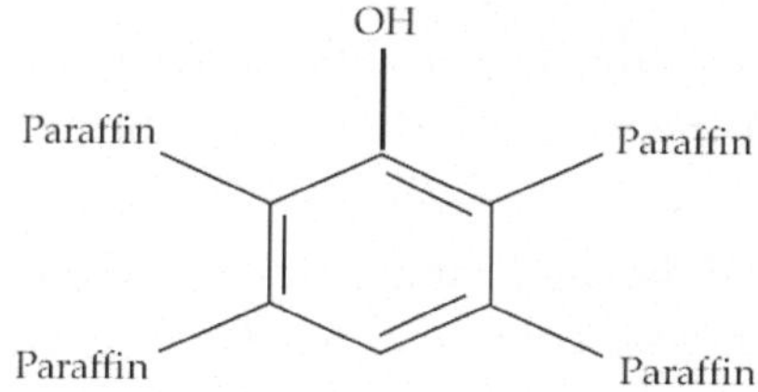

Condensation product of a chlorinated paraffin wax with naphthalene (107)

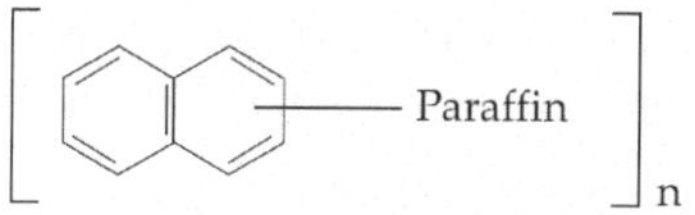

It has been suggested that these molecules adsorb into the wax crystals and redirect their growth, forming smaller and more isotropic crystals that interfere less with oil flow.

1.3.9 Multifunctional nature of additives (Rizvi, 2009)

A number of additives perform more than one function. Zinc dialkyl dithiophosphates, known mainly for their antiwear action, are also potent oxidation and corrosion inhibitors. Styrene-ester polymers and functionalized polymethacrylates and can act as viscosity modifiers, dispersants, and pour point depressants. Basic sulfonates, in addition to acting as detergents, perform as rust and corrosion inhibitors. They do so by forming protective surface films and by neutralizing acids that arise from fuel combustion, lubricant oxidation, and additive degradation.

2. Future work

- Using nanotechnology in preparation of lube oil additives, "synthesis of overbased nanodetergent". Production of stable, efficient nanodetergent system depends on development and new generation of surfactant. These nano-particles are relatively insensitive to temperature.
- In spite of the increasing temperature, loads and other requirements imposed on lubricants, mineral oils are likely to continue to be employed in the foreseeable future for the majority of automotive, industrial and marine applications. However, in the aviation field, synthetic lubricants are extensively used and there are a growing number of critical automotive, industrial and marine application where the use of synthetic lubricants van be justified on a cost / performance basis.

References

1. Alun L., Ken B.T., Randy C.B., and Joseph V.M., Large-scale dispersant leaching and effectiveness experiments with oils on calm water; *Marine Pollution Bulletin,* 60, 244– 254, (2010).
2. Battez A.H., Viesca J.L., González R., Blanco D., Asedegbega E., and Osorio A., Friction reduction properties of a CuO nanolubricant used as lubricant for a NiCrBSi coating; *Wear*, 268, 325–328, (2010).
3. Bharambe D.P., Designing maleic anhydride-α-olifin copolymeric combs as wax crystal growth nucleators; *Fuel Processing Technology,* 91, 997–1004, (2010).
4. Chen B., Sun Y., Fang J., Wang J., and Wu Jiang, Effect of cold flow improvers on flow properties of soybean biodiesel; *Biomass and bioenergy*, 34, 1309-1313, (2010).
5. Kyunghyun R., The characteristics of performance and exhaust emissions of a diesel engine using a bio-diesel with antioxidants; using a bio-diesel with antioxidants; *Bioresource Technology*, 101, 578–582, (2010).
6. Leslie R.R., Lubricant Additives, "Chemistry and Applications", Marcel Dekker, Inc., 293- 254, (2003).
7. Ludema K.C.; Friction, Wear, Lubrication, A Textbook in Tribology, CRC Press L.L.C., 124- 134, (1996).
8. Margareth J.S., Peter R.S., Carlos R.P.B., and José R.S., Lubricant viscosity and viscosity improver additive effects on diesel fuel economy; *Tribology International,* 43, 2298– 2302, (2010).
9. Masabumi M., Hiroyasu S., Akihito S.,and Osamu K., Prevention of oxidative degradation of ZnDTP by microcapsulation and verification of its antiwear performance; *Tribology International,* 41, 1097–1102, (2008).
10. Mel'nikov V.G., Tribological and Colloid-Chemical Aspects of the Action of Organic Fluorine Compounds as Friction Modifiers in Motor Oils; *Chemistry and Technology of Fuels and Oils,* 33, No. 5, 286-295, (1997).

11. Ming Z., Xiaobo W., Xisheng F., and Yanqiu X., Performance and anti-wear mechanism of $CaCO_3$ nanoparticles as a green additive in poly-alpha-olefin; *Tribology International,* 42, 1029–1039, (2009).
12. Rizvi, S.Q.A., A comprehensive review of lubricant chemistry, technology, selection, and design, ASTM International, West Conshohocken, PA., 100-112, (2009).

Part 3

Solid Lubricants and Coatings

11

Tribological Behaviour of Solid Lubricants in Hydrogen Environment

1. Introduction

In a future energy supply system based on renewable sources hydrogen technology will play a key role. Because the amount of energy from renewable sources, such as wind or solar power, differs seasonally and regionally, an energy storage method is necessary. Hydrogen, as an environmentally friendly energy carrier, can fill this gap in an ideal way, in particular for mobile applications (Wurster et al., 2009). Already today, in Germany the amount of hydrogen as a byproduct in chemical industry is enough for fuelling about 1 Mio passenger cars1. Excess electrical power can be used to produce hydrogen by electrolysis. On demand, this hydrogen can be used for mobile or stationary fuel cells. Beside this new developing technology, hydrogen is used as fuel for rocket engines and in chemical industry since a long time. Table 1 comprises some physical parameters of hydrogen. It can be seen that hydrogen gas has a very low density which makes storage at high pressure or in liquid form (LH_2) necessary.

With increasing utilisation of hydrogen it will be necessary to optimize components which are in contact with this medium. If these components contain tribosystems directly exposed to hydrogen they are critical in respect of excess wear, because of vanishing protective oxide layers in the presence of a chemically reducing environment. Furthermore, liquid lubricants are often not applicable, because of purity requirements, or very low temperatures in the case of liquid hydrogen. Thus, for numerous components in hydrogen technology, solid lubrication is the only possible method for reducing friction and wear.

Table 1. Physical parameters of hydrogen (Bulletin M 055, 1991 and Frey & Haefer, 1981)

Melting temperature		-259.35°C (13.80 K)
Boiling temperature (1.013 bar)		-252.87°C (20.28 K)
Gas densities	at 0°C, p = 1.013 bar	0.08989 kg/m3
	at boiling temperature density ratio H_2/air	1.338 kg/m3 0.0695 kg/m3
Density of the liquid at boiling temperature		0.07098 kg/l
evaporation enthalpy at boiling temperature h_l		0.915 kJ/mol
Heat conductivity (0°C; 1.013 bar)		0.1739 W/(m K)
Heat capacity of the liquid at boiling temperature		9.69 kJ/(kg K)
Specific heat (0°C; 1.013 bar)	c_p c_v	28.59 J/(mol K) 20.3 J/(mol K)
Critical temperature T_c		-240.17°C (32.98 K)
Critical pressure p_c		12.93 bar

Although the tribological behaviour of typical solid lubricants such as graphite, DLC, and MoS_2 has been characterized comprehensively (Landsdown, 1999; Donnet & Erdemir, 2004; Gradt et al., 2001), information about their suitability for hydrogen environment is very limited. Therefore, based on available literature and own measurements, an overview of solid lubricants and other materials for tribosystems in hydrogen environment is given in the following.

2. Tribosystems in hydrogen environment

Prominent examples for extremely stressed components in hydrogen environment are turbopumps for cryogenic propellants in rocket engines. They comprise numerous tribosystems, such as shaft seals or self-lubricated bearings, which have to work in hydrogen environment at low temperatures and high pressures. In the tribo-components of a turbopump various solid lubricants such gold, silver, silver-copper alloy, PTFE, graphite, and MoS_2 or wear resistant coatings as WC, Cr, Cr_2O_3, and TiN are applied (Nosaka, 2011). The LH_2-turbopump of the LE-7 engine for the Japanese H-2 rocket has a flow rate of 510 l/s, a shaft power of 19,700 kW, and a rotational speed of 42,000 rpm. All steel bearings made from AISI 440C with lubrication by PTFE transfer from the retainer to the raceways showed sufficient performance. These bearings were operated at 50,000 rpm without severe wear (Nosaka, 2011). For better performance and rotational speeds up to 100,000 rpm hybrid ceramic bearings with Si_3N_4 balls and steel rings are used. Such bearings were developed for the space shuttle (Gipson, 2001), the future VINCI launcher in Europe, and the Japanese LE- 7 rocket engine (Nosaka et al., 2010). Hybrid ball bearings with ceramic balls can be operated up to 120,000 rpm in liquid hydrogen.

Hydrogen environment influences the friction behaviour of materials such as transition metals and metals that react chemically with hydrogen by building stable hydrides (Fukuda & Sugimura, 2008). In the case of transition metals, chemisorption is the main mechanism. The influence of hydrogen on the tribological properties of steels cannot be derived directly from these mechanisms, although the main components in steels are transition metals (Fukuda et al., 2011).

A general problem for materials, especially metals, exposed to hydrogen is environmentally induced embrittlement, which is also active in tribologically stressed systems. One example is embrittlement of raceways in ball bearings. The effect of hydrogen on the fatigue behaviour of bearing steel AISI 52100 was studied by Fujita et al. (2010). He investigated samples of steel under cyclic loading. Samples precharged with hydrogen showed a significant shorter lifetime, which could be attributed to the occurrence of an increased number of cracks.

However, hydrogen embrittlement is a general materials problem and not specific to tribo- systems. Friction induced changes in the structure of steels that lead to embrittlement phenomena are treated in chapter 4. More specific to tribologically stressed surfaces is the fact that oxide layers, which protect many metals against wear and corrosion, are not renewed after they are worn away. In the special case of tribosystems running in liquid hydrogen, the environmental temperature is 20 K (-253°C) and far too low for any liquid lubricant. In such cases, solid lubricants can be employed for reducing friction and wear. Also, in applications such as fuel cells or semiconductor fabrication, gaseous hydrogen of high purity is required and high demands on the outgassing of the materials are made, which usually cannot be met by liquid lubricants.

Commercial hydrogen gas contains a certain amount of water and oxygen. The influence of residual water in hydrogen gas on the fretting wear behaviour of bearing steel SUJ2 (similar to AISI 52100) was investigated by Izumi et al. (2011). These tests were performed in hydrogen and nitrogen gas with water content between 2 and 70 ppm. In both gases friction decreases, but wear increases with increasing water content.

3. Test devices for friction tests in cryogenic hydrogen environment

In general, test equipment for hydrogen environment has to meet the safety standards for handling this medium. In particular, explosion safe electrical installations, proper venting, gas tight experimental chambers, filling, and venting tubes are necessary. For liquid hydrogen also cryogenic equipment has to be employed. As examples two cryotribometers which are available at BAM2 are shown in Figures 1 and 2 (Gradt, et al., 2001). Both cryotribometers are appropriate for liquid and gaseous hydrogen. The sample chambers are thermally insulated by vacuum superinsulation and cooled directly by a bath of liquid cryogen or by a heat exchanger.

In the case of CT 2 (Fig. 1) the liquid coolant is filled directly into the sample chamber (bath cryostat). The complete friction sample is immersed into the liquid cryogen and the environmental temperature is equal to the boiling temperature of the coolant (liquid nitrogen, LN_2: 77.3 K; liquid hydrogen, LH_2: 20.3 K; liquid helium, LHe: 4.2 K). The advantage of this method is a very effective cooling of the sample by making use of the heat of evaporation of the liquid.

Most of the tests are carried out by using the standard pin-on-disc configuration where a fixed flat pin or ball is continuously sliding against a rotating disc. The rotation is transmitted via a rotary vacuum feedthrough to a shaft with the sample disc at the lower end. In CT2 loading is performed by means of a gas bellow which acts on a frame with the fixed sample (pin) mounted on its lower beam. The mechanical stability of this assembly allows normal forces up to 500 N. The friction force is measured by means of a torque sensor on top of the motor journal or a beam force transducer integrated in the sample holder.

The sample chamber of the tribometer CT 3 is designed for pressures between 10-3 mbar and 20 bar and cooled by a heat exchanger (continuous flow cryostat). The coolant is pumped through the heat exchanger, evaporates and removes heat from the inner vessel. Thus, it is possible to adjust the temperature between 4.2 K (with LHe cooling) and room temperature independently from the pressure. There is not limitation to an equilibrium state of the boiling coolant as in a bath cryostat. In hydrogen environment, the behaviour of tribosystems in gaseous or liquid environment as well as in the vicinity of the critical point can be investigated, which is of importance for the design of high performance hydrogen pumps.

2 BAM Federal Institute for Materials Research and Testing, Berlin, Germany

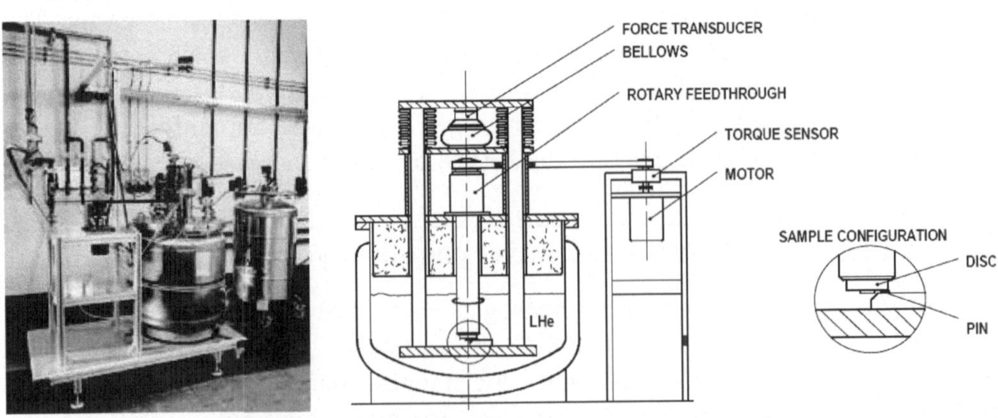

Fig. 1. Cryotribometer CT 2 (bath cryostat)

While in CT 2 the loading unit is located in the room temperature part of the apparatus, in CT 3 loading and force measurement is performed close to the friction couple in the cold part. Therefore, combined loading and measuring units are employed. The sample holder for the counterbody is directly mounted on a two dimensional beam force transducer for measuring normal and friction forces. Loading is accomplished by pressurized He-gas which acts on a piston that moves the beam with the sample holder upwards and presses the counterbody against the lower face of the rotating disk.

To remove any residual gases and condensed liquids, the sample chamber is evacuated to a pressure below 10-3 mbar and filled with pure He-gas. The pump-down-refill cycle is repeated three times. During the experiment, the sliding force and the displacement of the pin are measured. After the measurement, the wear scars of both bodies can be examined by profilometry, light, electron, or atomic force microscopy.

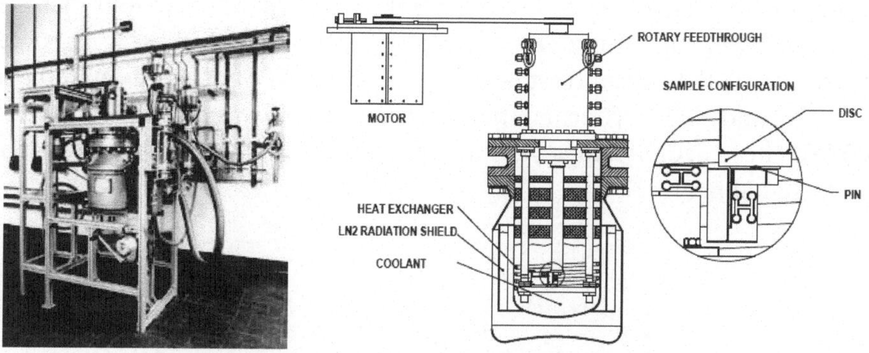

Fig. 2. Cryotribometer CT 3 (flow cryostat)

4. Tribological behaviour of metals in hydrogen environment

4.1 Soft metals as solid lubricants

Soft Metals such as gold, silver, lead, and indium can serve as solid lubricants. Thin films with good adhesion can be applied by ion-plating with an optimum thickness of about 1 μm. The tribological properties of soft metals are similar in ambient air and vacuum environment with friction coefficients of about 0.1 and remain unchanged during cooling down to cryogenic temperatures. Furthermore, as they have a f.c.c. crystal structure, they are not affected by hydrogen embrittlement (Moulder & Hust, 1983) and therefore, applicable for tribosystems in gaseous and liquid hydrogen. However, in sliding friction in vacuum these materials have higher friction and wear than lamellar solids (Roberts, 1990, Subramonian et al., 2005).

4.2 Properties of steels in cryogenic hydrogen environment

A large number of ferrous alloys are employed for tribosystems, including those running in hydrogen environment. As many of these materials suffer from hydrogen embrittlement, they are treated in this chapter, although they are no solid lubricants. In particular, ferritic and martensitic steels with b.c.c. lattice are strongly affected by hydrogen. Austenitic FeCrNi alloys with f.c.c. structure don't show hydrogen embrittlement, and therefore, these alloys are the favoured materials in hydrogen technology. As these steels have good mechanical properties even at cryogenic temperatures they are also appropriate for components in contact with liquid hydrogen. However, in highly stressed tribosystems deformation-induced generation of martensite is possible, and the danger of embrittlement in these regions arises. Furthermore, an uptake of hydrogen can intensify the deterioration of the material. In an austenitic lattice solute hydrogen decreases the stacking fault energy (SFE) (Holzworth & Louthan, 1968). As a consequence, the deformation behaviour changes and the martensite generation is facilitated. In Fig. 3 (Butakova, 1973) the generation of martensite in tensile testing in dependence of the SFE for various FeCrNi-alloys is shown.

Therefore, it is necessary to investigate the tribological behaviour of austenitic steels in hydrogen-containing environments. The friction and wear behaviour in liquid hydrogen of the austenitic steels 1.4301 (AISI 304), 1.4439 (comparable to AISI 316), 1.4876, and 1.4591 (German materials numbers) was studied by Huebner, et al. (2003a). These FeCrNi alloys have different stability of their austenitic structure and are included in Fig. 3.

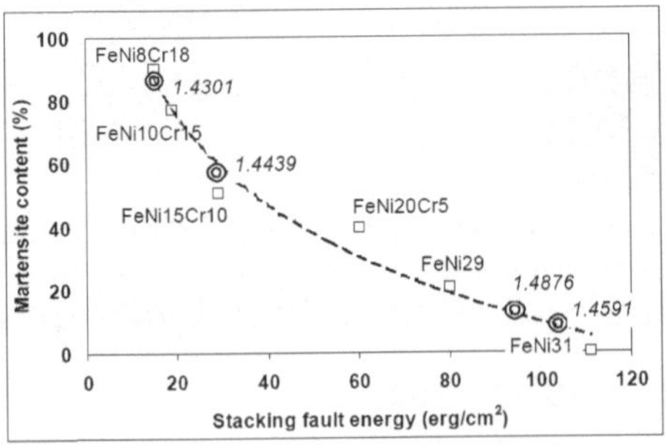

Fig. 3. Influence of the SFE of austenitic FeNiCr alloys on the martensite volume fraction after 80% plastic deformation in tensile testing (according to Butakova, 1973)

Steel 1.4301 is a metastable austenite. Its SFE is very low and thus, deformation-induced structure transformation is possible, even at room temperature. Steel 1.4439 is a so-called stable austenitic steel. Transformation is impeded because of its increased SFE. Finally, in materials 1.4876 and 1.4591 with very high contents of Ni, the SFE is rather high, and the generation of martensite should be impossible. As counterbodies Al_2O_3 ceramic balls were used to avoid metal transfer to the steels samples. The austenitic steels were tested in inert environments at low temperatures and in LH_2. After the friction experiments, the transformation to martensite in the wear scars was detected by changes of the materials magnetic properties (magneto-inductive single-pole probe). This method has been shown to be sensitive enough to describe the transformation at a crack tip (Bowe et al., 1979).

The amount of martensite vs. temperature for 1.4301 is shown in Fig. 4. The amount of martensite strongly depends on the temperature with a maximum at about 30 K. Below this temperature the generation of martensite decreases. For this metastable steel, hydrogen environment was without any influence on the amount of austenite transformed into martensite (symbol ′).

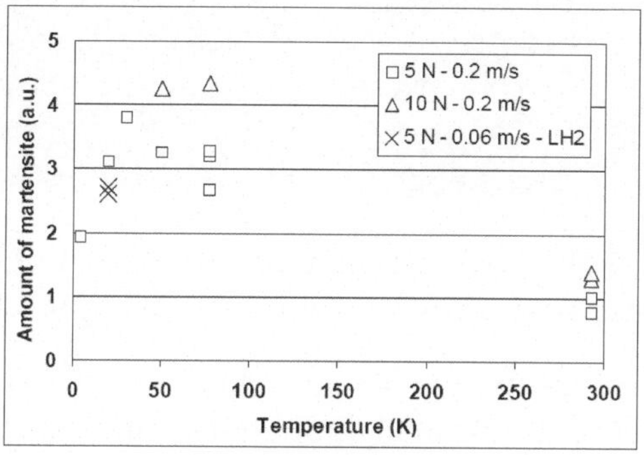

Fig. 4. Steel 1.4301, Temperature-dependence of friction-induced generation of martensite

Contrary to steel 1.4301, the transformation behaviour of the steel 1.4439 showed a distinct influence of the environment (Fig. 5). In LN_2 and at 20 K in gaseous He, only local magnetisation was detected in the wear scars (symbols: □, D). It could be shown by scanning electron microscopy that locations with magnetic signals correspond to extremely deformed transfer particles (Hübner, 2001). After a test in liquid hydrogen (symbol: +), magnetic changes were observed in the entire circular wear track.

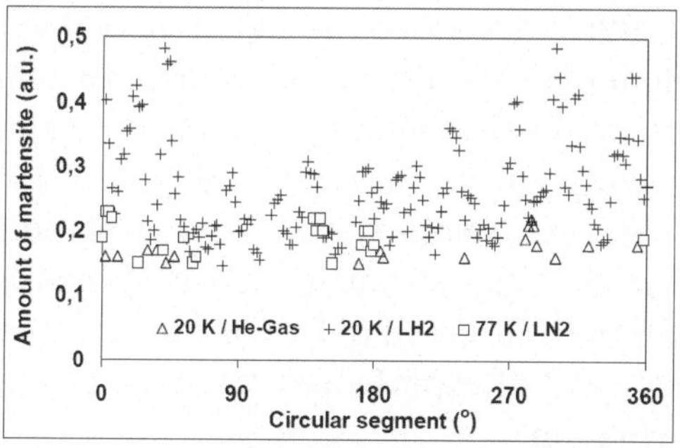

Fig. 5. Steel 1.4439, Influence of hydrogen on the generation of martensite during friction

After the tests in inert environment, extremely deformed wear debris was found all over the wear track. However, these particles did not show any embrittlement. After sliding in hydrogen, the surface showed completely different features. The wear scar exhibits a net of microcracks (Fig. 6). This topography was detected for all austenitic alloys chosen for these experiments, even for the highly alloyed materials 1.4876 and 1.4591. This is clear indication for the occurrence of hydrogen induced embrittlement, even in LH_2. These findings could be confirmed by measurements of residual stresses in the deformed zone (Hübner et al., 2003b).

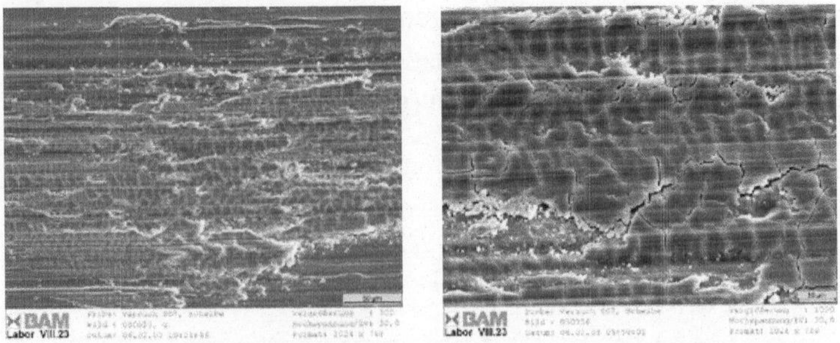

Fig. 6. Steel 1.4591, SEM images of the wear track; net of brittle cracks in the wear scar after frictional stressing in LH_2

For influencing the deformation behaviour, it is necessary that atomic hydrogen exists in the material. In LH_2 thermally initiated dissociation is not possible. Thus, the dissociation process could only be activated by mechanical energy from sliding.

The influence of hydrogen on the deformation mechanisms is also visible in the shape of the X-ray diffraction line profiles. Fig. 7 shows the γ_{311} reflection of the austenitic steel 1.4876 after sliding in air, LHe, and LH_2. The reflection profiles of the tests in air and LHe are symmetrical. They exhibit only deformation-induced broadening. However, in LH_2 an asymmetry occurs, which is a clear indication for hydrogen uptake. Hydrogen lowers the stacking fault energy of the austenite lattice, which enhances the building of the epsilon phase (Whiteman & Troiano, 1984, Pontini & Hermida, 1997).

Gavriljuk et al. (1995) described in detail how hydrogen influences the transformation behaviour of unstable as well as stable austenitic steels. In so-called unstable steels, already cold working induces phase transformation. Stable steels may be subject to structure changes after charging with hydrogen, which causes a decrease in SFE. These explanations are in good agreement with the results shown in Figures 4 and 5. A significant influence of hydrogen on the austenite-martensite transformation is observed only in the stable steel (Fig. 5), because the metastable steel 1.4301 (Fig. 4) experiences structure changes already during the low-temperature deformation.

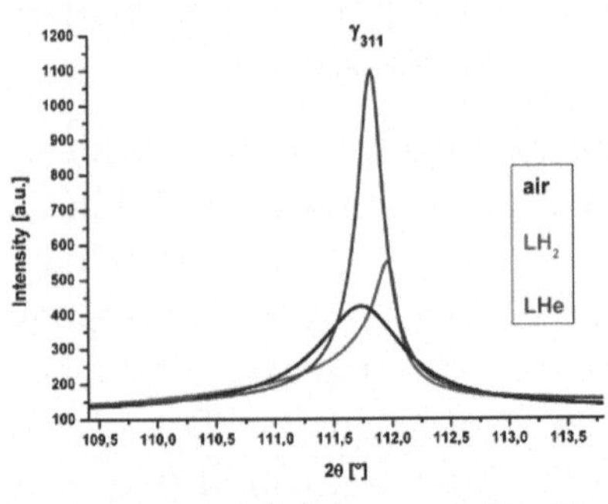

Fig. 7. Steel 1.4876, Asymmetry of the γ_{311} reflection of the after frictional stressing in LH_2

Beside deformation enhanced creation of martensite, also other mechanisms can lead to increased wear in austenitic stainless steel. Kubota et al. (2011) reported a reduction of the fretting fatigue limit in hydrogen gas for steel AISI 304. He found that small cracks which were stable in air propagated in hydrogen gas. The reason for this effect was an increased local adhesion in hydrogen environment.

4.3 Other metals

The tribological properties of Zr and Nb in hydrogen environment were investigated by Murakami et al. (2010). Coatings of Zr-alloys on high strength steels are considered as a diffusion-barrier for hydrogen. Furthermore, Zr forms hydrides which have the same structure as CaF_2, which can be used as solid lubricant. As NbH_2 has a lattice structure similar to CaF_2 it may also have lubricating properties. Pure Zr and Nb were tested as self- mated pairs in pin-on-disc tests with a sliding speed of 3.49 x 10-2 ms-1 and loads of 25 and 70 N. Both materials showed lower friction coefficients in H_2 gas atmosphere than in air, He gas, and vacuum. In H_2-gas atmosphere the friction coefficients of the Nb specimens were much higher than those of the Zr specimens. X-ray diffraction analysis showed that the wear particles, which were formed by sliding Zr and Nb specimens in the H_2 gas atmosphere, consisted mainly of the ZrH_2 phase (ε phase) and NbH phase (β phase), respectively. X-ray diffraction analysis also showed that the wear particles, which were formed by sliding in air, consisted mainly of the αZr and Nb phases, respectively.

5. Solid lubricant coatings

An amorphous carbon (DLC-), a MoS_2-coating, prepared by physical vapour deposition (PVD), and two types of anti-friction coatings (AFC 1 and AFC 2) were tested in dry and humid N_2-, H_2-, and CH_4-environment at BAM (Gradt & Theiler, 2010). In dry gas, the residual water content was in the ppm-range. In humid environment the relative humidity was close to 100%. The solid lubricant in both AF- coatings was PTFE. The tests were performed in ball-on-flat configuration in reciprocating sliding at room temperature. The test parameters are given in Table 2.

Table 2. Test parameters, solid lubricant coatings

Parameter	Value
Substrate	100Cr6 (AISI 52100)
Coatings	DLC, MoS_2, AF-Coatings (lubricant: PTFE)
Counterbody	uncoated ball, d = 4 mm. X90CrMoV18 (AISI 440B)
Environment	N_2-, H_2-, CH_4-gas, dry/humid
Gas pressure, bar	3
F_N, N	5; 10
Stroke, µm	200
Frequency, Hz	20
Test duration	2 h (144,000 Cycles)

Fig. 8 summarizes the measured friction coefficients in the miscellaneous environments. The carbon coating shows a distinct sensitivity to the humidity. While in dry gases the friction coefficient is about 0.15, it rises to 0.19 to 0.25 in humid environment. Also the wear of this type of DLC-coating rises under high humidity, as can be seen in Figures 9 and 10. Fig. 9 shows an SEM-image of a wear scar after a test in hydrogen of high humidity. The complete coating is worn away, and abrasive wear of the substrate is visible. Fig. 10 shows an image of a wear scar after the same sliding distance in dry hydrogen. It can be seen that the coating is still intact, and the wear track has a very smooth, polished-like surface.

Both AF-coatings showed friction coefficients around 0.15 in dry hydrogen and nitrogen. They have not been tested in these gases with high humidity. Such comparative measurements were done in methane gas. It can be seen that AFC 1 reaches a low friction coefficient of 0.08 in dry and humid CH_4. Thus, this coating is sensitive to the particular type of gas and not to general chemical reactivity or humidity. AFC 2 also changes its frictional behaviour in CH_4. However, while methane seems to have a beneficial effect on AFC 1, it causes an increasing friction of AFC 2.

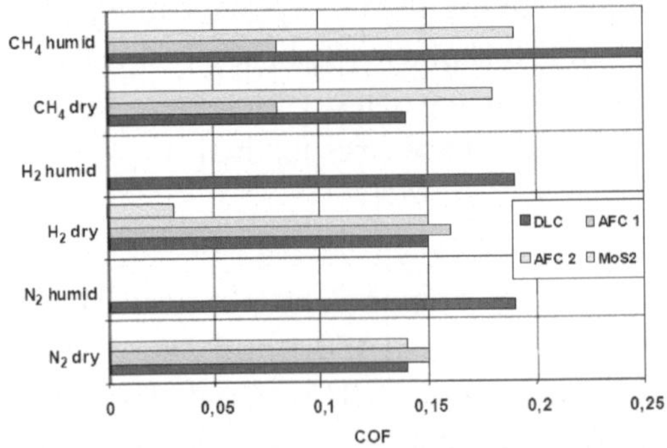

Fig. 8. Friction coefficients of several solid lubricants in inert and reactive gaseous environment

Fig. 9. Coating failure of a DLC-coating after a reciprocating sliding test in humid H_2- environment (SEM-image of the wear scar)

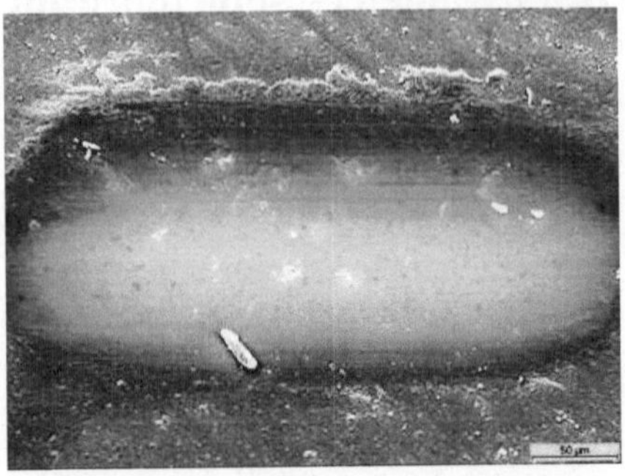

Fig. 10. Smooth wear track of the a DLC-coating after a reciprocating sliding test in dry H_2- environment (SEM-image of the wear scar)

The lowest coefficient of friction (COF = 0.03) was observed for MoS_2. This coating showed a very smooth sliding behaviour with nearly no running-in. Fig. 11 shows the development of the COF of the three tested MoS_2- coatings in comparison to DLC. The DLC-coatings showed a higher COF and a pronounced running-in behaviour. However, the lifetime of the MoS_2-coatings was much shorter than that of DLC and not sufficient in the scope of this test series, where more than 100,000 cycles were necessary. Therefore, no further tests in other environments were carried out. Nevertheless, for dry sliding tribosystems, where a lifetime of 10,000 friction cycles is sufficient, MoS_2-lubrication seems to be applicable in hydrogen environment.

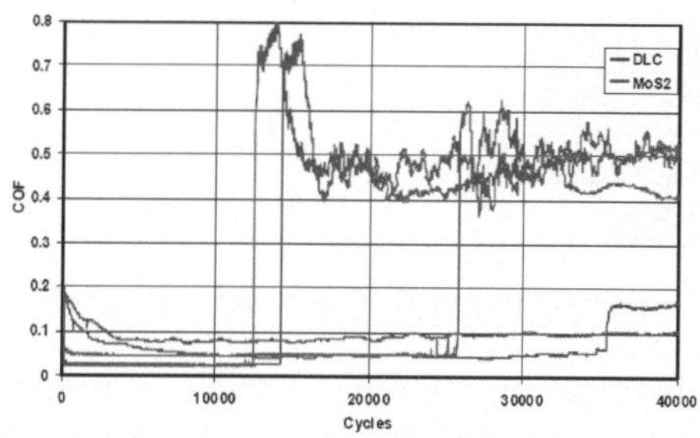

Fig. 11. Oscillating friction of DLC- and MoS_2-coatings in gaseous hydrogen

6. Solid lubricants in polymer composites

Table 3. Materials and test parameters, polymer composites

Polymer matrix	PTFE: polytetrafluoroethylene PEEK: polyetheretherketone PI: polyimide PA: Polyamide PEI: polyetherimide EP: epoxy
Fibers	CF: carbon fibers
Fillers	PEEK, PPS bronze TiO_2
Lubricants	PTFE, MoS_2, graphite
Normal load, N	16; 50 N
Sliding speed, m/s	0.2
Sliding distance, m	2000

Polymers and polymer composites are widely used as dry sliding materials in friction assemblies where external supply of lubricants is impossible, or not recommended. The field of application of self-lubricating materials in tribological systems is considerably extending also to extreme environments (Gardos, 1986). Over the years, composite materials have replaced many traditional metallic materials in sliding components. They offer not only low weight and corrosion resistance, but also excellent tribological properties. In view of hydrogen technology, numerous polymer composites containing PTFE, MoS_2, and graphite respectively have been tested in hydrogen and inert media such as nitrogen and helium (Theiler & Gradt, 2007). Some of these materials were also tested in liquid hydrogen. Fig. 12 shows the test configuration, and Table 3 summarizes the materials and test parameters. The material compositions are given in the figures of the test results below.

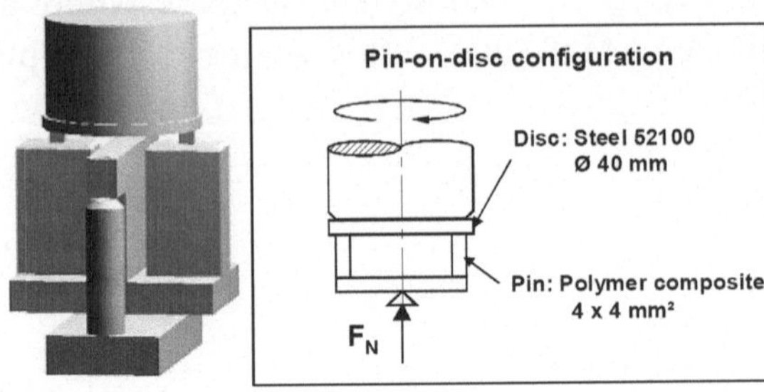

Fig. 12. Sample configuration for tests of polymer composites

Fig. 13 shows the friction coefficient of various polymer composites against steel in air and liquid hydrogen (Theiler & Gradt, 2007). Except the first one, all tested composites have lower friction in LH_2 than in air at room temperature. A decrease of friction at lower

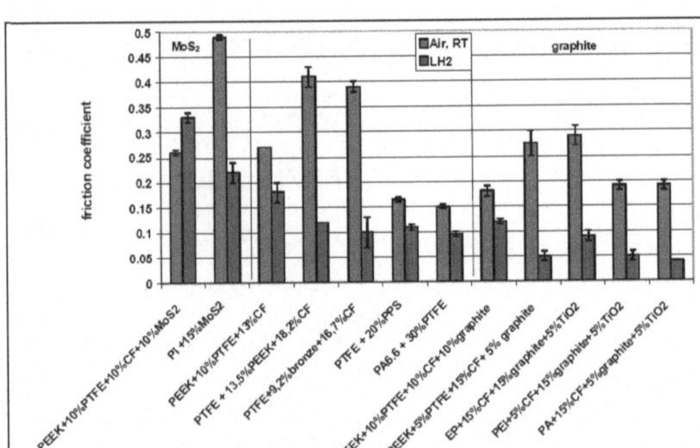

Fig. 13. Sliding friction of polymer composites against steel (Theiler & Gradt, 2007)

temperatures is observed for many polymers and is due to the fact that hardness and Young's modulus of the polymers increase with decreasing temperature. Both lead to lower deformation and a smaller real area of contact. This causes a lower shearing force at the interface and thus a lower friction (Theiler et al., 2004).

Another tendency is that graphite containing composites have the lowest friction coefficients in liquid hydrogen, in one case even lower than 0.05. On the other hand, composites containing MoS_2 don't reach values below 0.2. Thus, for hydrogen applications graphite seems to be a much more efficient component for improving the lubricating properties of polymers. The friction coefficients of the composites without graphite or MoS_2 are between 0.1 and 0.2 in LH_2 which is sufficient for many applications. All materials of this group contain PTFE, which also acts as a solid lubricant. In some cases, the large difference between ambient air and LH_2 is a possible drawback for practical application.

A comparison of the friction coefficients in liquid hydrogen, hydrogen gas, and ambient air at room temperature for two composites with PTFE- and two with PEEK-matrix is shown in Fig. 14. The materials with PTFE-matrix show a large difference in COF between normal air and hydrogen environment and no significant influence of the temperature. This difference is much smaller for the PEEK materials with additions of PTFE. Additional admixture of graphite leads to a COF of about 0.15, which depends only very little on the environment. Although the other composites exhibit lower friction under certain conditions, this low dependence on the environment makes the graphite containing composite a most suitable material for hydrogen applications.

The wear behaviour of the PTFE- and PEEK-composites follows a similar tendency. As shown in Fig. 15, the wear rate of the two materials without graphite is much smaller in hydrogen environment than in air. The wear of the graphite containing material is not significantly influenced by the environment. Furthermore, a wear rate below 10-6 makes this material suitable for application in sliding bearings or in cages for roller bearings.

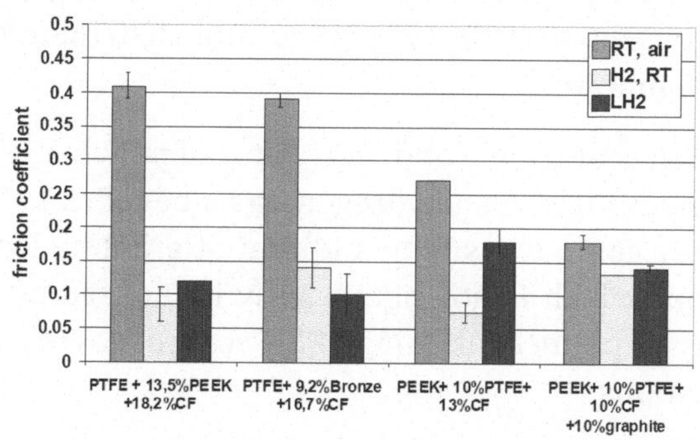

Fig. 14. Friction of polymer composites in air and H_2

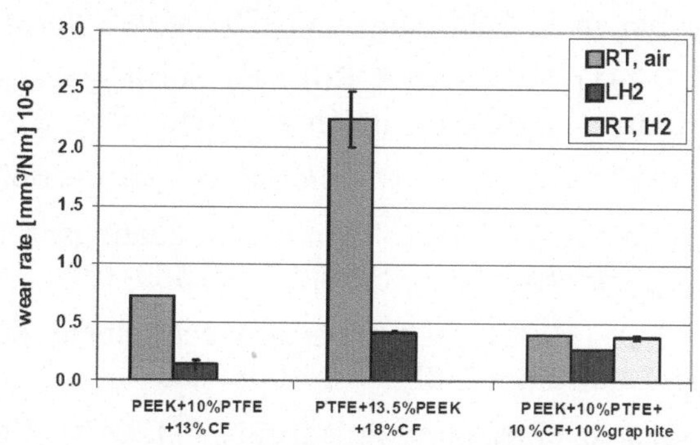

Fig. 15. Wear of polymer composites in air and H_2

7. Conclusion

Tribosystems directly exposed to hydrogen are critical in respect of excess wear, because they may experience hydrogen embrittlement, chemical reactions to hydrides, and vanishing protective oxide layers respectively. Furthermore, liquid lubricants are often not applicable, because of purity requirements, or very low temperatures in the case of liquid hydrogen.

Hydrogen uptake and material deterioration influences wear processes also in austenitic stainless steels. Hydrogen lowers the stacking fault energy of the austenite lattice, which enhances the building of deformation induced martensite that is prone to hydrogen embrittlement.

For numerous components in hydrogen technology solid lubrication is the only possible method for reducing friction and wear. Solid lubricants such as PTFE, graphite, DLC, and MoS_2 applied as coatings, or as components in polymer composites, in general are able to reduce friction and wear in gaseous as well as in liquid hydrogen.

MoS_2-coatings have low friction, but a very short lifetime in hydrogen environment. The tested carbon coating showed higher friction, but a much longer lifetime in dry environment. In humid environment this type of coating fails rapidly.

PTFE-based anti friction (AF-) coatings exhibit low friction and a negligible sensitivity to humidity. However, the type of gas influences their frictional behaviour, independent of the humidity.

In general, friction coefficients and wear rates of polymer composites decrease with decreasing temperature. Also hydrogen has a beneficial effect on the friction behaviour of polymer composites. The addition of graphite leads to a favourable tribological behaviour which is not significantly influenced by the environmental medium. This makes graphite- containing PEEK-composites most suitable materials for hydrogen applications.

References

1. Bowe, K.H.; Hornbogen, E.; Wittkamp, I. (1979). *Materialprüfung*, Vol. 21, pp. 74
2. Bulletin M 055 (1991). Wasserstoff, BGI 612, Berufsgenossenschaft Rohstoffe und chemische Industrie, Jedermann Verlag, Heidelberg, Germany
3. Butakova, E.D. (1973). *Fizika Met. Metalloved.* Vol. 35, pp. 662
4. Donnet, C. & Erdemir, A. (2004). Solid lubricant coatings: recent developments and future trends, *Tribology Letters,* Vol. 17, pp. 389-397
5. Frey, H.; Haefer, R.A. (1981). Tieftemperaturtechnologie, Eder, F.X. (Ed.), VDI-Verlag, Düsseldorf, Germany, ISBN 3-18-400503-8
6. Fujita, S.; Matsuoka, S.; Murakami, Y.; Marquis, G. (2010). Effect of hydrogen on

Mode II fatigue crack behaviour of tempered bearing steel and microstructural changes. *International Journal of Fatigue.* Vol. 32, pp. 934-951

7. Fukuda, K.; Sugimura, J. (2008). Sliding Properties of Pure Metals in Hydrogen Environment, *Proc. STLE/ASME,* IJTC2008-71210
8. Fukuda, K.; Hashimoto, M.; Sugimura, J. (2011). Friction and Wear of Ferrous Materials in a Hydrogen Gas Environment, *Tribology Online,* Vol. 6, pp. 142-147
9. Gardos, M.N. (1986). Self lubricating composites for extreme environmental conditions, In: *Friction and Wear of Polymer Composites, Composite Materials Series 1,* K. Friedrich (Ed.), pp. 397-447
10. Gavriljuk, V.G.; Hänninen, H.; Tarasenko, A.V.; Tereshchenko, A.S.; Ullakko, K.. (1995). Phase transformations and relaxation phenomena caused by hydrogen in stable austenitic stainless steels, *Acta metal. mater.*, Vol. 43, pp. 559-568
11. Gipson, H. (2001). Lubrication of Space Shuttle Main Engine Turbopump Bearings, *Lubrication Engineering,* August 2001, pp. 10-12
12. Gradt T., Börner H., Schneider T. (2001). Low Temperature Tribometers and the Behaviour of ADLC Coatings in Cryogenic Environment, *Tribology Intern.*, Vol. 34, pp. 225-230
13. Gradt, Th.; Theiler, G. (2010). Tribological Behaviour of Solid Lubricants in Hydrogen Environment, *Tribology Online*, Vol. 6, pp. 117-122, ISSN 1881-2198
14. Holzworth, M.L.; Louthan jr., M.R. (1968). Hydrogen-induced phase transformations in type 304L stainless steels, *Corrosion*, Vol. 244, pp. 110-124
15. Hübner, W. (2001). Phase transformations in austenitic stainless steels during low temperature tribological stressing. *Tribology Intern.*, Vol. 34, 231-236
16. Huebner, W.; Gradt, Th.; Assmus, K.; Pyzalla, A.; Pinto, H.; Stuke, S. (2003a). Hydrogen Absorption by Steels during Friction, *Proc. Hydrogen and Fuel Cells Conf. and Trade Show*, Vancouver, Canada
17. Hübner, W.; Pyzalla, A.; Assmus, K.; Wild, E.; Wroblewski, T. (2003b). Phase stability of AISI 304 stainless steel during sliding wear at extremely low temperatures, *Wear*, Vol. 255, pp. 476-480
18. Izumi, N.; Morita, T.; Sugimura, J. (2011). Fretting Wear of a Bearing Steel in Hydrogen Gas Environment Containing a Trace of Water, *Tribology Online*, Vol. 6, pp. 148-154
19. Kubota, M.; Kuwada, K.; Tanaka, Y.; Kondo, Y. (2011). Mechanism of reduction of fretting fatigue limit caused by hydrogen gas in SUS304 austenitic stainless steel, *Tribol. Int., in press*
20. Landsdown, A. R. (1999). Molybdenium Disulphide Lubrication, *Tribology Series*, Vol. 35, D. Dowson (Ed.), Elsevier, Amsterdam,
21. Moulder, J.C.; Hust, J.G. (1983). Compatibility of Materials with Cryogens, in: *Materials at low temperatures*, Reed, R.P.; Clark, A.F. (Ed.), ASME, USA, ISBN: 0-87170-146-4

22. Murakami, T.; Mano, H.; Kaneda, K.; Hata, M.; Sasaki, S.; Sugimura, J. (2010). Friction and wear properties of zirconium and niobium in a hydrogen Environment, *Wear*, Vol. 268, pp. 721–729
23. Nosaka, N; Takata, S.; Yoshida, M.; Kikuchi, M; Sudo, T.; Nakamura, S. (2010). Improvement of Durability of Hybrid Ceramic Ball Bearings in Liquid Hydrogen at 3 Million DN (120,000 rpm), *Tribology Online*, Vol. 5, pp. 60-70
24. Nosaka, M. (2011), Cryogenic Tribology of High-speed Bearings and Shaft Seals in Liquid Hydrogen, *Tribology Online*, Vol. 6, 133-141
25. Pontini, A.E.; Hermida, J.D. (1997). X-Ray Diffraction Measurement of the Stacking Fault energy Reduction Induced by Hydrogen in an AISI 304 Steel, *Scripta Materialia*, Vol. 37, pp. 1831-1837
26. Roberts, E.W. (1990). Thin Solid Lubricant Films in Space, *Tribol. Int.*, Vol. 23, pp. 95-104 Subramonian, B.; Kato, K.; Adachi, K.; Basu, B. (2005). Experimental Evaluation of Friction and Wear Properties of Solid Lubricant Coatings on SUS440C Steel in Liquid Nitrogen, *Tribol. Lett.*, Vol. 20, pp. 263-272
27. Theiler, G.; Hübner, W.; Gradt, T.; Klein, P. (2004). Friction and Wear of Carbon Fibre Filled Polymer Composites at Room and Low Temperatures", *Materialwissenschaft und Werkstofftechnik*, Vol. 35, pp. 683-689
28. Theiler, G. & Gradt, Th. (2007). Polymer Composites for Tribological Applications in Hydrogen Environment, *Proceedings of the 2nd International Conference on Hydrogen Safety*, San Sebastian, Spain
29. Whiteman, M.B.; Troiano, A.R. (1984). *Phys. Status Solidi (a)*, Vol. 85, pp. 85-89
30. Wurster, R.; Zerta, M.; Stiller, C.; Wolf, J. (2009). Energy Infrastructure 21 – Role of Hydrogen in Addressing the Challenges in the new Global Energy System. *German Hydrogen and Fuell Cell Association*, EHA/DWV publication

12

Alternative Cr+6-Free Coatings Sliding Against NBR Elastomer

1. Introduction

Hexavalent chromium compounds result attractive primarily for industrial activity because they provide manufactured products with enhanced hardness, shininess, durability, color, corrosion resistance, heat resistance, decay resistance and tribological properties. On the other hand, it poses far more health hazards than trivalent chromium. It is a hazardous substance that increases the risk of developing lung cancer if itis inhaled. Ingestion or even simple skin exposure of chromic acid could increase the risk of cancer formation. In this situation, hexavalent chromium is classified by the International Agency for Research on Cancer (IARC) as a known human carcinogen (Group 1) (Working Group on the Evaluation of Carcinogenic Risks to Humans, 1987), where workers have the highest risk of adverse health effects from hexavalent chromium exposure.

Hexavalent chromium has been deeply used in tribological applications being friction and wear reduction also one of the main objectives in sliding mechanical parts for minimizing loss of energy and improving systems performance (Flitney, 2007) (Monaghan, 2008). In the last years, in fact, attention in maintenance costs saving also grow up, therefore a key question is to achieve low levels of friction as well as high wear resistance. In the field of elastomeric materials in 1978 A. N. Gent et al. (Gent, 1978) studied wear of metal by rubber attributing those phenomena at the direct attack upon metals of free radical species generated by mechanical rupture of elastomer molecules during abrasion. It suggested that such studies might lead to new metal texturing processes and surface treatment that can have the double effect of improving the tribological performances and protecting from external agents. Furthermore, coating technology is gaining ground thanks to new available technologies and focusing in particular to the need of using new alternative non toxic surface treatment with equivalent functionality of Cr+6.

The availability of new coating technologies like High Velocity Oxy-Fuel (HVOF) permits to have a wide range of hard coatings, but a deep study of their mechanical and tribological characteristics is needed due to the strong influence of their roughness, hardness, finishing and resistance to wear and corrosion.

HVOF thermal spray technique allows depositing variety of materials (alloys and ceramics). The powdered feedstock of deposition material is heated and accelerated to high velocities in oxygen fuel. The material hits and solidifies as high density well adherent coating material on the sample/component. HVOF coatings are also strong and show low residual tensile stress or in some cases compressive stress, which enable very much thicker coatings to be applied than previously possible with the other processes.

An investigation is herein proposed considering NBR (Nitrile butadiene rubber) material sliding against HVOF coated steel rod in order to clarify the influence of the surface characteristics (hardness, roughness and texture) on the tribological measurements. Many times, in addition, the metallic parts need to have good corrosion resistance for protecting them from external hostile atmospheres. A study of the corrosion resistance of the HVOF coatings is then presented, in comparison with the reference Hard Chromium Plating (HCP) treatment.

In this situation, the main objective of this work was to investigate and compare the tribological and corrosion behavior of a reference tribopair NBR/HCP versus some alternatives based on NBR/HVOF coatings. These materials combinations simulates contact occurring in sealing systems, where polymer and metallic parts are rubbed each other (Conte, 2006).

2. Materials and methodology

2.1 Rod coatings

Fig. 1. Coated rod samples. Image corresponds to rods with Grinding process

Three different material powders were sprayed by HVOF on a 15-5PH steel rod (diameter 19 mm, length 33 mm): AlBronze, NiCrBSi and WCCoCr. After the HVOF coating process, the cylinders were subjected to different surface modification processes identified as Grinding (G), Superfinishing (F) and Shot Peening (SP). Shot peening was performed with glass balls of diameter in the range of 90- 150 mm, which were injected on the surface of the rods at a pressure of 7 bar and at approximately a distance of 20 mm from the rod. By combining grinding, finishing and shot peening processes it was possible to create different textures on the surface of the coated rods. In addition, reference surface treatment, Hard Chromium Plating (HCP), was also investigated. Coated rods are shown in Fig. 1 where it can be seen that the coatings have been homogeneously deposited on the surface of the rods.

Table 1 shows some information about hardness and roughness of the tested coated rods. In all the materials "Grinding" and "Grinding + Superfinishing" processes modified the surface of the rods to an averaged roughness of approximately 0.20 mm and 0.04 mm, respectively. However, differences were observed when shot peening was applied. WCCoCr material had a high hardness so impacts of microballs did not modify its surface and hence the final surface was very similar to the roughness achieved with the "Grinding" process, that is, 0.28. However, the other two materials (AlBronze and NiCrBSi) were strongly affected by the shots, so final roughnesses were 1.36 and 2.06 mm, respectively. These two last high values have to be considered as rough figures, since the surface of the shot peened coatings was very irregular, so high dispersion of values was obtained.

Table 1. Tested coated rods

Rod identification	Coating	Hardness (HV)	Surface texture process	Ra (mm)
HCP + G *(reference)*	Hard Chromium Plating	850 ±11	Grinding	0.20
AlBronze+G+F	AlBronze (HVOF)	260±10	Grinding + Superfinishing	0.04
AlBronze+G			Grinding	0.22
AlBronze+SP+G			Shot peening+ Grinding	1.36
NiCrBSi+G+F	NiCrBSi (HVOF)	745±15	Grinding +Superfinishing	0.04
NiCrBSi+G			Grinding	0.16
NiCrBSi+SP+G			Shot peening+ Grinding	2.06
WCCoCr+G+F	WCCoCr (HVOF)	1115±92	Grinding +Superfinishing	0.03
WCCoCr+G			Grinding	0.23
WCCoCr+SP+G			Shot peening+ Grinding	0.28

Fig. 2 to Fig. 4 show the cross section of the HVOF coated rods where structure can be examined. For this characterization, rods with shot peening process where selected in order to analyze the deformation suffered by the coating after the glass impacts. The thickness of the coatings was in the range of 120-150 mm. Neither pores nor cracks in the interface of the coating where found in the coatings, which improves corrosion resistance and facilitates proper bonding, respectively. However, the analysis of the SEM images evidences the presence of some irregularities in the coatings which were analyzed in detail.

In the WCCoCr coating (Fig. 2) Nickel traps form some clusters of material. These clusters could come from previous processes were Nickel was deposited (for example in the preparation of the NiCrBSi coating). It was also identified alumina particles between the substrate and the coating (darker area in Fig. 2) which could come from the machining process. No evidence of craters was present on the surface of the coatings. It seemed that the hard nature of this coating (1115±92 HV) made difficult the creation of craters on its surface. The NiCrBSi coating (Fig. 3) had many clusters of material particles. The pale clusters corresponded to Molybdenum, also detected in the surface of this rod; the dark polygonal clusters corresponded again to alumina. The alumina was detected not only between the substrate and the coating, but also in the matrix of the coating. In this case, shots of the glass balls did perform craters on the coating, increasing then the roughness of the coating till 2.06 mm. In some areas of the surface of the coating it was appreciated flakes-like irregularities which could had been provoked during the finishing process. These non homogeneous features under severe working conditions could accelerate the fail of the coating.

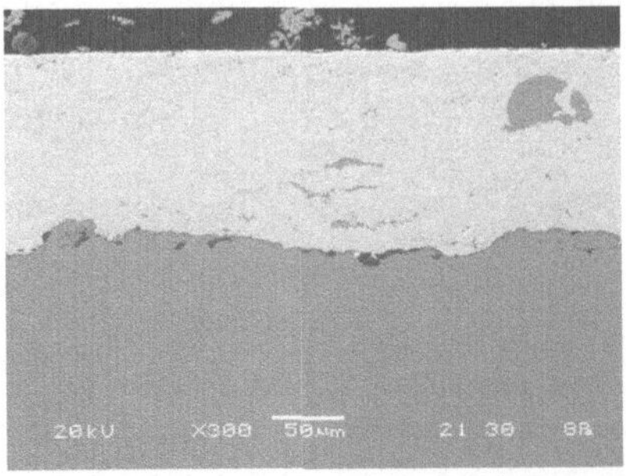

Fig. 2. WCCoCr + SP+ G coating

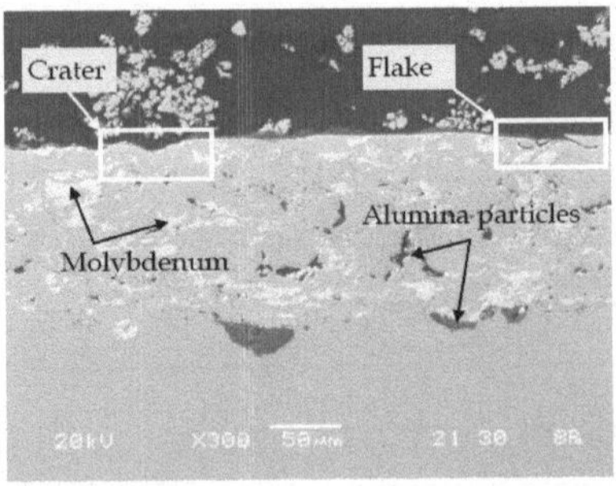

Fig. 3. NiCrBSi + SP+ G coating

Fig. 4. Al Bronze + SP+ G coating

The superficial appearance of the AlBronze coating (Fig. 4) was similar to the NiCrBSi coating. It showed high roughness (Ra=1.36 mm) because of the combination of its relatively low hardness (260 HV) and the craters performed during the shot peening; flake-like cracks an alumina clusters were again found within the coating.

2.2 NBR elastomer

NBR elastomer samples were obtained from real seals, and had a hardness of 85±1 ShA. The material was analyzed by Thermogravimetry Analysis (TGA) and Scanning Electro Microscopy with Energy Dispersive X-ray Spectroscopy (SEM-EDS) techniques. The composition of the tested NBR is shown in Table 2. The analysis of the inorganic part revealed the presence of Magnesium Silicate (talc), Sulphur and Zinc Oxide. Magnesium Silicate is used as compounding material, Sulphur acts as vulcanization agent and Zinc Oxide is used for activating this process.

Table 2. Composition of the NBR rubber

Component	Quantity (% in weigth)
Elastomer and plasticizers	49
Carbon black	46
Inorganic filler	5

2.3 Tribological tests

Friction and wear tests were carried out using the cylinder on plate configuration (Fig. 5). Coated rods were put in contact against flat sample of NBR under sliding linear reciprocating conditions. Contacting surfaces were lubricated using AeroShell Fluid 41 hydraulic mineral oil.

During the test, the coated rod was linearly reciprocated at a maximum linear speed of 100 mm/s with a stroke of 2 mm. Testing normal load was applied gradually in order to soften the contact between the metallic rod and the rubber sample: during the first 30 s it was set a normal load of 50 N and then a ramp of load was applied to reach 100 N, the testing normal load. Tests had a duration of 30 min.

Specimens were located in a climate chamber to set temperature and relative humidity at 25 ºC and 50 %RH, respectively. Each material combination was tested at least twice in order to evaluate the dispersion of the results.

It was recorded the evolution of the coefficient of friction through time and, after the tests, surface damage on the specimens was analyzed by optical microscopy. It was also considered the evaluation of the mass loss but no significant results were obtained, so it was not reported.

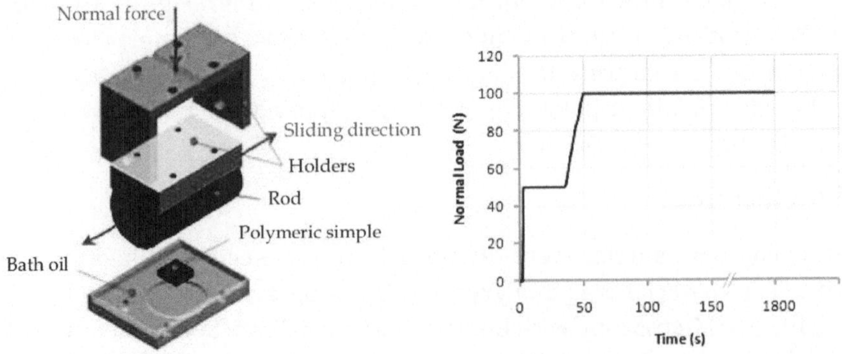

Fig. 5. Scheme of the testing arrangement (Cylinder on Plate configuration) (a) and load history (b)

2.4 Corrosion tests

Corrosion tests were performed in a conventional electrochemical cell of three electrodes. The reference electrode used for these measurements was a silver/silver chloride electrode (SSC, 0.207V vs SHE), the counter electrode was a platinum wire and the working electrode was the studied surface in each case. The exposed area of the samples was 1.47 cm2. Tests were done at room temperature and under aerated conditions. The aggressive media used was NaCl 0.06M. The electrochemical techniques applied for the corrosion behaviour study were electrochemical impedance spectroscopy in function of immersion time (4 and 24 hours of immersion) and potentiodynamic polarization.

On the other hand, impedance measurements were performed at a frequency range between 100 kHz and 10 mHz (10 freq/decade) with a signal amplitude of 10 mV. Polarization curves were registered from -0.4V versus open circuit potential (OCP) and 0.8 V vs OCP at a scan rate of 0.5mV/s.

3. Friction and wear behaviour of hard coatings and rubber material

The evolution of friction coefficient through time for the different rods is shown in Fig. 6. The steady-state of the coefficient of friction was reached from the beginning of the tests, that is, the running-in phase is really short. The high values during the first seconds corresponded to the loading phase since the setting of the testing normal load was reached after 50 s.

Considering the mean values of the friction curves it was found that in general, for the three HVOF coatings, the lower the averaged roughness, the higher the mean friction coefficient, independently of the material of the coating (Fig. 7). The effect of reducing roughness by mechanical surface treatments revealed that lowering rod roughness did not promote the formation of the lubrication film in the interphase rod/rubber, resulting in friction force increment. This general tendency was not followed by the AlBronze coating. This material had the lowest hardness so it was very affected by the shot peening process, which generated a very irregular surface with unbalanced tribological effect.

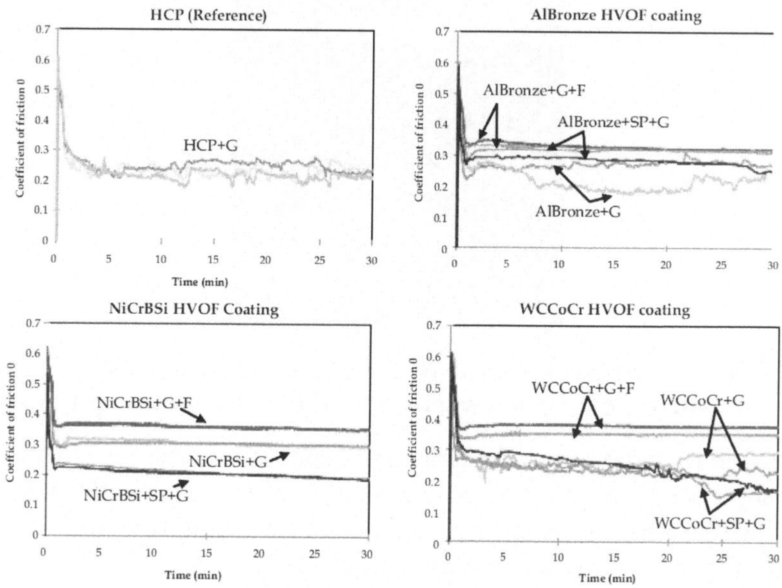

Fig. 6. Friction curves

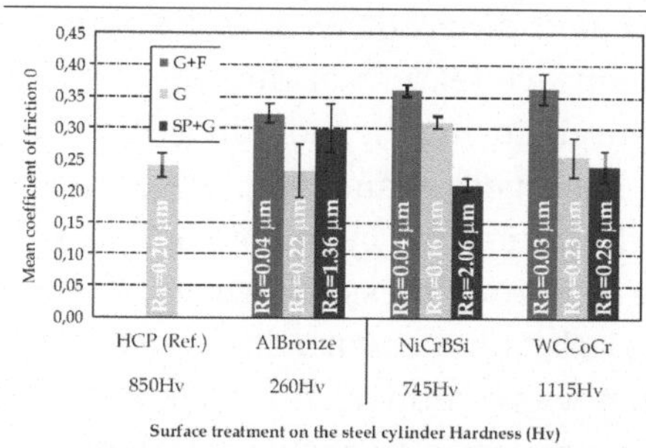

Fig. 7. Mean coefficient of friction, averaged roughness and hardness

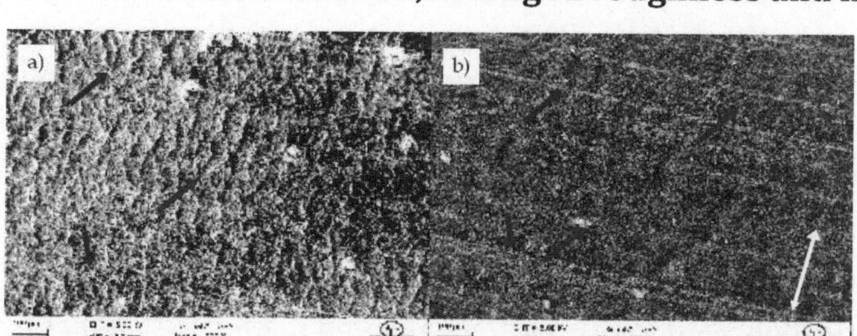

Fig. 8. Not tested area on the NBR elastomeric samples (a) and worn area after tests againts HCP+G reference material (b). White arrow indicates sliding direction. Blue arrows indicate straigth marks from the mould. Red arrows indicate points where X-Ray analysis was done

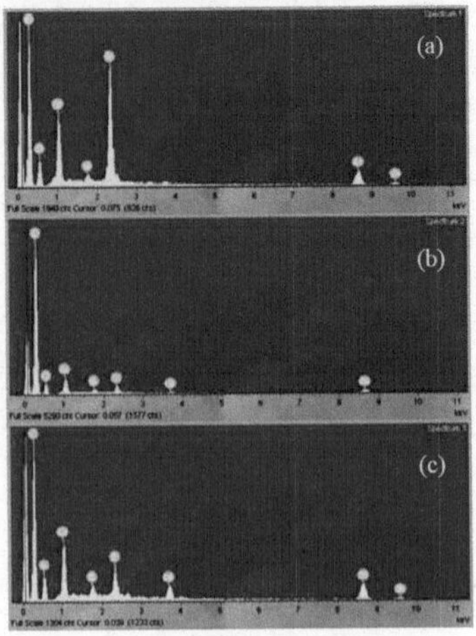

Fig. 9. X-Ray microanalysis on the NBR sample: not tested surface (a), plain worn area (b) and particle on the worn surface (c)

The coated rods did not suffer damage as consequence of the contact with the relatively soft rubber sample; the lubrication film protected effectively the metallic surfaces. On the other hand, strong influence of the counterbody was observed when analyzing the wear behaviour of the NBR elastomers.

An overview of the SEM images showing the surface damage on the surface of the NBR samples revealed different wear behaviour depending on the tested counterbody. The initial surface texture of the NBR sample had a flake-like shape (Fig. 8 (a)), a texture acquired during the moulding phase of the elastomeric sample. Straight lines were also observed, again a replica of the texture of the mould. As observed in Fig. 8 (b) the reference cylinder coating HCP softened this texture by reducing the microscopic roughness. However, straight lines from the mould remained still visible. Particles on the worn area were analyzed by X-Ray. Spectrum of Fig. 9 (c) indicated they were rubber with a significant amount of Sulphur and Zinc. These elements corresponded to the components used in the vulcanization process of the rubber. They tend to emigrate to surface of the NBR sample and thus, they remain within the matrix of the detached wear particles. Important presence of these two elements was found on the untested area ((Fig. 9 (a)); contrary, the plain worn area had less quantity of these elements as observed in Fig. 9 (b), since the successive cycles removed the upper film of the NBR sample.

In relation to the tests with the HVOF coated rods, the intensity of the surface damage on the NBR sample was very influenced by the surface texture of the rod. Rods with high roughness (AlBronze+SP+G and NiCrBSi+SP+G) produced important abrasion marks in the sliding direction as observed in Fig. 10 (c) and Fig. 11 (c). With rods of lower roughness this phenomenon was still present, but with lower intensity (Fig. 10 (b) and Fig. 12 (c)). Schallamach waves (Schallamach, 1971) perpendicular to the sliding direction were observed on the NBR after the test with the AlBronze+G (Fig. 10 (b)), which indicated that micro-bonding between contacting surfaces occurred. This material produced light surface damage on the NBR when the surface roughness was low according to the Superfinishing process (Fig. 10 (a)). There is still present the flake-like shape of the texture of the untested rubber, as well as the straight lines from the mould. The same behaviour was observed with the WCCoCr+G+F rod as shown in Fig. 12 (a). On the other hand, the NiCrBSi alloy with the G+F and G processes roughened the NBR surface in very similar way; the rubber failed by cracking and fatigue phenomena.

4. Corrosion resistance of coatings

Open circuit measurements registered during the initial 5000 s of immersion in the electrolyte appear in Fig. 13. The potential in case of reference chromed sample differs from the rest of coatings showing a more stable and noble open circuit potential.

After the first 4 hours of immersion an electrochemical impedance spectroscopy was performed on each surface to evaluate the electrochemical response of the coatings to the selected aggressive media. In this study, EIS (Electrochemical Impedance Spectroscopy) was employed to detect the pinholes in the coatings proposed and assessed their effect on the system corrosion behaviour over longer immersion times. Because of that, a second EIS was additionally measured on each sample after 24 hours of exposure to the aggressive electrolyte. Fig. 14 shows the impedance diagrams registered at 4 h and 24 h of immersion for each coating.

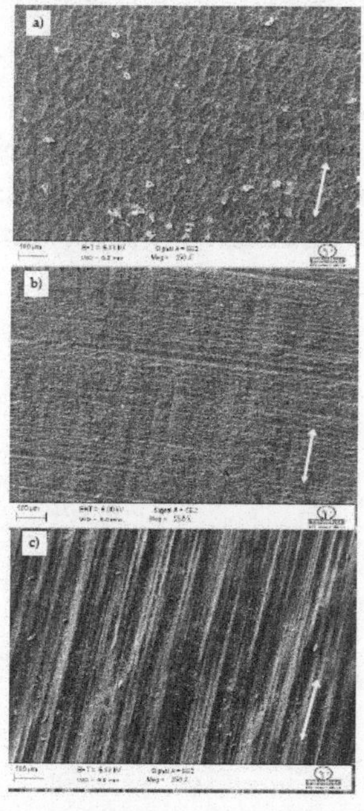

Fig. 10. Worn areas on NBR elastomeric samples against AlBronze coatings: G+F (a), G (b) and SP+G (c). White arrows indicate sliding direction

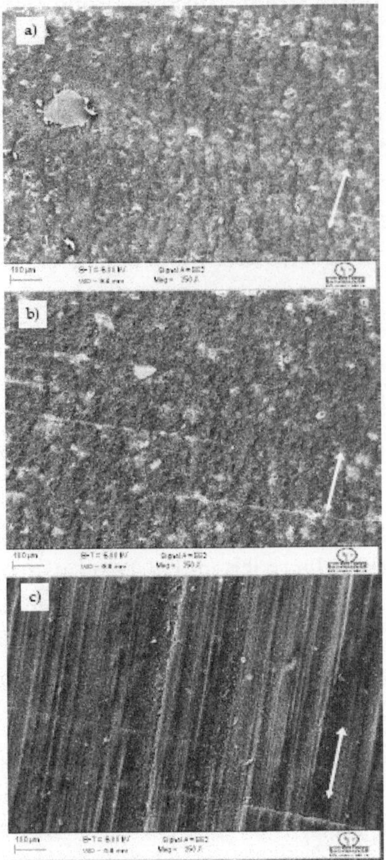

Fig. 11. Worn areas on NBR elastomeric samples against NiCrBSi coatings: G+F (a), G (b) and SP+G (c). White arrows indicate sliding direction

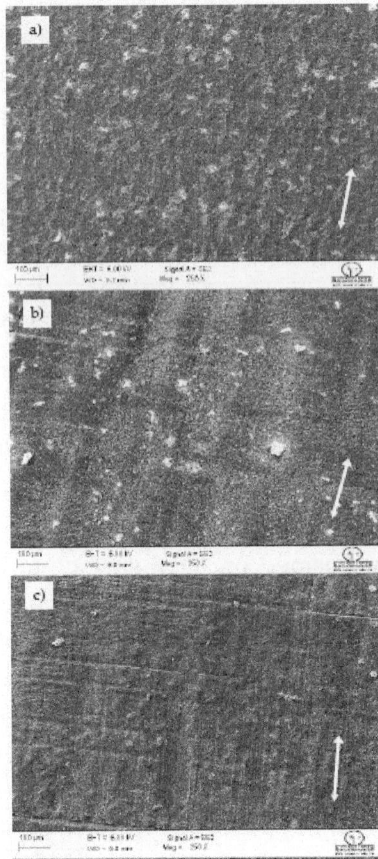

Fig. 12. Worn areas on NBR elastomeric samples against WCCoCr coatings: G+F (a), G (b) and SP+G (c). White arrows indicate sliding direction

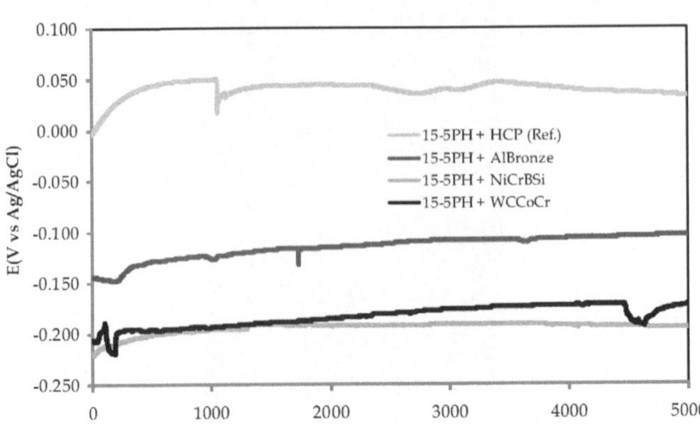

Fig. 13. Open circuit potential measurements of coated rods in NaCl 0.06M

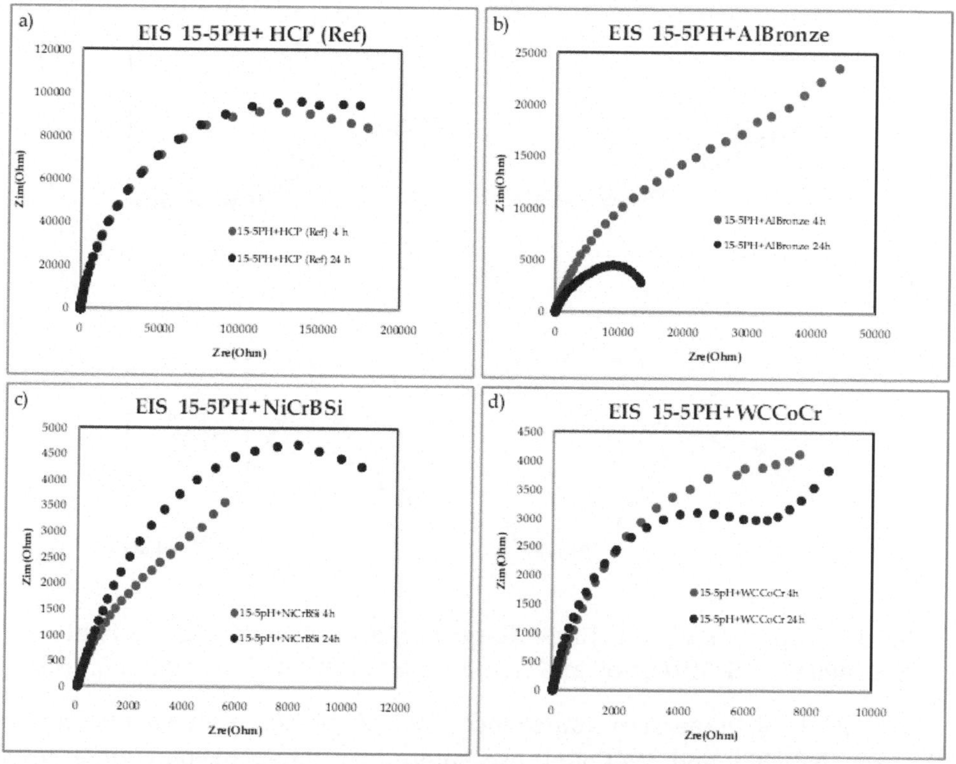

Fig. 14. Impedance diagrams at 4 h and 24 h of immersion in NaCl 0.6M; a) chromed reference, b) AlBronze coating; c) NiCrBSi coating and d) WCCoCr coating

Fig. 15 gives the Bode plots from the coated samples over the two immersion times in NaCl. According to the impedance diagram, after 4 h immersion, only one semi-circle was shown in all cases, corresponding to the coatings time constant. Low immersion periods were too short to reveal any contribution of the 15-5PH substrate. When the immersion period was increased to 24 h, the phase shift was different to that of 4 h in all alternative coatings, except in case of reference HCP film, whose Bode spectra remains stable and very similar to the first one registered at 4 h of exposure time.

At 4 h of immersion time, all coatings showed diffusion processes in the low frequency range and the experimental data could be fitted by using the equivalent circuit (A) drawn in Fig. 16. The electrochemical parameters obtained using this circuit are listened in Table 3. In this case, CPE1 is the constant phase element of the coating (CPE-c) which impedance can be written as ZCPE=1/Yo(iw)n. R1 is the charge transfer resistance (Rct)in the interface coating/electrolyte and W is the diffusion element (Zw).

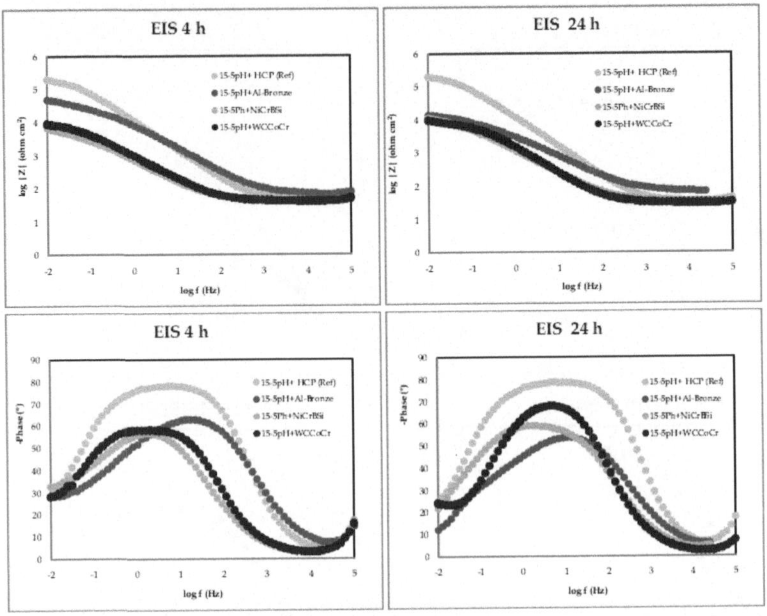

Fig. 15. Impedance data (Bode diagrams) of reference and alternative coatings for 15-5PH alloy at 4 h and 24 h of immersion in NaCl 0.06M

After 24 h of immersion, impedance data of the three alternative coatings (AlBronze, NiCrBSi and WCCoCr) presented two time constants due to the contribution of the substrate through the coatings micropores or defects. In this case, the experimental data could be fitted with the equivalent circuit (B) where CPE-c corresponds to CPE1, the constant phase element of the coating, R2 is Rpo, the resistance through the coating pores, CPE-s is CPE-2, the constant phase element of the substrate and Rct corresponds to R2, the charge transfer resistance in the interface substrate/electrolyte.

Table 3. Electrochemical parameters obtained from EIS tests using the equivalent circuits of Fig. 16 in NaCl 0.06M

	HCP		AlBronze		NiCrBSi		WCCrCr	
Time (h)	4	24	4	24	4	24	4	24
Eoc (V)	0.025	0.050	-0.087	-0.183	-0.192	-0.258	-0.171	-0.174
Rs (W.cm2)	68.2	46.6	89.3	88.2	56.9	42.9	52.6	39.1
R1 (KW.cm2)	238.0	242.1	38.2	11.26	6.7	16	12.1	24.9
Y0-CPE-1 (10-4F/cm2)	0.127	0.123	0.203	0.566	2.751	3.337	1.973	10.21
N1	0.885	0.888	0.742	0.687	0.716	0.668	0.73	0.691
Zw (10-3 W-1.cm-2.s1/2)	0.039	0.049	8.769	/	0.701	/	0.845	/
R2 (KW.cm2)	/	/	/	9.9	/	5.9	/	8
Y0-CPE-2 (10-4F/cm2)	/	/	/	1.646	/	3.681	/	1.165
n2	/	/	/	0.762	/	0.758	/	0.843

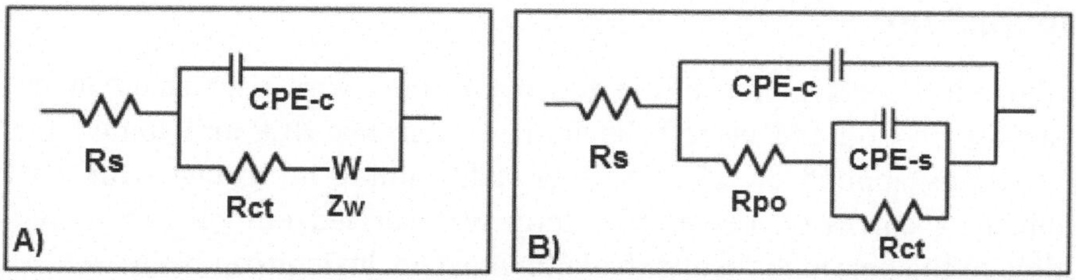

Fig. 16. Equivalent circuits used to simulate impedance experimental data. Circuit A) used in all cases at 4 hours of immersion time, and at 24h in case of chromed reference sample. Circuit B) used at 24h of immersion time for the three alternative coatings: AlBronze, NiCrBSi and WCCoCr

According to this results, it was seen that the HCP coating was a very good reference for corrosion protection in chloride media since it showed the most constant and stable behaviour after 24 hours of immersion time, as well as high corrosion resistance in comparison to the other alternative coatings.

After 24 hours of exposure, a potentiodynamic polarization curve was performed on the different coated rods. The potential-current curves are exposed in Fig. 17. The results of polarization tests were in agreement with impedance measurements. Chromed rod showed the lowest corrosion current over the whole potential range analyzed, whereas in the case of AlBronze and NiCrBSi coatings the current progressively increased when potential went to more anodic values which involved a more active behaviour in these cases. WCCoCr coating showed more stable and lower corrosion current than the other two alternatives but the corrosion resistances were worst than those measured in case of reference coating (Table 4).

Table 4. Tafel analysis of potential-current curves

	Ecorr (V)	Icorr (10-6A)	Rp (KW)
15-5PH+HCP (Ref)	-0.095	0.13	417
15-5PH+AlBronze	-0.209	12.50	7
15-5PH+NiCrBSi	-0.269	1.79	21
15-5PH+WCCoCr	-0.271	1.40	38

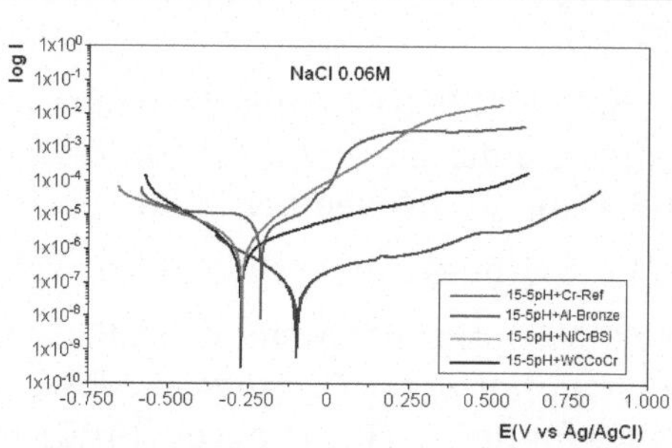

Fig. 17. Potentiodynamic polarization curves of coated 15-5PH samples after 24 hours of immersion in NaCl 0.06M

5. Conclusions

Tribological tests under lubricated conditions were performed in order to compare the friction and wear behaviour of reference HCP and some alternative HVOF coatings applied on 15-5PH steel rods, sliding in against NBR elastomer. Additionally a corrosion resistance study was carried out on the coated rods. According to the obtained results the following conclusions can be drawn:

- In terms of friction, in general it was seen that for the studied HVOF coatings, the lower the averaged roughness, the higher the mean friction coefficient, independently of the material of the coating. In addition, wear suffered by the NBR elastomer was very sensitive to the surface texture on the rod, and, rods with elevated roughness generated not acceptable surface damage on the rubber. So, surprisingly, those NBR samples with lower surface damage did not corresponded with tests with low coefficients of friction. This phenomenon suggested significant temperature rise in the contact.
- The corrosion tests revealed that the reference HCP surface coating was a very good reference for corrosion protection and had better behaviour that the proposed HVOF coatings. However, it must be pointed out that obtained values indicated good behaviour of these coatings.
- Considering the tribological and corrosion results, it can be said that the AlBronze+G HVOF coating could be considered as good alternative to replace the reference HCP treatment since it generated a equivalent friction and produced an acceptable damage on the surface of the elastomeric material. Additionally, its corrosion response was good enough for protecting the substrate material.

References

1. Conte, M. (2006), Interaction between seals and counterparts in pneumatic and hydraulic components. *PhD Thesis* (June 2009)
2. Flitney, B. (2007). Alternatives to chrome for hydraulic actuators. *Sealing Technology*, Vol 2007, Issue 10, (October 2007), pp.8-12
3. Gent A.N., Pulford C.T.R. (1978). Wear of steel by rubber. *Wear*, Vol. 49, Issue 1, (July 1978), pp. 135-139
4. Monaghan, K. J. & Straub, A. (2008). Comparison of seal friction on chrome and HVOF coated rods under conditions of short stroke reciprocating motion. *Sealing Technology*, Vol 2008, Issue 11, (November 2008), pp. 9-14
5. Schallamach, A. (1971), How does rubber slide?, *Wear*, Vol. 17, Issue 4, pp.301–312
6. Working Group on the Evaluation of Carcinogenic Risks to Humans (1987), *IARC Monographs on the evaluation of the Carcinogenic Risks to Humans*, Supplement 7, International Agency for Research on Cancer (IARC), ISBN 9283214110, Lyon

13

The New Methods for Scuffing and Pitting Investigation of Coated Materials for Heavy Loaded, Lubricated Elements

1. Introduction

In modern technology due to the increase of the unit pressure, velocities, and hence temperatures in the tribosystems of machines, a risk of two very dangerous forms of wear exists. These forms are scuffing and pitting.

Scuffing is a form of wear typical of highly-loaded surfaces working at high relative speeds. Scuffing is considered to be a localised damage caused by the occurrence of solid-phase welding between sliding gear flanks, due to excessive heat generated by friction, and it is characterised by the transfer of material between sliding surfaces. This condition occurs during metal-to-metal contact and due to the removal of the protective oxide layer of the metal surfaces (Burakowski et. al., 2004).

A typical scuffing zone of gear teeth (Michalczewski et al., 2010) is illustrated in Fig. 1.

Fig. 1. A typical scuffing wear of gear teeth

Another form of wear is rolling contact fatigue (pitting). Pitting is a form of wear typical of highly-loaded surfaces working at a sliding-rolling and rolling contact, e.g. such components in transmissions like toothed gears and rolling bearings (Torrance et al, 1996). It is caused by the cyclic contact stress, which leads to cracks initiation (Libera et al., 2005). The lubricant is pressed into the cracks at a very high pressure (elastohydrodynamic lubrication), making them propagate. Finally, cyclic stress results in breaking a piece of material off the surface. Examples of a gear and a race worn due to pitting (Michalczewski et al., 2010) are presented in Fig. 2.

(a) (b)

Fig. 2. The pitting wear: a) on a pinion gear, b) on a bearing race

For many engineering materials, further improvement of their properties through a modification of their microstructure, chemical composition, and phase composition, is practically impossible. In this situation the most effective way of improving mechanical properties of various engineering components is the modification of surface properties by the deposition of PVD/CVD coatings (Michalczewski, 2008). One of the most important characteristics of these coatings is the fact that its thickness, usually in the range from 1 to 5 μm, is located in the field of dimensional tolerances of typical machine elements.

There are many successful applications of thin hard PVD/CVD coatings in various technical devices like engines, pumps, compressors. However the problem of application of such coatings for heavy-loaded friction parts (e.g. gears, bearings) is still open - the share of mechanical components that are coated is extremely small (less than 2%). Why? The service life of heavy-loaded machine parts is essentially determined by two types of tribological failures: scuffing which is a severe form of mechanical wear, and pitting which is a surface fatigue phenomenon. Up to now, there was a lack of verified laboratory test methods intended for correlated determination of coating material and lubricating media on scuffing and pitting resistance of heavy-loaded system. So, the selection of coating material and technology was realised mainly basing on very expensive and long-term practical component research and the results are frequently contradictory (Szczerek, Michalczewski, & Piekoszewski, 2009).

The evaluation of friction and wear characteristics of PVD/CVD coatings is only possible on the way of experimental research. The experimental research of friction and wear of interacting surfaces is realised by means of a special device called tribotester. The new test methods and testing machines have been developed based on the achievements of the System for Tribological Research (SBT) implemented in the Tribology Department at the Institute for Sustainable Technologies - National Research Institute (ITeE-PIB), Radom, Poland (Szczerek, 1996). The SBT system was developed on the basis of the combinatorics that enables to reduce the tendency which is widely known as "testing rush".

For the purpose of the tribological research in the areas mentioned above, two tribological devices have been developed:

- The T-02U Universal Four-Ball Testing Machine,
- The T-12U Universal Back-to-back Gear Test Rig.

The set of methods and devices intended for the comprehensive tribological evaluation of PVD/CVD coatings is presented in Fig. 3.

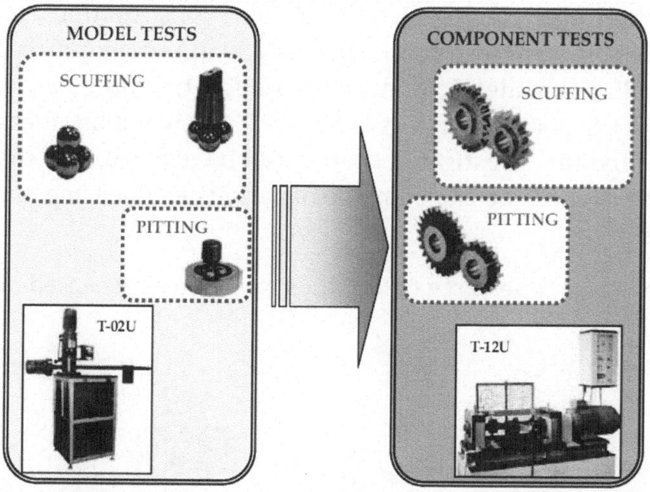

Fig. 3. The tribological methods and devices developed intended for comprehensive tribological evaluation of elements covered with PCD/CVD coatings

By means of this set, low-friction and antiwear PVD/CVD coatings can be evaluated from micro to macroscale in model and component tests (Antonov et al., 2009).

Using new devices, five test methods, giving the possibility of comprehensive testing of various low-friction and antiwear PVD/CVD coatings intended for machine elements, were developed. They are as follows:

- The model method for the evaluation of scuffing in the four-ball tribosystem,
- The model method for the evaluation of scuffing in the cone-three balls tribosystem,

- The model method for the evaluation of pitting wear in the cone-three balls tribosystem,
- The component gear method for the evaluation of scuffing resistance of gears,
- The component gear method for the evaluation of pitting wear of gears.

The new test methods and the new devices for the experimental evaluation of friction and wear of low-friction and antiwear PVD/CVD coatings are described below.

2. Model methods and T-02U Universal Four-Ball Testing Machine for evaluation of scuffing and pitting resistance of PVD/CVD coatings

2.1 Model scuffing tests in four-ball and cone-three balls tribosystems

For evaluation of scuffing resistance of lubricants, coatings, and engineering materials two tribosystems were employed: four-ball and cone-three balls. In typical four-ball test balls are made of chrome alloy 100Cr6 bearing steel, with diameter of 12.7 mm (0.5 in.). Surface roughness is Ra = 0.032 μm and hardness 60 to 65 HRC. In the new method the investigated coating can be deposited on the ball or on the cone. Furthermore the cone can be made of various engineering material, not only of bearing steel.

The four-ball and cone-three balls tribosystems are presented in Fig. 4.

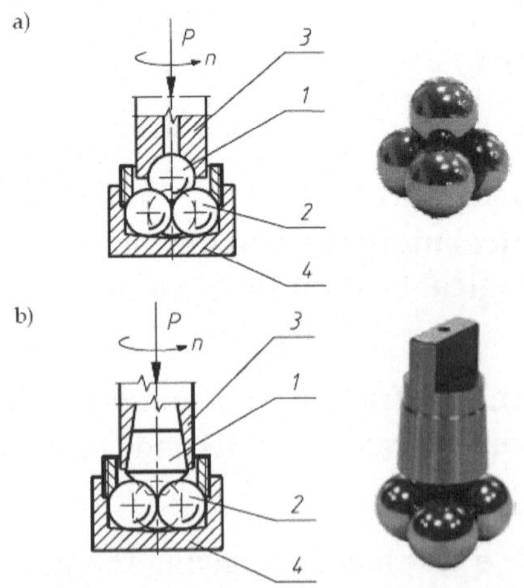

Fig. 4. Model tribosystems for testing scuffing: a) four-ball tribosystem: 1- top ball, 2- lower balls, 3- ball chuck, 4 – ball pot, b) cone-three balls tribosystem; 1 – top cone, 2 – bottom balls, 3 – ball chuck, 4-ball pot

The three stationary, bottom balls (2), having a diameter of 0.5 in., are fixed in the ball pot (4) and pressed against the top ball or cone (1) at the continuously increasing load P. The top ball/cone is fixed in the ball chuck (3) and rotates at the constant speed n. The tribosystem is immersed in the tested lubricant. During the run the friction torque is observed until seizure occurs.

The test conditions are as follows: rotational speed: 500 rpm, speed of continuous load increase: 409 N/s, initial applied load: 0 N, maximum load: 7200 ± 100 N.

The methods are described in detail in works (Szczerek & Tuszynski, 2002) and patented (Polish Patent No. 179123 - B1 – G01N 33/30). A friction torque curve (M_t) obtained at the continuously increasing load (P) is shown in Fig. 5.

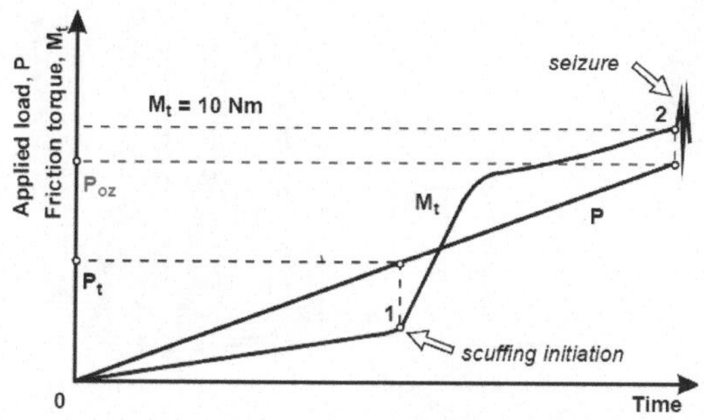

Fig. 5. Simplified friction torque curve (M_t) obtained at continuously increasing load (P); 1 – scuffing initiation, 1-2 – scuffing propagation, 2 – seizure (exceeding 10 Nm friction torque)

Scuffing initiation occurs at the time of a sudden increase in the friction torque – point 1. The load at this moment is called the *scuffing load* and denoted P_t.

According to the new test method, the load still increases (over a value of P_t) until seizure occurs (i.e. friction torque exceeds 10 N m – point 2). The load at this moment will be called the *seizure load* and denoted P_{oz}. If 10 Nm is not reached, maximum load (c.a. 7200 N) is considered to be the seizure load (although in such a case there is no seizure). For every tested lubricant the so-called *limiting pressure of seizure* (denoted p_{oz}) should be calculated. This value reflects the lubricant behaviour under scuffing conditions and is equal to the nominal pressure exerted on the wear scar surface at the moment of seizure or at the end of the run (when seizure has not appeared). The limiting pressure of seizure is calculated from the equation (1):

$$p_{oz} = 0.52 \frac{P_{oz}}{d^2} \tag{1}$$

where:

p_{oz} – limiting pressure of seizure, N/mm2,

P_{oz} – seizure load [N],

d – average wear scar diameter measured on the stationary balls, mm.

The 0.52 coefficient results from the force distribution in the four-ball tribosystem. The higher p_{oz} value, the better action of the tested lubricant under scuffing conditions is.

The developed test methods were successfully used for testing the scuffing resistance of components with thin hard coatings (thickness of 2 μm) deposited by PVD/CVD method. The example of their application (Michalczewski et al., 2010) is presented in Fig. 6.

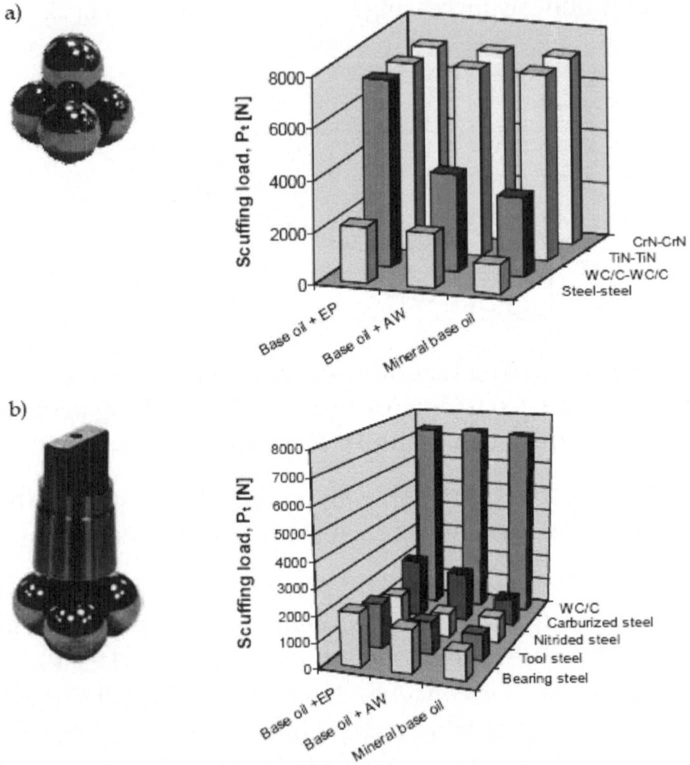

Fig. 6. Results from scuffing tests for lubricants, engineering materials and thin hard coatings: a) modified four-ball scuffing test, b) cone-three balls scuffing test

Wear scars images on lower balls from scuffing tests for steel-steel and CrN-CrN tribosystems are presented in Fig. 7.

The developed test methods have the resolution, not achieved by the other methods, good enough to differentiate between coatings, engineering materials and lubricants (Piekoszewski, Szczerek & Tuszynski, 2001). What is more, they are fast and inexpensive. So, these test methods can be effectively used to select the optimum substrate-coating-lubricant combinations best suited for highly loaded machine components (Michalczewski et al., 2009a).

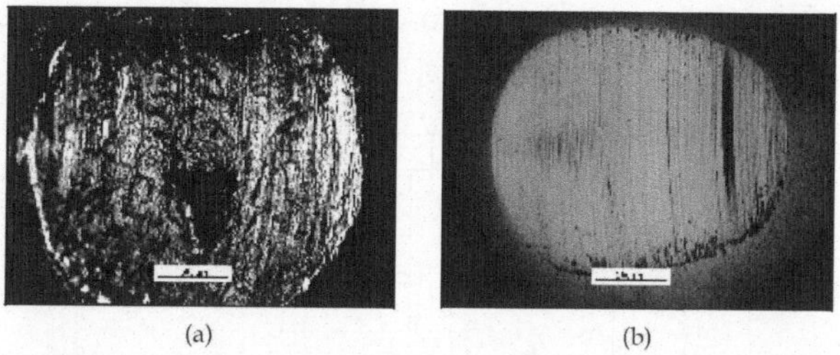

Fig. 7. Wear scars images on lower balls from scuffing tests for: a) steel-steel, b) CrN-CrN (four ball test, mineral base oil without lubricating additives)

2.2 Model method for evaluation of pitting wear in cone-three balls tribosystem

The cone-three balls test method is generally based on IP 300 standard (Rolling contact fatigue tests for fluids in a modified four-ball machine). The main change is the geometry of the contact of the rolling elements. The upper ball was replaced with a special cone (Michalczewski & Piekoszewski, 2006). The cone can be made of any material. The cone- three balls tribosystem is presented in Fig. 8.

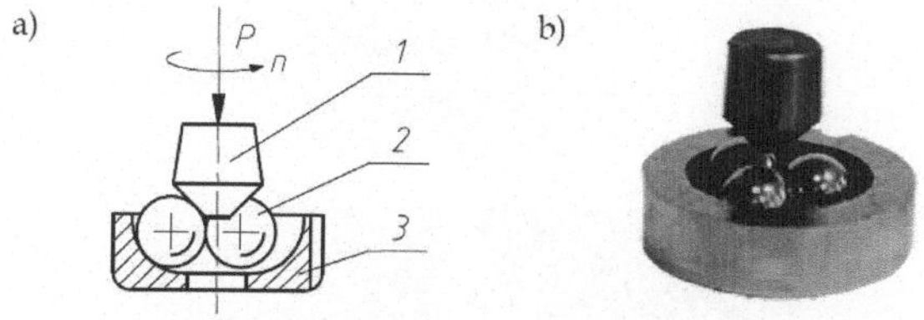

Fig. 8. Cone-three balls tribosystem: a) scheme, b) photograph; 1- cone, 2 - balls, 3 – race

The tribosystem consists of a rotating cone (1) loaded against three balls (2) which are able to rotate in the race (3). The specimens are immersed in the tested lubricant. During the run the vibration level is monitored until pitting occurs.

The tested cones are made of the tested material. The test balls are made of 100Cr6 chrome alloy bearing steel. For each test the new set of balls should be used. According to the method the test conditions are 3924 N (400 kg) load and 1450 rpm top cone speed. 24 top cone failures are necessary to assess the performance of the lubricant and the material. The tested materials can be compared on the basis of L_{10} or L_{50} values as well as scatter factor K. The value of L_{10} represents the life at which 10% of a large number of cones made of the tested material would be expected to have failed. The value of L_{50} relates in a corresponding manner to the failure of 50% of tested cones. The higher L_{10} and L_{50} value, the better the resistance of the tested material to pitting is.

The developed test method was successfully used for testing the fatigue life of components with thin hard coatings deposited by PVD/CVD method and presented in.

The results from pitting tests for uncoated steel and steel coated with single and low-friction coatings are presented in Fig. 9.

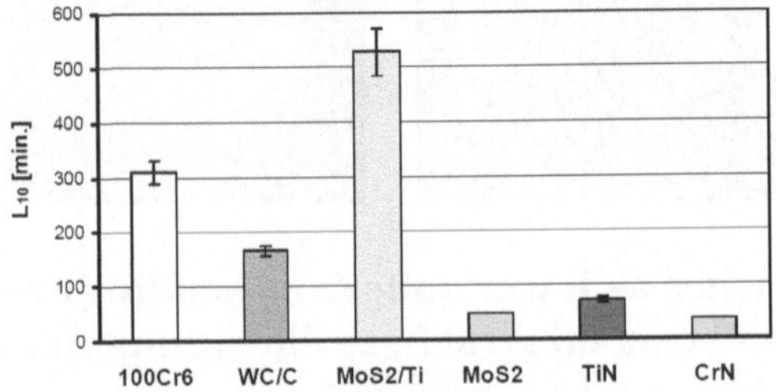

Fig. 9. Results from pitting tests for 100Cr6 steel covered with thin, hard coating

The SEM images of wear on the test cone from pitting tests for 100Cr6 steel covered with WC/C coating are presented in Fig. 10.

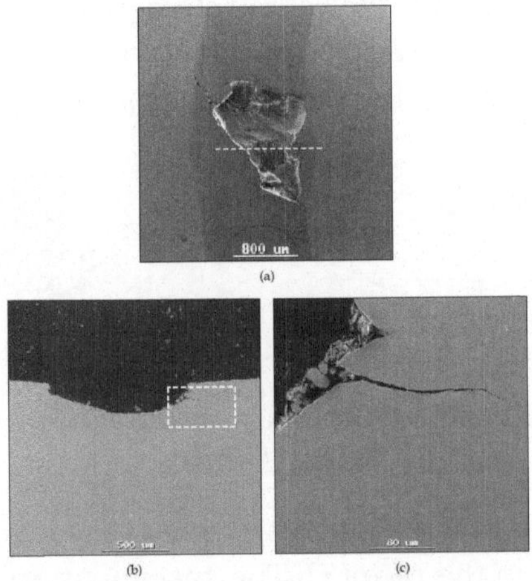

Fig. 10. The pitting wear on the test cone: a) upper view, b) cross-section, c) enlargement of selected fragment (WC/C coated cone, RL-144/4 mineral oil)

The results indicate beneficial impact of low friction coatings on pitting wear (e.g. MoS_2/Ti coating).

The presented method for testing pitting in cone-three balls tribosystem can be applied to testing fatigue wear of various materials, surface coatings as well as various lubricants. In comparison to other existing methods the new method gives better resolution and is time- and cost-effective.

2.3 T-02U Universal Four-Ball Testing Machine

The methods for evaluation of pitting and scuffing resistance of PVD/CVD coatings is realised by means of T-02U Universal Four-Ball Testing Machine (Michalczewski et al., 2009b). The photo of the machine is presented in Fig. 11. The tribotester is equipped with a computer-aided system of control and measurements.

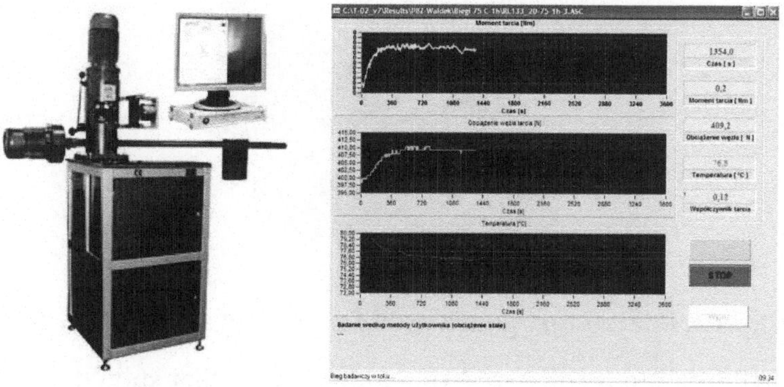

(a) (b)

Fig. 11. T-02U Universal Four-Ball Testing Machine: a) photograph, b) computer screen during data acquisition

A very wide range of lubricants can be tested using the T-02U Machine, e.g.: gear oils, hydraulic-gear oils, motor oils, eco-lubricants, non-toxic lubricants, new EP additives, cutting fluids, and greases. Many test methods described in international and national standards can be performed - ISO 20623, ASTM D 2783, D 2596, D 4172, D 2266, D 5183, DIN 51350, IP 239, IP 300, PN-76/C-04147. They concern the determination of the influence of the tested lubricants on scuffing, pitting, friction coefficient, and sliding wear, at ambient and elevated temperatures.

3. Component methods and T-12U Universal Back-to-back Gear Test Rig for evaluation of scuffing resistance and rolling contact fatigue of PVD/CVD coated gears

In research where high reliability is at stake, there is a tendency to use such test specimens that are similar to real machine components. The gear testing is incomparably more expensive and time consuming than tests carried out on simple specimens. But the main advantage is better reliability of the results obtained.

Concerning the most dangerous kinds of wear of gear wheels, two types can be specified: scuffing and pitting. These forms have been described previously in this study.

3.1 Component method for evaluation of scuffing resistance of gears

The test method for the evaluation of scuffing resistance of gears has been originally developed by FZG (Gear Research Centre) at the Technical University of Munich. This method was adapted for investigation of PVD/CVD coated gears at ITeE-PIB.

All test gears are case carburised, with HRC 60 to 62 surface hardness and case depth of 0.6 to 0.9 mm. "A" test gears are cross-Maag's ground, and their tips are especially shaped to achieve high sliding velocities, hence the tendency to scuffing. The tested PVD/CVD coating can be deposited on one or both gears – Fig. 12.

Fig. 12. Coated test gears used for testing scuffing - type A

The only limitation is the deposition temperature that should be below 180°C, which is connected with thermal stability of gear material.

Special coated gears (e.g. A20 type) are run in the test lubricant, at constant speed for a fixed time, in dip lubrication system. From load stage 4 the initial temperature is controlled. The oil is heated up to 90°C. Loading of the gear teeth is raised in stages. During the running time of each load stage the oil temperature is allowed to rise freely. After load stage 4 the pinion gear teeth flanks are inspected for damage and any changes in tooth appearance are noted. The maximum load stage is 12. If the summed total width of the damaged areas on all the pinion gear teeth faces is estimated to equal or exceed one gear tooth width then this load stage should be taken as the failure load stage (FLS). Additionally the oil temperature, vibration level and motor load during the test can be measured.

The main advantage of the method is the possibility of scuffing testing of various materials, surface coatings as well as various lubricants intended for heavy-loaded friction joints. Furthermore the test can be realised by means of the worldwide popular back-to-back gear test rig manufactured by many producers.

Table 1. The teeth failure at load stage for various DLC coatings (gears lubricated with eco-oil)

Load stage	Hertzian stress at pitch point p_{max}	The type for tooth failure			
	[MPa]	uncoated steel	a-C:H:W	a-C:Cr	a-C:H
4	621	light grooves	light scars	face polished	none
5	773	light grooves	light scars	face polished	none
6	927	light grooves	light scars	face polished	none
7	1080	light grooves	light scars	face polished	light scars
8	1232	grooves	light scars	face polished	light scars
9	1386	scuffing strips	light scars	face polished	light scars
10	1538	wide scuffing areas	light scars	wide scuffing areas	light scars
11	1691		light scars		numerous scars
12	1841		light scars		numerous scars

The test method has been successfully used for extensive research to determine the effect of ecological gear oils on scuffing resistance of coated gears and for the selection of coating types for gear applications. An example of the research on gear oils is presented below.

The method has been applied for selecting a proper DLC coating for increasing the scuffing resistance of gears. The results from gear tests are presented in Table 1 and Fig. 13.

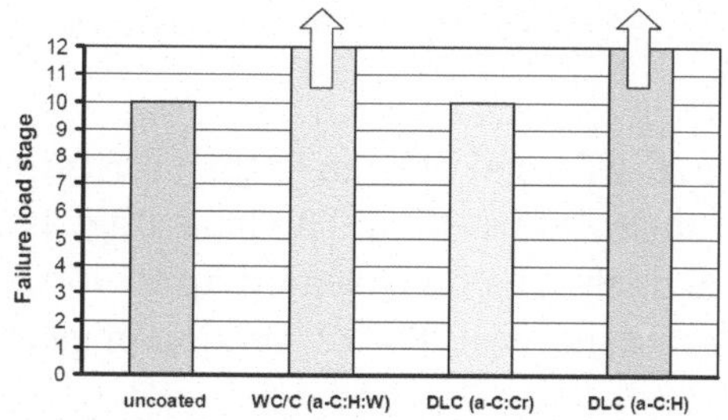

Fig. 13. Failure load stages for uncoated steel gears and for teeth coated with DLC coatings lubricated with eco-oil (A/8.3/90 method)

For uncoated gears lubricated eco-oil the 10th failure load stage only was achieved. The application of the coating (a-C:H:W or a-C:H) increased the scuffing resistance of gears. They passed the maximum 12th stage without scuffing. Only a-C:Cr coating did not improve the scuffing resistance of the tested gears.

The photographs of teeth surfaces after tests for tested DLC coatings (gears lubricated with eco-oil) are presented in Fig. 14.

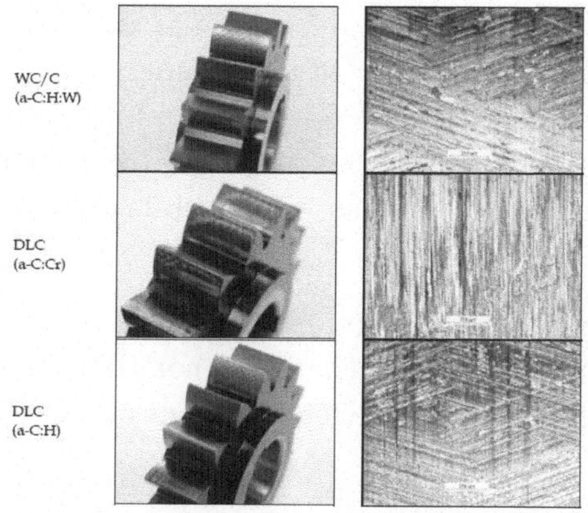

Fig. 14. The photographs of teeth surfaces after tests for various DLC coatings (gears lubricated with eco-oil)

The presented component method for evaluation of scuffing resistance of gears have been applied for developing a new solution for manufacturing steel heavy-loaded machine components covered with low friction coatings that enables increase service life of components and allows lubricating with environmentally friendly oils. This will increase the reliability of machines and reduce pollution of the environment by oil.

3.2 Component gear method for evaluation of pitting wear of gears

Similarly to scuffing gear tests, the method for evaluation of pitting wear of gears has been originally developed by FZG (Gear Research Centre) in the Technical University of Munich. This method was also adapted for the investigation of PVD/CVD coated gears at ITeE-PIB. The experiments are performed using the single-stage pitting test procedure (PT C/10/90) in an FZG type gear test rig, using C-PT gears – Fig. 15.

Special coated gears (C-PT type) are run in the lubricant test, at constant speed for a fixed time, in dip lubrication system. The load stage is 9 or 10 giving 302 Nm and 372 of torque respectively. The oil is heated up to 90°C. The oil temperature is controlled and kept at constant level. The inspection of gears is performed every 7 or 14 hours.

Fig. 15. Coated test gears used for testing pitting – type C-PT

The result of the tests is the LC_{50} fatigue life, related to 50% probability of failure. LC_{50} is defined as the number of load cycles when the damage area of the most damaged tooth flanks exceeds 4% (about 5 mm2). The total test time of each run is limited to 40 millions load cycles at pinion (300 operating hours). In some cases other criteria can be used. At least three valid runs are necessary to calculate the LC_{50} parameter.

The main advantage of the method is the possibility of comprehensive testing on various low-friction and antiwear PVD/CVD coatings intended for heavy-loaded machine elements. The method is realised by means of the worldwide popular back-to-back gear test rig.

The test method has been successfully used for extensive research to determine the effect of low-friction and antiwear coatings on pitting wear. An example of the research on gear oils is presented below.

The results indicate that for the coated/coated pair (pinion and wheel coated) and coated pinion/steel wheel pair a significant decrease in the fatigue life compared to the uncoated gears was obtained – Fig. 16.

The best results were obtained in the case of the steel pinion/W-DLC coated wheel – even fourfold increase in the fatigue life was observed. This shows a very high potential of the application of DLC coatings for gears.

Thanks to the component gear method for the evaluation of pitting wear of gears, it was possible to overcome the main factor hampering application of thin coatings on heavy loaded elements for many years i.e. their poor behaviour under cyclic stress conditions. This new method will allow for selection of low-friction and antiwear PVD/CVD coatings intended for manufacturing of steel heavy-loaded machine components. This will increase the service life of components and allow for the application of environmentally friendly oils. This will increase the reliability of machines and reduce environmental pollution.

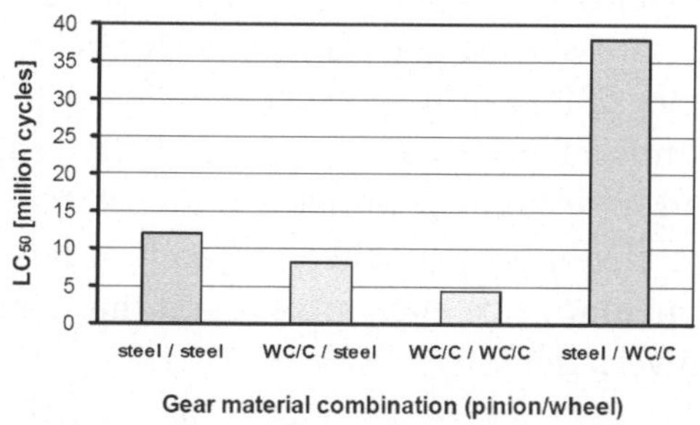

Fig. 16. Fatigue life LC_{50} for various pinion/wheel gear material

3.3 T-12U Universal Back-to-back Gear Test Rig

The T-12U Universal Back-to-back Gear Test Rig makes it possible to investigate both aforementioned forms of wear. The photo of the tester is presented in Fig. 17.

Fig. 17. T-12U Universal Back-to-back Gear Test Rig

The tribotester is equipped with a microprocessor-aided controller and as an option, it may also be equipped with a computer-aided measuring system.

A very wide range of lubricants can be tested using the T-12U Test Rig, e.g.: gear oils, hydraulic-gear oils, eco-oils, non-toxic oils, and new EP additives. What is more, there is a possibility of testing modern engineering materials and surface coatings intended for gear manufacturing. Many test methods described in international and national standards can be performed - ISO 14635-1, 14635-2, 14635-3, CEC L-07-A-95, L-84-02, DIN 51354, IP 334, ASTM D 5182, D 4998, PN-78/C-04169, FVA information sheets: 2/IV (1997), 54/7 (1993), 243 (2000). For the last few years, the T-12U Rig has been successfully used at ITeE-PIB for the extensive research to determine an effect of modern gear oils (including ecological oils) on different forms of gear tooth wear, as well as possibility of improving the gear life by the deposition of low-friction coatings.

4. Conclusion

Presented methods give the possibility of comprehensive testing on various low-friction and antiwear PVD/CVD coatings intended for machine elements. All the presented methods and both tribotesters i.e. T-02U Universal Four-Ball Testing Machine, T-12U Universal Back- to-back Gear Test Rig have been implemented at the Tribology Laboratory of ITeE-PIB and successfully verified. They are employed to perform various kinds of projects e.g. grants, R&D projects, ordered by the Polish government and international projects (COST Actions, 6th EU Framework Programme). They are also used to realise research orders from Polish industry (especially small and medium size enterprises) and the scientific sector (research institutes, technical universities).

The new methods exhibit very good resolution and precision comparable to standardised test methods and are time and cost effective. Furthermore the cone-three ball method gives the possibility of testing fatigue wear of any coating and substrate material. Basing on the elaborated methods the optimal selection and development of PVD/CVD technologies applied for extension of the life of the heavy-loaded friction joints as well as the elimination of toxic lubricating additives have been obtained.

The further development of tribological devices is performed in the frame of Strategic Programme "Innovative Systems of Technical Support for Sustainable Development of Economy," which is currently realised at the Institute for Sustainable Technologies-National Research Institute (ITeE-PIB) in Radom, in Poland. The Programme is realised within the framework of the Innovative Economy Operational Programme co-funded from European structural funds. The greatest emphasis is put on the development of advanced machines for testing spur gears and rolling bearings under extreme conditions.

References

1. Antonov, M., Michalczewski, R., Pasaribu, R. & Piekoszewski W. (2009). Assessment of the potential of lubricated contact conditions laboratory testing and surface analysis for improving the performance of machine elements. Comparision of model and real components test methods. *Estonian Journal of Engineering*, Vol. 15. No. 4, pp. 349- 358, ISSN 1736-6038

2. Burakowski, T.; Szczerek, M. & Tuszynski, W. (2004). Scuffing and seizure - characterization and investigation, In: Mechanical tribology. Materials, characterization, and applications, Totten, G.E. & Liang, H., (Ed.), pp. 185-234, Marcel Dekker, Inc., ISBN 0-8247-4873-5, New York-Basel
3. Libera, M., Piekoszewski, W. & Waligóra W. (2005). The influence of operational conditions of rolling bearings elements on surface fatigue scatter. *Tribologia*. 2005, No. 3, pp. 205–215, ISSN 0208-7774
4. Michalczewski, R. (2008). Chemomechanical synergy of PVD/CVD coatings and environmentally friendly lubricants in rolling and sliding contacts. In. Triboscience and tribotechnology superior friction and wear control in engines and transmissions., K. Holmberg (Ed.), pp. 191-199, ISBN 978-92-989-0040-2, COST Office Belgium
5. Michalczewski, R. & Piekoszewski, W. (2006). The method for assessment of rolling contact fatigue of PVD/CVD coated elements in lubricated contacts. *Tribologia. Finish Journal of Tribology*, Vol. 25 (4), pp. 34-43, ISSN 0780-2285
6. Michalczewski, R., Piekoszewski, W., Szczerek, M. & Tuszynski W. (2009a) The lubricant- coating interaction in rolling and sliding contacts. *Tribology International*, Vol. 42, pp. 554– 560, ISSN 0301-679X
7. Michalczewski, R., Piekoszewski, W., Szczerek, M., Tuszyński, W. & Wulczyński, J. (2010). Development of methods and devices for evaluation of low-friction and antiwear PVD/CVD coatings. In: Innovative Technological Solutions for Sustainable Development, A. Mazurkiewicz (Ed.), pp. 63-82, ISBN 978-83-7204-955-1, ITeE-PIB, Radom
8. Michalczewski, R., Szczerek, M., Tuszynski, W. & Wulczyński, J. (2009b). A four-ball machine for testing antiwear, extreme-pressure properties, and surface fatigue life with a possibility to increase the lubricant temperature. *Tribologia*, No. 1, pp. 113- 127, ISSN 0208-7774
9. Piekoszewski, W.; Szczerek, M. & Tuszynski, W. (2001). The action of lubricants under extreme pressure conditions in a modified four-ball tester. Wear, Vol. 249, pp. 188- 193, ISSN 0043-1648
10. Szczerek, M. (1996) Metodologiczne problemy systematyzacji eksperymentalnych badañ tribologicznych, ITeE, Radom (in Polish), ISBN 83-87039-42-X
11. Szczerek, K., Michalczewski, R., & Piekoszewski, W. (2009). The Correlated Selection of PVD/CVD Coatings and Eco-Lubricants for Heavy-Loaded Machine Components – A New Approach. In. IV World Tribology Congress, Kyoto, Japan, 6-11 Sept., 2009, p. 338
12. Szczerek, M. & Tuszynski, W. (2002). A method for testing lubricants under conditions of scuffing. Part I. Presentation of the method. *Tribotest*, Vol. 8, No. 4, pp. 273-284, ISSN 1354-4063
13. Torrance, A.A., Morgan, J.E. & Wan, G.T.Y. (1996). An additive's influence on the pitting and wear of ball bearing steel. Wear, Vol. 192, pp. 66-73, ISSN 0043-1648

Index

9 789360 840938

Printed at Repro India Ltd.